TC 3-04.4

Fundamentals of Flight

DECEMBER 2016

DISTRIBUTION RESTRICTION. Approved for public release; distribution is unlimited.
This publication supersedes FM 3-04.203, dated 1 May 2007.

Headquarters, Department of the Army

This publication is available at the Army Publishing
Directorate site (http://www.apd.army.mil),
and the Central Army Registry site
(https://atiam.train.army.mil/catalog/dashboard)

*TC 3-04.4 (FM 3-04.203)

Training Circular
TC 3-04.4

Headquarters
Department of the Army
Washington, D.C., 22 December 2016

Fundamentals of Flight

Contents

	Page
Preface	xv
Introduction	xvii

Chapter 1	**Aerodynamics of Flight**	**1-1**
	Section I – Physical Laws and Principles of Airflow	1-1
	Newton's Laws of Motion	1-1
	Bernoulli's Principle of Differential Pressure	1-2
	Vectors and Scalars	1-4
	Section II – Flight Mechanics	1-6
	Airfoil Characteristics	1-6
	Airflow and Reactions in the Rotor System	1-8
	Rotor Blade Angles	1-10
	Rotor Blade Actions	1-11
	Helicopter Design and Control	1-16
	Section III – In-Flight Forces	1-25
	Total Aerodynamic Force	1-25
	Lift and Lift Equation	1-25
	Drag	1-26
	Centrifugal Force and Coning	1-28
	Torque Reaction and Antitorque Rotor (Tail Rotor)	1-30
	Balance of Forces	1-31
	Section IV – Hovering	1-33
	Airflow in Hovering Flight	1-33
	Ground Effect	1-33
	Translating Tendency	1-35
	Section V – Rotor in Translation	1-36
	Airflow in Forward Flight	1-36
	Translational Lift	1-40
	Transverse Flow Effect	1-41
	Effective Translational Lift	1-42
	Autorotation	1-42
	Section VI – Maneuvering Flight	1-50

Distribution Restriction: Approved for public release; distribution is unlimited.

*This publication supersedes FM 3-04.203, dated 1 May 2007.

Contents

	Aerodynamics	1-50
	Guidelines	1-55
	Section VII – Factors Affecting Performance	**1-55**
	Density Altitude	1-55
	High and Low Density Altitude Conditions	1-56
	Weight	1-57
	Winds	1-58
	Performance Charts	1-58
	Hovering Performance	1-58
	Climb Performance	1-58
	Section VIII – Emergencies	**1-59**
	Settling with Power	1-59
	Dynamic Rollover	1-61
	Retreating Blade Stall	1-64
	Ground Resonance	1-65
	Compressibility Effects	1-66
Chapter 2	**Weight, Balance, and Loads**	**2-1**
	Section I – Weight	**2-1**
	Weight Definitions	2-1
	Weight Versus Aircraft Performance	2-2
	Section II – Balance	**2-2**
	Center of Gravity	2-2
	Lateral Balance	2-2
	Balance Definitions	2-3
	Principle of Moments	2-5
	Section III – Weight and Balance Calculations	**2-5**
	Weight and Balance Methods	2-6
	Center of Gravity Limits	2-7
	Section IV – Loads	**2-8**
	Planning	2-8
	Internal Loads	2-10
	External Loads	2-19
	Hazardous Materials	2-22
Chapter 3	**Rotary-Wing Environmental Flight**	**3-1**
	Section I – Cold Weather Operations	**3-1**
	Environmental Factors	3-1
	Taxiing and Takeoff	3-6
	Maintenance	3-10
	Training	3-12
	Section II – Desert Operations	**3-12**
	Environmental Factors	3-12
	Flying Techniques	3-16
	Maintenance	3-18
	Training	3-19
	Section III – Jungle Operations	**3-20**
	Environmental Factors	3-20

	Flying Techniques	3-22
	Maintenance	3-23
	Training	3-23
	Section IV – Mountain Operations	**3-24**
	Environmental Factors	3-24
	Flying Techniques	3-31
	Maintenance	3-43
	Training	3-43
	Section V – Overwater Operations	**3-44**
	Environmental Factors	3-44
	Flying Techniques	3-45
	Maintenance	3-45
	Training	3-45
Chapter 4	**Rotary-Wing Night Flight**	**4-1**
	Section I – Night Vision	**4-1**
	Night Vision Capability	4-1
	Types of Vision	4-1
	Combat Visual Impairments	4-1
	Aircraft Design	4-2
	Section II – Hemispheric Illumination and Meteorological Conditions	**4-3**
	Light Sources	4-3
	Other Considerations	4-4
	Section III – Terrain Interpretation	**4-5**
	Factors	4-8
	Other Considerations	4-12
	Section IV – Night Vision Sensors	**4-13**
	Electromagnetic Spectrum	4-13
	Night Vision Devices	4-15
	Forward-looking infrared Systems	4-20
	Section V – Night Operations	**4-27**
	Premission Planning	4-27
	Night Flight Techniques	4-29
	Emergency and Safety Procedures	4-35
Chapter 5	**Rotary-Wing Terrain Flight**	**5-1**
	Section I – Terrain Flight Operations	**5-1**
	Mission Planning and Preparation	5-1
	Aviation Mission Planning System	5-1
	Terrain Flight Limitations	5-2
	Terrain Flight Modes	5-2
	Selection of Terrain Flight Modes	5-4
	Pickup Zone/Landing Zone Selection	5-4
	Route-Planning Considerations	5-6
	Map Selection and Preparation	5-8
	Charts, Photographs, and Objective Cards	5-11
	Route Planning Card Preparation	5-11
	Hazards to Terrain Flight	5-14

Contents

	Terrain Flight Performance	5-15
	Section II - Training	**5-16**
	Command Responsibility	5-17
	Identification of Unit/Individual Needs	5-17
	Training Considerations	5-17
	Training Safety	5-17
Chapter 6	**Multi-Aircraft Operations**	**6-1**
	Section I – Formation Flight	**6-1**
	Formation Discipline	6-1
	Crew Coordination	6-1
	Crew Responsibilities	6-1
	Considerations	6-3
	Formation Breakup	6-9
	Rendezvous and Join-Up Procedures	6-12
	Lost Visual Contact Procedures	6-13
	Communication During Formation Flight	6-13
	Section II – Formation Types	**6-13**
	Two-Helicopter Team	6-14
	Fixed Formations	6-14
	Maneuvering Formations	6-18
	Section III – Basic Combat Maneuvers	**6-21**
	Maneuvering Flight Communications	6-22
	Basic Combat Maneuvers	6-22
	Section IV – Planning Considerations and Responsibilities	**6-28**
	Planning Considerations	6-28
	Planning Responsibilities	6-29
	Section V – Wake Turbulence	**6-30**
	In-Flight Hazard	6-30
	Ground Hazard	6-30
	Vortex Generation	6-30
	Induced Roll and Counter Control	6-31
	Operational Problem Areas	6-31
	Vortex Avoidance Techniques	6-32
Chapter 7	**Fixed-Wing Aerodynamics and Performance**	**7-1**
	Section I – Fixed-Wing Stability	**7-1**
	Motion Sign Principles	7-1
	Static Stability	7-1
	Dynamic Stability	7-2
	Pitch Stability	7-3
	Lateral Stability	7-11
	Cross-Effects and Stability	7-13
	Section II – High-Lift Devices	**7-16**
	Purpose	7-16
	Increasing the Coefficient of Lift	7-17
	Types of High-Lift Devices	7-19
	Section III – Stalls	**7-22**

Aerodynamic Stall	7-22
Stall Warning and Stall Warning Devices	7-24
Stall Recovery	7-26
Spins	7-26
Section IV – Maneuvering Flight	**7-29**
Climbing Flight	7-29
Angle of Climb	7-31
Rate of Climb	7-32
Aircraft Performance in a Climb or Dive	7-32
Turns	7-34
Slow Flight	7-37
Descents	7-38
Section V – Takeoff and Landing Performance	**7-40**
Procedures and Techniques	7-40
Takeoff	7-40
Section VI – Flight Control	**7-45**
Development	7-45
Control Surface and Operation Theory	7-45
Longitudinal Control	7-47
Directional Control	7-49
Lateral Control	7-49
Control Forces	7-49
Control Systems	7-52
Propellers	7-54
Section VII – Multiengine Operations	**7-56**
Twin-Engine Aircraft Performance	7-56
Asymmetric Thrust	7-57
Critical Engine	7-57
Minimum Single-Engine Control Speed	7-58
Single-Engine Climbs	7-60
Single-Engine Level Flight	7-62
Single-Engine Descents	7-62
Single-Engine Approach and Landing	7-62
Propeller Feathering	7-63
Accelerate-Stop Distance	7-63
Accelerate-Go Distance	7-65

Chapter 8	**Fixed-Wing Environmental Flight**	**8-1**
	Section I – Cold Weather/Icing Operations	**8-1**
	Environmental Factors	8-1
	Aircraft Equipment	8-7
	Flying Techniques	8-10
	Training	8-14
	Section II – Mountain Operations	**8-15**
	Environmental Factors	8-15
	Flying Techniques	8-16
	Section III – Overwater Operations	**8-17**

Contents

	Oceanographic Terminology	8-17
	Ditching	8-17
	Section IV – Thunderstorm Operations	**8-22**
	Wind Shear	8-22
Chapter 9	**Fixed-Wing Night Flight**	**9-1**
	Section I – Preparation and Preflight	**9-1**
	Equipment	9-1
	Lighting	9-1
	Parking Ramp Check	9-2
	Preflight	9-2
	Section II – Taxi, Takeoff, and Departure Climb	**9-3**
	Taxi	9-3
	Takeoff and Climb	9-3
	Section III – Orientation and Navigation	**9-4**
	Visibility	9-4
	Maneuvers	9-5
	Disorientation and Reorientation	9-5
	Cross-Country Flights	9-5
	Overwater Flights	9-5
	Illusions	9-5
	Section IV – Approaches and Landings	**9-5**
	Distance	9-5
	Airspeed	9-5
	Depth Perception	9-6
	Approaching Airports	9-6
	Entering Traffic	9-6
	Final Approach	9-6
	Executing Roundout	9-7
	Section V – Night Emergencies	**9-9**
	Glossary	**1**
	References	**1**
	Index	**1**

Figures

Figure 1-1. Aerodynamic lift–explained by Newton's Law of Motion	1-2
Figure 1-2. Water flow through a tube	1-2
Figure 1-3. Venturi effect	1-3
Figure 1-4. Venturi flow	1-3
Figure 1-5. Resultant by parallelogram method	1-4
Figure 1-6. Resultant by the polygon method	1-5
Figure 1-7. Resultant by the triangulation method	1-5
Figure 1-8. Force vectors on an airfoil segment	1-6
Figure 1-9. Force vectors on aircraft in flight	1-6

Figure 1-10. Symmetrical airfoil section 1-8
Figure 1-11. Nonsymmetrical (cambered) airfoil section 1-8
Figure 1-12. Relative wind 1-9
Figure 1-13. Rotational relative wind 1-9
Figure 1-14. Induced flow (downwash) 1-10
Figure 1-15. Resultant relative wind 1-10
Figure 1-16. Angle of incidence and angle of attack 1-11
Figure 1-17. Blade rotation and blade speed 1-12
Figure 1-18. Feathering 1-12
Figure 1-19. Flapping in directional flight 1-13
Figure 1-20. Flapping (advancing blade 3 o'clock position) 1-13
Figure 1-21. Flapping (retreating blade 9-o'clock position) 1-14
Figure 1-22. Flapping (blade over the aircraft nose) 1-14
Figure 1-23. Flapping (blade over the aircraft tail) 1-14
Figure 1-24. Lead and lag 1-15
Figure 1-25. Under slung design of semirigid rotor system 1-16
Figure 1-26. Gyroscopic precession 1-17
Figure 1-27. Rotor head control systems 1-18
Figure 1-28. Stationary and rotating swashplates tilted by cyclic control 1-18
Figure 1-29. Stationary and rotating swashplates tilted in relation to mast 1-19
Figure 1-30. Pitch-change arm rate of movement over 90 degrees of travel 1-19
Figure 1-31. Rotor flapping in response to cyclic input 1-20
Figure 1-32. Cyclic feathering 1-21
Figure 1-33. Input servo and pitch-change horn offset 1-21
Figure 1-34. Cyclic pitch variation–full forward, low pitch 1-22
Figure 1-35. Fully articulated rotor system 1-23
Figure 1-36. Semirigid rotor system 1-23
Figure 1-37. Effect of tail-low attitude on lateral hover attitude 1-24
Figure 1-38. Cyclic control response around the lateral and longitudinal axes 1-25
Figure 1-39. Total aerodynamic force 1-25
Figure 1-40. Forces acting on an airfoil 1-26
Figure 1-41. Drag and airspeed relationship 1-27
Figure 1-42. Effects of centrifugal force and lift 1-29
Figure 1-43. Decreased disk area (loss of lift caused by coning) 1-29
Figure 1-44. Torque reaction 1-30
Figure 1-45. Balanced forces; hovering with no wind 1-31
Figure 1-46. Unbalanced forces causing acceleration 1-32
Figure 1-47. Balanced forces; steady-state flight 1-32
Figure 1-48. Unbalanced forces causing deceleration 1-32
Figure 1-49. Airflow in hovering flight 1-33
Figure 1-50. In ground effect hover 1-34
Figure 1-51. Out of ground effect hover 1-35
Figure 1-52. Translating tendency 1-36
Figure 1-53. Differential velocities on the rotor system caused by forward airspeed 1-37

Figure 1-54. Blade areas in forward flight 1-38
Figure 1-55. Flapping (advancing blade, 3-o'clock position) 1-39
Figure 1-56. Flapping (retreating blade, 9-o'clock position) 1-39
Figure 1-57. Blade pitch angles 1-40
Figure 1-58. Translational lift (1 to 5 knots) 1-41
Figure 1-59. Translational lift (10 to 15 knots) 1-41
Figure 1-60. Transverse flow effect 1-42
Figure 1-61. Effective translational lift 1-42
Figure 1-62. Blade regions in vertical autorotation descent....................................... 1-43
Figure 1-63. Force vectors in vertical autorotative descent 1-45
Figure 1-64. Autorotative regions in forward flight 1-46
Figure 1-65. Force vectors in level-powered flight at high speed 1-46
Figure 1-66. Force vectors after power loss–reduced collective 1-47
Figure 1-67. Force vectors in autorotative steady-state descent 1-47
Figure 1-68. Autorotative deceleration 1-48
Figure 1-69. Drag and airspeed relationship 1-49
Figure 1-70. Counterclockwise blade rotation 1-50
Figure 1-71. Lift to weight 1-53
Figure 1-72. Aft cyclic results 1-54
Figure 1-73. Density altitude computation 1-57
Figure 1-74. Induced flow velocity during hovering flight 1-59
Figure 1-75. Induced flow velocity before vortex ring state 1-59
Figure 1-76. Vortex ring state 1-60
Figure 1-77. Settling with power region 1-61
Figure 1-78. Downslope rolling motion 1-62
Figure 1-79. Upslope rolling motion 1-63
Figure 1-80. Retreating blade stall (normal hovering lift pattern) 1-64
Figure 1-81. Retreating blade stall (normal cruise lift pattern) 1-64
Figure 1-82. Retreating blade stall (lift pattern at critical airspeed–retreating blade stall) 1-65
Figure 1-83. Ground resonance 1-66
Figure 1-84. Compressible and incompressible flow comparison 1-68
Figure 1-85. Normal shock wave formation 1-69
Figure 2-1. Helicopter station diagram 2-4
Figure 2-2. Aircraft balance point 2-5
Figure 2-3. Locating aircraft center of gravity 2-6
Figure 2-4. Fuel moments 2-7
Figure 2-5. Center of gravity limits chart 2-8
Figure 2-6. Weight-spreading effect of shoring 2-11
Figure 2-7. Load contact pressure 2-11
Figure 2-8. Formulas for load pressure calculations 2-12
Figure 2-9. Determining general cargo center of gravity 2-13
Figure 2-10. Determining center of gravity of wheeled vehicle 2-13
Figure 2-11. Compartment method steps 2-14
Figure 2-12. Station method steps 2-15

Figure 2-13. Effectiveness of tie-down devices 2-17
Figure 2-14. Calculating tie-down requirements 2-19
Figure 3-1. Weather conditions conducive to icing 3-3
Figure 3-2. Ambient light conditions 3-5
Figure 3-3. Depth perception 3-9
Figure 3-4. Desert areas of the world 3-13
Figure 3-5. Sandy desert terrain 3-14
Figure 3-6. Rocky plateau desert terrain 3-15
Figure 3-7. Mountain desert terrain 3-15
Figure 3-8. Jungle areas of the world 3-20
Figure 3-9. Types of wind 3-25
Figure 3-10. Light wind 3-25
Figure 3-11. Moderate wind 3-26
Figure 3-12. Strong wind 3-26
Figure 3-13. Mountain (standing) wave 3-27
Figure 3-14. Cloud formations associated with mountain wave 3-28
Figure 3-15. Rotor streaming turbulence 3-28
Figure 3-16. Wind across a ridge 3-29
Figure 3-17. Snake ridge 3-30
Figure 3-18. Wind across a crown 3-30
Figure 3-19. Shoulder wind 3-31
Figure 3-20. Wind across a canyon 3-31
Figure 3-21. Mountain takeoff 3-32
Figure 3-22. High reconnaissance flight patterns 3-35
Figure 3-23. Computing wind direction between two points 3-36
Figure 3-24. Computing wind direction using the circle maneuver 3-37
Figure 3-25. Approach paths and areas to avoid 3-38
Figure 3-26. Nap-of-the-earth or contour takeoff (terrain flight) 3-40
Figure 3-27. Ridge crossing at a 45-degree angle (terrain flight) 3-41
Figure 3-28. Steep turns or climbs at terrain flight altitudes 3-41
Figure 3-29. Flight along a valley (terrain flight) 3-42
Figure 3-30. Nap-of-the-earth or contour approach (terrain flight) 3-43
Figure 4-1. Identification by object size 4-6
Figure 4-2. Identification by object shape 4-7
Figure 4-3. Identification by object contrast 4-7
Figure 4-4. Identification by object viewing distance 4-8
Figure 4-5. Electromagnetic Spectrum 4-13
Figure 4-6. IR energy 4-15
Figure 4-7. Image intensifier 4-15
Figure 4-8. AN/AVS-6 in operational position 4-16
Figure 4-9. MPNVS/MTADS 4-20
Figure 4-10. Atmospheric effects on infrared radiation 4-22
Figure 4-11. Infrared energy crossover 4-23
Figure 4-12. Parallax effect 4-24

Figure 4-13. Polarity 4-26
Figure 4-14. Stop-turn scanning 4-29
Figure 4-15. Scanning with ten degree circular overlap 4-29
Figure 4-16. Night visual meteorological conditions takeoff 4-31
Figure 4-17. Approach to a lighted inverted Y 4-33
Figure 4-18. Approach to a lighted T 4-34
Figure 5-1. Modes of flight 5-3
Figure 5-2. Route planning map symbols 5-10
Figure 5-3. Sample–joint operations graphic map preparation 5-11
Figure 5-4. Example of an en route card 5-12
Figure 5-5. Example of an objective card 5-13
Figure 6-1. Horizontal distance 6-5
Figure 6-2. Stepped-up vertical separation 6-6
Figure 6-3. Echelon formation before breakup 6-9
Figure 6-4. Left break with 10-second interval for landing 6-9
Figure 6-5. Breakup into two elements 6-10
Figure 6-6. Formation breakup–inadvertent instrument meteorological conditions ... 6-11
Figure 6-7. Two-helicopter section/element 6-14
Figure 6-8. Staggered right and left formation 6-15
Figure 6-9. Echelon right and left formation 6-16
Figure 6-10. Trail formation 6-17
Figure 6-11. V-formation 6-18
Figure 6-12. Team combat cruise 6-19
Figure 6-13. Flight combat cruise 6-19
Figure 6-14. Combat cruise right 6-20
Figure 6-15. Combat cruise left 6-20
Figure 6-16. Combat trail 6-21
Figure 6-17. Combat spread 6-21
Figure 6-18. Basic combat maneuver circle 6-22
Figure 6-19. Tactical turn away 6-23
Figure 6-20. Tactical turn to 6-24
Figure 6-21. Dig and pinch maneuvers 6-24
Figure 6-22. Split turn maneuver 6-25
Figure 6-23. In-place turn 6-25
Figure 6-24. Cross turn in or out 6-26
Figure 6-25. Cross turn cover (high/low) 6-26
Figure 6-26. Break turn left/right 6-27
Figure 6-27. Break turn left/right (high/low) 6-27
Figure 6-28. Shackle turn 6-28
Figure 6-29. Wake vortex generation 6-30
Figure 7-1. Stability nomenclature 7-1
Figure 7-2. Nonoscillatory motion 7-2
Figure 7-3. Oscillatory motion 7-3
Figure 7-4. C_M versus C_L 7-4

Figure 7-5. Fixed-wing aircraft center of gravity and aerodynamic center 7-5
Figure 7-6. Wing contribution to longitudinal stability 7-6
Figure 7-7. Negative pitching moment about the aerodynamic center of a positive-cambered airfoil.. 7-6
Figure 7-8. Positive longitudinal stability of a positive-cambered airfoil 7-7
Figure 7-9. Negative longitudinal stability of a positive-cambered airfoil 7-7
Figure 7-10. Lift as a stabilizing moment to the horizontal stabilizer 7-8
Figure 7-11. Thrust axis about center of gravity.. 7-9
Figure 7-12. Positive sideslip angle.. 7-9
Figure 7-13. Directional stability (β versus C_N).. 7-10
Figure 7-14. Dorsal fin decreases drag 7-11
Figure 7-15. Fixed-wing aircraft configuration positive yawing moment 7-11
Figure 7-16. Horizontal lift component produces sideslip 7-12
Figure 7-17. Positive static lateral stability 7-12
Figure 7-18. Dihedral angle... 7-12
Figure 7-19. Dihedral stability... 7-13
Figure 7-20. Adverse yaw.. 7-14
Figure 7-21. Slipstream and yaw.. 7-16
Figure 7-22. Asymmetric loading (propeller-factor)... 7-16
Figure 7-23. Increasing camber with trailing-edge flap 7-17
Figure 7-24. Suction boundary-layer control 7-18
Figure 7-25. Blowing boundary-layer control... 7-19
Figure 7-26. Vortex generators... 7-19
Figure 7-27. Angle of incidence change with flap deflection 7-20
Figure 7-28. Types of high-lift devices 7-20
Figure 7-29. Coefficient of lift maximum increase with slotted flap 7-22
Figure 7-30. Coefficient of lift curve.. 7-23
Figure 7-31. Various airfoil angles of attack 7-23
Figure 7-32. Boundary-layer separation 7-24
Figure 7-33. C_L curves for cambered and symmetrical airfoils 7-25
Figure 7-34. Stall strip 7-26
Figure 7-35. Flapper switch 7-26
Figure 7-36. Spins 7-27
Figure 7-37. Climb angle and rate 7-29
Figure 7-38. Force-vector diagram for climbing flight... 7-30
Figure 7-39. Wind effect on maximum climb angle 7-31
Figure 7-40. Full-power polar diagram 7-33
Figure 7-41. Polar curve 7-34
Figure 7-42. Effect of turning flight 7-35
Figure 7-43. Effect of load factor on stalling speed 7-36
Figure 7-44. Best glide speed... 7-40
Figure 7-45. Net accelerating force 7-41
Figure 7-46. Landing roll velocity 7-43
Figure 7-47. Using flaps to increase camber... 7-45
Figure 7-48. Operation of aileron in a turn 7-46

Contents

Figure 7-49. Effect of elevator and rudder on moments 7-46
Figure 7-50. Effect of center of gravity location on longitudinal control 7-47
Figure 7-51. Adverse moments during takeoff 7-48
Figure 7-52. Hinge moment 7-50
Figure 7-53. Aerodynamic balancing using horns 7-50
Figure 7-54. Aerodynamic balancing using a balance board 7-51
Figure 7-55. Aerodynamic balancing using a servo tab 7-51
Figure 7-56. Spoiler used as control surface 7-53
Figure 7-57. Wing flap control 7-54
Figure 7-58. Blade angle affected by revolutions per minute 7-55
Figure 7-59. Forces created during single-engine operation 7-58
Figure 7-60. Sideslip 7-60
Figure 7-61. One-engine inoperative flight path 7-61
Figure 7-62. Windmilling propeller creating drag 7-63
Figure 7-63. Required takeoff runway lengths 7-64
Figure 7-64. Balanced field length 7-65
Figure 8-1. Lift curve 8-3
Figure 8-2. Drag curve 8-3
Figure 8-3. Tail stall pitchover 8-6
Figure 8-4. Pneumatic boots 8-9
Figure 8-5. Propeller ice control 8-10
Figure 8-6. Wind swell ditch heading 8-18
Figure 8-7. Single swell 8-19
Figure 8-8. Double swell (15 knot wind) 8-19
Figure 8-9. Double swell (30 knot wind) 8-20
Figure 8-10. Swell (50 knot wind) 8-20
Figure 8-11. Effect of microburst 8-23
Figure 9-1. Positive climb 9-4
Figure 9-2. Typical light pattern for airport identification 9-6
Figure 9-3. Visual approach slope indicator 9-7
Figure 9-4. Roundout (when tire marks are visible) 9-8

Tables

Table 1-1. Airfoil terminology 1-7
Table 1-2. Aircraft reaction to forces 1-17
Table 1-3. Bank angle versus torque 1-53
Table 1-4. Speed of sound variation with temperature and altitude 1-67
Table 2-1. Responsibilities 2-9
Table 2-2. Internal loading considerations 2-10
Table 2-3. Percentage restraint chart 2-18
Table 4-1. Design eye point reference distance 4-3
Table 5-1. Mission, enemy, terrain and weather, troops and support available, time available, civil considerations and terrain flight modes 5-4
Table 5-2. Pickup zone selection considerations 5-5

Table 5-3. Pickup zone selection considerations ... 5-6
Table 5-4. Route planning considerations ... 5-7
Table 5-5. Example of a navigation card ... 5-12
Table 6-1. Sample lighting conditions ... 6-7
Table 8-1. Temperature ranges for ice formation ... 8-2
Table 8-2. Oceanographic terminology ... 8-17

This page intentionally left blank.

Preface

Training Circular (TC) 3-04.4 presents the basic physics of flight, the dynamics associated with rotary and FW aircraft, and covers basic tactical flight profiles, formation flight, and maneuvering flight techniques. It contains theoretical and practical concepts which Army Aviators and crewmembers apply to tactical and operational expertise technical base from which Army Aviation executes its core competencies.

The principal audience for TC 3-04.4 is all Army Aviators and crewmembers. Trainers and educators throughout the Army will also use this publication.

Commanders, staffs, and subordinates ensure that their decisions and actions comply with applicable United States, international, and in some cases host-nation laws and regulations. Commanders at all levels ensure that their Soldiers operate in accordance with the law of war and the rules of engagement. (See FM 27-10.)

TC 3-04.4 uses joint terms where applicable. Selected joint and Army terms and definitions appear in both the glossary and the text. Terms for which TC 3-04.4 is the proponent publication (the authority) are italicized in the text and are marked with an asterisk (*) in the glossary. Terms and definitions for which TC 3-04.4 is the proponent publication are boldfaced in the text. For other definitions shown in the text, the term is italicized and the number of the proponent publication follows the definition.

The proponent of this publication is Headquarters, United States Army Training and Doctrine Command (TRADOC). Send comments and recommendations on Department of the Army (DA) Form 2028 (Recommended Changes to Publications and Blank Forms) or automated link (http://www.apd.army.mil) to Commander, United States Army Aviation Center of Excellence (USAACE) ATTN: ATZQ-TD-D, Fort Rucker, Alabama 36362-5263. Comments may be e-mailed to the Directorate of Training and Doctrine at usarmy.rucker.avncoe.mbx.doctrine-branch@mail.mil. Other doctrinal information can be found on the Internet at Army Knowledge Online (AKO).

TC 3-04.4 applies to the Active Army, Army National Guard/Army National Guard of the United States and United States Army Reserve unless otherwise stated.

This publication has been reviewed for operations security considerations.

This page intentionally left blank.

Introduction

One of the underlying premises of Army Aviation is if crewmembers understand 'why' they will be better prepared to 'do' when confronted with the unexpected. This publication is an excellent reference for Army crewmembers; however, it is not expected that this training circular is all inclusive or a full comprehension of the information will be obtained by simply reading the text. A firm understanding will begin to occur as crewmembers become more experienced in their particular aircraft; study the tactics, techniques, and procedures (TTP) of their units; and study other sources of information. Crewmembers seeking to hone their skills should review this document periodically to gain new insights.

This publication ensures that crewmembers understand the basic physics of flight, and the dynamics associated with rotary and FW aircraft. A comprehensive understanding of these principles will better prepare aviators for flight, transition training, and tactical flight operations. The fundamentals of flight outlined in this TC form the technical base from which Army Aviation executes its core competencies.

Because the United States Army prepares its Soldiers to operate anywhere in the world, this publication describes the unique requirements and flying techniques crewmembers will use to successfully operate in various environments. The environments may not always be encountered during home station training, but may be replicated in constructive, gaming, and virtual training.

Army Aviation leverages superior night operation tactics and technologies to gain and maintain advantage over our adversaries. To that end, Army crewmembers must be familiar and capable of performing their missions proficiently at night. The information on night vision systems (NVSs) and night operations in this circular provides the basis for mastering these skills.

Every aviator understands that they must operate the aircraft safely. Every crewmember must perform the mission effectively and decisively in training and combat. TC 3-04.4 also covers basic tactical flight profiles, formation flight, and maneuvering flight techniques.

This publication incorporates the following changes:
- Removes references to the OH 58-D.
- Adds the rigid rotor system.
- Removes thermal imaging systems.
- Adds forward-looking infrared radar (FLIR).

This page intentionally left blank.

Chapter 1

Aerodynamics of Flight

This chapter presents aerodynamic fundamentals and principles of rotary-wing flight. Understanding these principles will facilitate an aviator's ability to maximize performance of the aircraft. The content relates to flight operations and performance of normal mission tasks. It covers theory and application of aerodynamics for the aviator. Chapter 7 presents additional information on FW flight.

SECTION I – PHYSICAL LAWS AND PRINCIPLES OF AIRFLOW

1-1. There are several ways to explain how an airfoil generates lift. For the purpose of this manual, we will discuss Newton's laws of motion and Bernoulli's principal. The "Newton" position is that lift is the reaction force on a body caused by deflecting a flow of gas, and the "Bernoulli" position is that lift is generated by a pressure difference across the wing. Both Bernoulli and Newton are correct, and we can use equations developed by each of them to determine the magnitude and direction of the aerodynamic force.

NEWTON'S LAWS OF MOTION

1-2. Newton's three laws of motion are inertia, acceleration, and action/reaction. These laws apply to flight of any aircraft. A working knowledge of the laws and their applications will assist in understanding aerodynamic principles discussed in this chapter. Interaction between the laws of motion and aircraft mechanical actions causes the aircraft to fly and allows aviators to control such flight.

FIRST LAW: INERTIA

1-3. A body at rest will remain at rest, and a body in motion will remain in motion at the same speed and in the same direction unless acted upon by an external force. Nothing starts or stops without an outside force to bring about or prevent motion. Inertia is a body's resistance to a change in its state of motion. For a constant mass, force (F) equals mass (M) times acceleration (A), expressed in the formula ($F=MA$). Mass is then the property of matter that manifests itself as inertia.

SECOND LAW: ACCELERATION

1-4. The force required to produce a change in motion of a body is directly proportional to its mass and rate of change in its velocity. Acceleration is a change in velocity with respect to time, it is directly proportional to force and inversely proportional to mass. This takes into account the factors involved in overcoming Newton's First Law. It covers both changes in direction and speed, including starting up from rest (positive acceleration) and coming to a stop (negative acceleration or deceleration). Expressed in the equation $A=F/M$.

THIRD LAW: ACTION/REACTION

1-5. For every action, there is an equal and opposite reaction. When an interaction occurs between two bodies, equal forces in opposite directions are imparted to each body. In a helicopter, the rotor blades move air downward; consequently, the air pushes the rotor blades (and thus the helicopter) in the opposite direction (figure 1-1, page 1-2).

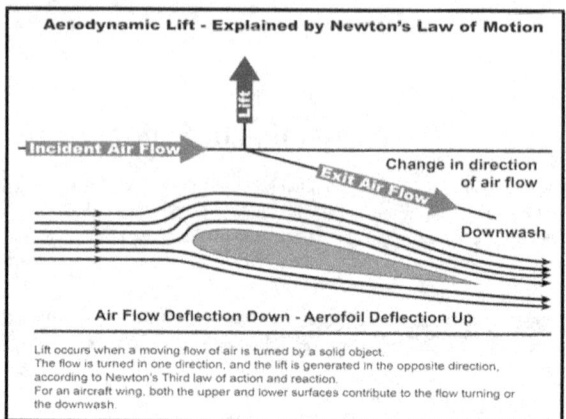

Figure 1-1. Aerodynamic lift—explained by Newton's Law of Motion

BERNOULLI'S PRINCIPLE OF DIFFERENTIAL PRESSURE

1-6. This principle describes the relationship between internal fluid pressure and fluid velocity. It is a statement of the law of conservation of energy and helps explain why an airfoil develops an aerodynamic force. The concept of conservation of energy states energy cannot be created or destroyed and the amount of energy entering a system must also exit. A simple tube with a constricted portion near the center of its length illustrates this principle. An example is using water through a garden hose (figure 1-2). The mass of flow per unit area (cross sectional area of tube) is the mass flow rate. In figure 1-2, the flow into the tube is constant, neither accelerating nor decelerating; thus, the mass flow rate through the tube must be the same at stations 1, 2, or 3. If the cross sectional area at any one of these stations—or any given point—in the tube is reduced, the fluid velocity must increase to maintain a constant mass flow rate to move the same amount of fluid through a smaller area. Fluid speeds up in direct proportion to the reduction in area. Venturi effect is the term used to describe this phenomenon. Figure 1-3, page 1-3, illustrates what happens to mass flow rate in the constricted tube as the dimensions of the tube change.

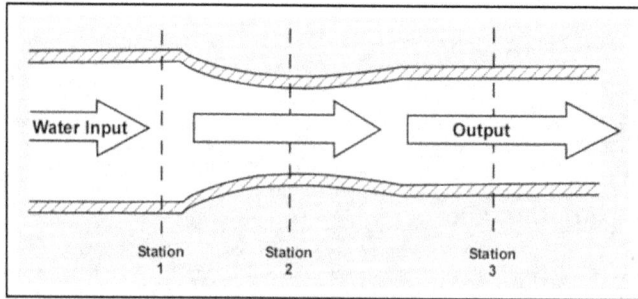

Figure 1-2. Water flow through a tube

Aerodynamics of Flight

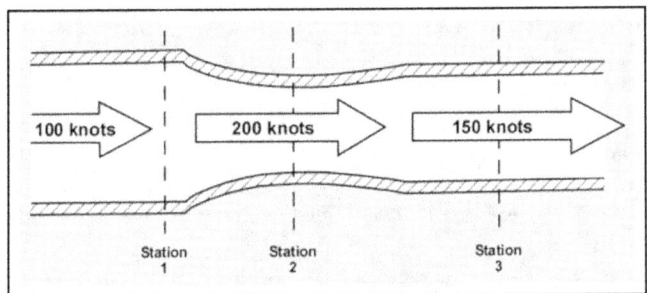

Figure 1-3. Venturi effect

VENTURI FLOW

1-7. While the amount of total energy within a closed system (the tube) does not change, the form of the energy may be altered. Pressure of flowing air may be compared to energy in that the total pressure of flowing air will always remain constant unless energy is added or removed. Fluid flow pressure has two components—static and dynamic pressure. Static pressure is the pressure component measured in the flow but not moving with the flow as pressure is measured. Static pressure is also known as the force per unit area acting on a surface. Dynamic pressure of flow is that component existing as a result of movement of the air. The sum of these two pressures is total pressure. As air flows through the constriction, static pressure decreases as velocity increases. This increases dynamic pressure. Figure 1-4 depicts the bottom half of the constricted area of the tube, which resembles the top half of an airfoil. Even with the top half of the tube removed, the air still accelerates over the curved area because the upper air layers restrict the flow—just as the top half of the constricted tube did. This acceleration causes decreased static pressure above the curved portion and creates a pressure differential caused by the variation of static and dynamic pressures.

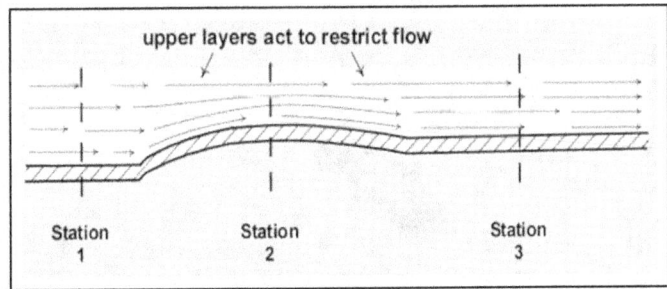

Figure 1-4. Venturi flow

AIRFLOW AND THE AIRFOIL

1-8. Airflow around an airfoil performs similar to airflow through a constriction. As velocity of the airflow increases, static pressure decreases above and below the airfoil. The air usually has to travel a greater distance over the upper surface; thus, there is a greater velocity increase and static pressure decrease over the upper surface than the lower surface. The static pressure differential on the upper and lower surfaces produces about 75 percent of the aerodynamic force, called lift. The remaining 25 percent of the force is produced as a result of action/reaction from the downward deflection of air as it leaves the trailing edge of the airfoil and by the downward deflection of air impacting the exposed lower surface of the airfoil.

Chapter 1

VECTORS AND SCALARS

1-9. Vectors and scalars are useful tools for the illustration of aerodynamic forces at work. Vectors are quantities with a magnitude and direction. Scalars are quantities described by size alone such as area, volume, time, and mass.

VECTOR QUANTITIES

1-10. Velocity, acceleration, weight, lift, and drag are examples of vector quantities. The direction of vector quantities is as important as the size or magnitude. When two or more forces act upon an object, the combined effect may be represented by the use of vectors. Vectors are illustrated by a line drawn at a particular angle with an arrow at the end. The arrow indicates the direction in which the force is acting. The length of the line (compared to a scale) represents the magnitude of the force.

VECTOR SOLUTIONS

1-11. Individual force vectors are useful in analyzing conditions of flight. The chief concern is with combined, or resultant, effects of forces acting on an airfoil or aircraft. The following three methods of solving for the resultant are most commonly used.

Parallelogram Method

1-12. This is the most commonly used vector solution in aerodynamics. Using two vectors, lines are drawn parallel to the vectors determining the resultant. If two tugboats push a barge with equal force, the barge will move forward in a direction that is the mean of the direction of both tugboats (figure 1-5).

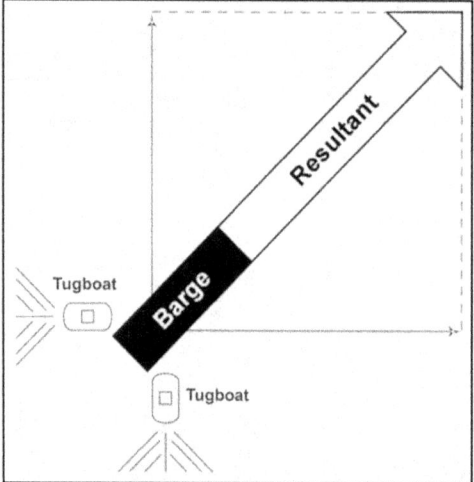

Figure 1-5. Resultant by parallelogram method

Polygon Method

1-13. When more than two forces are acting in different directions, the resultant may be found by using a polygon vector solution. Figure 1-6, page 1-5, shows an example in which one force is acting at 90 degrees with a force of 180 pounds (vector A), a second force acting at 45 degrees with a force of 90 pounds (vector B), and a third force acting at 315 degrees with a force of 120 pounds (vector C). To determine the resultant, draw the first vector beginning at point 0 (the origin) with remaining vectors drawn consecutively. The resultant is drawn from point of origin (0) to the end of the final vector (C).

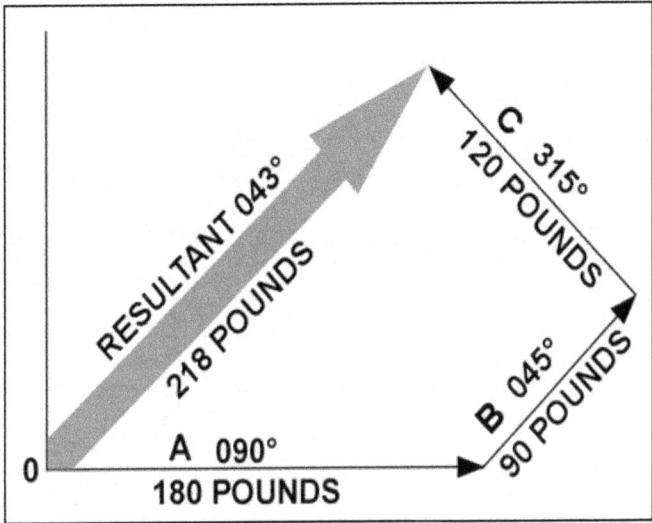

Figure 1-6. Resultant by the polygon method

Triangulation Method

1-14. This is a simplified form of a polygon vector solution using only two vectors and connecting them with a resultant vector line. Figure 1-7 shows an example of this solution. By drawing a vector for each of these known velocities and drawing a connecting line between the ends, a resultant velocity and direction can be determined.

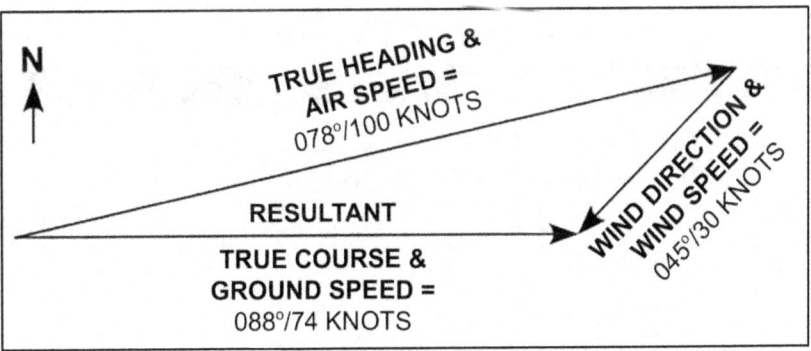

Figure 1-7. Resultant by the triangulation method

Chapter 1

VECTORS USED

1-15. Figures 1-8 and 1-9 show examples of vectors used to depict forces acting on an airfoil segment and aircraft in flight.

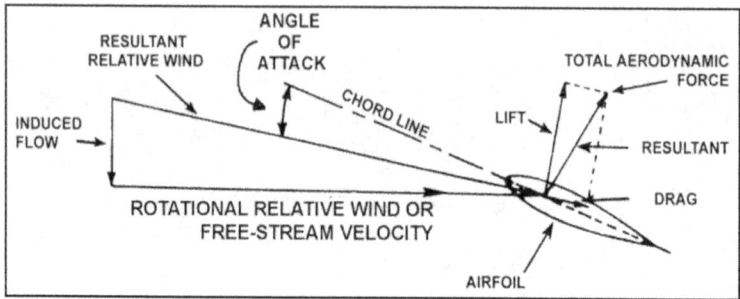

Figure 1-8. Force vectors on an airfoil segment

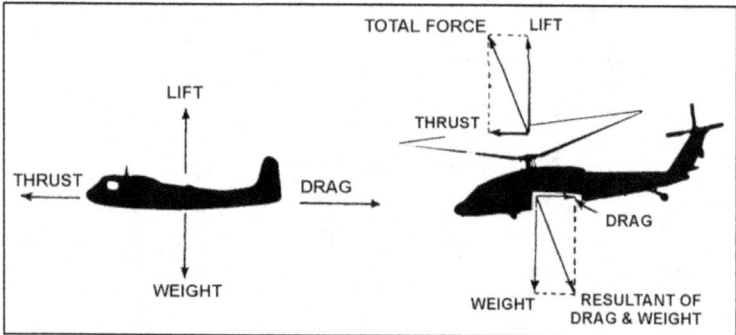

Figure 1-9. Force vectors on aircraft in flight

SECTION II – FLIGHT MECHANICS

AIRFOIL CHARACTERISTICS

1-16. Helicopters and conventional aircraft are able to fly due to aerodynamic forces produced when air passes around the airfoil. An airfoil is a structure or body designed to produce a reaction by its motion through the air. Airfoils are most often associated with production of lift. Airfoils are also used for stability (fin), control (elevator), and thrust or propulsion (propeller or rotor). Certain airfoils, such as rotor blades, combine some of these functions. Airfoils are carefully structured to accommodate a specific set of flight characteristics.

AIRFOIL TERMINOLOGY

1-17. Table 1-1, page 1-7, provides airfoil terms and their definitions common to all aircraft. The first four terms describe the shape of an airfoil. The remaining terms describe development of aerodynamic properties.

Aerodynamics of Flight

Table 1-1. Airfoil terminology

Terms	Definitions
Blade Span	The length of the rotor blade from point of rotation to tip of the blade.
Wing Span	The length of the wing from tip to tip.
Chord Line	A straight line intersecting leading and trailing edges of the airfoil.
Chord	The length of the chord line from leading edge to trailing edge; it is the characteristic longitudinal dimension of the airfoil section.
Mean Camber Line	A line drawn halfway between the upper and lower surfaces. The chord line connects the ends of the mean camber line. Camber refers to curvature of the airfoil and may be considered curvature of the mean camber line. The shape of the mean camber is important for determining aerodynamic characteristics of an airfoil section. Maximum camber (displacement of the mean camber line from the chord line) and its location help to define the shape of the mean camber line. The location of maximum camber and its displacement from the chord line are expressed as fractions or percentages of the basic chord length. By varying the point of maximum camber, the manufacturer can tailor an airfoil for a specific purpose. The profile thickness and thickness distribution are important properties of an airfoil section.
Leading-Edge Radius	The radius of curvature given the leading edge shape.
Flight-Path Velocity	The speed and direction of the airfoil passing through the air. For FW airfoils, flight-path velocity is equal to true airspeed (TAS). For helicopter rotor blades, flight-path velocity is equal to rotational velocity, plus or minus a component of directional airspeed.
Relative Wind	Air in motion equal to and opposite the flight-path velocity of the airfoil. This is rotational relative wind for rotary-wing aircraft and will be covered in detail later. As an induced airflow may modify flight-path velocity, relative wind experienced by the airfoil may not be exactly opposite its direction of travel.
Induced Flow	The downward flow of air (more distinct in rotary-wing).
Resultant Relative Wind	Relative wind modified by induced flow.
Angle of Attack (AOA)	The angle measured between the resultant relative wind and chord line.
Angle of Incidence (FW [fixed-wing] Aircraft)	The angle between the airfoil chord line and longitudinal axis or other selected reference plane of the airplane.
Angle of Incidence (Rotary-Wing Aircraft)	The angle between the chord line of a main or tail-rotor blade and rotational relative wind (tip-path plane). It is usually referred to as blade pitch angle. For fixed airfoils, such as vertical fins or elevators, angle of incidence is the angle between the chord line of the airfoil and a selected reference plane of the helicopter.
Center of Pressure	The point along the chord line of an airfoil through which all aerodynamic forces are considered to act. Since pressures vary on the surface of an airfoil, an average location of pressure variation is needed. As the AOA changes, these pressures change and center of pressure moves along the chord line.
Aerodynamic Center	The point along the chord line where all changes to lift effectively take place. If the center of pressure is located behind the aerodynamic center, the airfoil experiences a nose-down pitching moment. Use of this point by engineers eliminates the problem of center of pressure movement during AOA aerodynamic analysis.

AIRFOIL TYPES

1-18. The two basic types of airfoils are symmetrical and nonsymmetrical.

Symmetrical

1-19. The symmetrical airfoil (figure 1-10, page 1-8) is distinguished by having identical upper and lower surface designs, the mean camber line and chord line being coincident and producing zero lift at zero angle of attack (AOA). A symmetrical design has advantages and disadvantages. One advantage is the center-of-pressure remains relatively constant under varying angles of attack (reducing the twisting force exerted on the airfoil). Another advantage is it affords ease of construction and reduced cost. The disadvantages are less lift production at a given AOA than a nonsymmetrical design and undesirable stall characteristics.

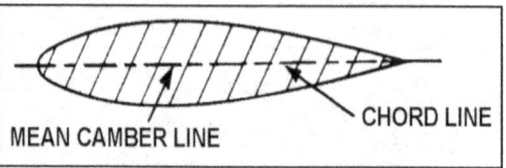

Figure 1-10. Symmetrical airfoil section

Nonsymmetrical (Cambered)

1-20. The nonsymmetrical airfoil (figure 1-11) has different upper and lower surface designs, with a greater curvature of the airfoil above the chord line than below. The mean camber line and chord line are not coincident. The nonsymmetrical airfoil design produces useful lift even at negative angles of attack. A nonsymmetrical design has advantages and disadvantages. The advantages are more lift production at a given AOA than a symmetrical design, an improved lift to drag ratio, and better stall characteristics. The disadvantages are the center-of-pressure travel can move up to 20 percent of the chord line (creating undesirable torque on the airfoil structure) and greater production costs.

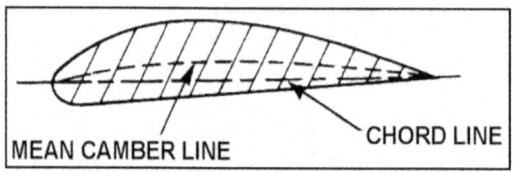

Figure 1-11. Nonsymmetrical (cambered) airfoil section

BLADE TWIST (ROTARY-WING AIRCRAFT)

1-21. Because of lift differential along the blade, it should be designed with a twist to alleviate internal blade stress and distribute the lifting force more evenly along the blade. Blade twist provides higher pitch angles at the root where velocity is low and lower pitch angles nearer the tip where velocity is higher. This increases the induced air velocity and blade loading near the inboard section of the blade.

AIRFLOW AND REACTIONS IN THE ROTOR SYSTEM

RELATIVE WIND

1-22. Knowledge of relative wind (figure 1-12, page 1-9) is essential for an understanding of aerodynamics and its practical flight application for the aviator. Relative wind is airflow relative to an airfoil. Movement of an airfoil through the air creates relative wind. Relative wind moves in a parallel but opposite direction to movement of the airfoil.

Aerodynamics of Flight

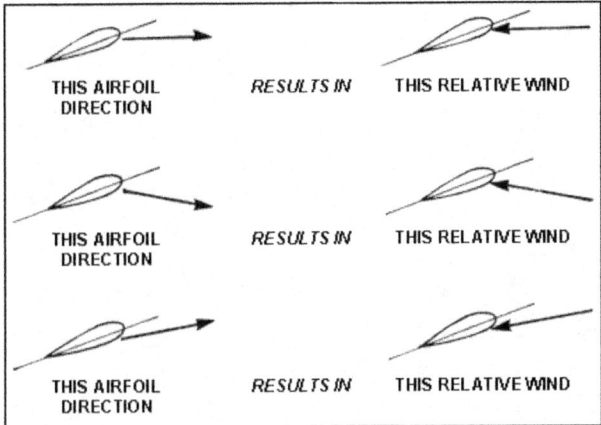

Figure 1-12. Relative wind

ROTATIONAL RELATIVE WIND

1-23. The rotation of rotor blades as they turn about the mast produces rotational relative wind (figure 1-13). The term rotational refers to the method of producing relative wind. Rotational relative wind flows opposite the physical flight path of the airfoil, striking the blade at 90 degrees to the leading edge and parallel to the plane of rotation, and is constantly changing in direction during rotation. Rotational relative wind velocity is highest at blade tips, decreasing uniformly to zero at axis of rotation (center of the mast).

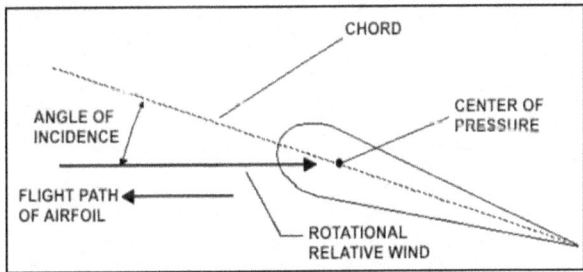

Figure 1-13. Rotational relative wind

INDUCED FLOW (DOWNWASH)

1-24. At flat pitch, air leaves the trailing edge of the rotor blade in the same direction it moved across the leading edge; no lift or induced flow is being produced. As blade pitch angle is increased, the rotor system induces a downward flow of air through the rotor blades creating a downward component of air that is added to the rotational relative wind. Because the blades are moving horizontally, some of the air is displaced downward. The blades travel along the same path and pass a given point in rapid succession. Rotor blade action changes the still air to a column of descending air. This downward flow of air is called induced flow (downwash). It is most pronounced at a hover under no-wind conditions (figure 1-14, page 1-10).

Chapter 1

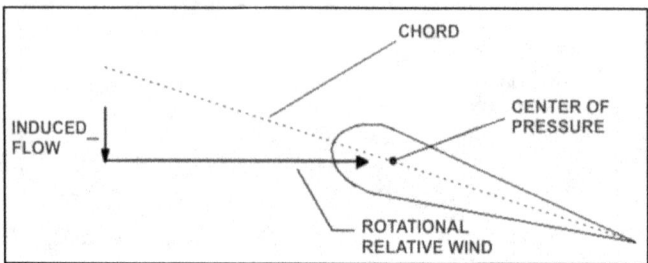

Figure 1-14. Induced flow (downwash)

RESULTANT RELATIVE WIND

1-25. The resultant relative wind (figure 1-15) at a hover is rotational relative wind modified by induced flow. This is inclined downward at some angle and opposite the effective flight path of the airfoil, rather than the physical flight path (rotational relative wind). The resultant relative wind also serves as the reference plane for development of lift, drag, and total aerodynamic force (TAF) vectors on the airfoil. When the helicopter has horizontal motion, airspeed further modifies the resultant relative wind. The airspeed component of relative wind results from the helicopter moving through the air. This airspeed component is added to, or subtracted from, the rotational relative wind, depending on whether the blade is advancing or retreating in relation to helicopter movement. Introduction of airspeed relative wind also modifies induced flow. Generally, the downward velocity of induced flow is reduced. The pattern of air circulation through the disk changes when the aircraft has horizontal motion. As the helicopter gains airspeed, the addition of forward velocity results in decreased induced flow velocity. This change results in an improved efficiency (additional lift) being produced from a given blade pitch setting. Section V further covers this process.

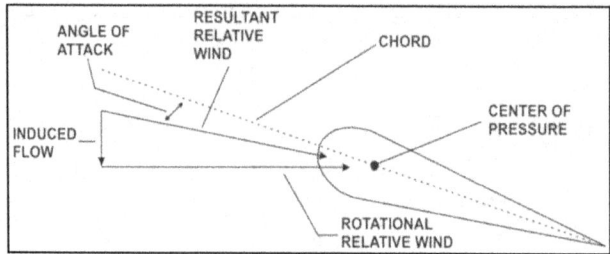

Figure 1-15. Resultant relative wind

UP FLOW (INFLOW)

1-26. Up flow (inflow) is airflow approaching the rotor disk from below as the result of some rate of descent. Up flow also occurs as a result of blades flapping down or an updraft, which alter the AOA.

ROTOR BLADE ANGLES

ANGLE OF INCIDENCE

1-27. Angle of incidence (figure 1-16, page 1-11) is the angle between the chord line of a main or tail rotor blade and the rotational relative wind of the rotor system (tip-path plane). It is a mechanical angle rather than an aerodynamic angle and is sometimes referred to as blade pitch angle. In the absence of induced flow, AOA and angle of incidence are the same. Whenever induced flow, up flow (inflow), or airspeed modifies relative

Aerodynamics of Flight

wind, then AOA is different from angle of incidence. Collective input and cyclic feathering change angle of incidence. A change in angle of incidence changes AOA, which changes the coefficient of lift, thereby changing the lift produced by the airfoil.

Angle of Attack

1-28. AOA (figure 1-16) is the angle between the airfoil chord line and resultant relative wind. AOA is an aerodynamic angle. It can change with no change in angle of incidence. Several factors may change the rotor blade AOA. Aviators control some of those factors; others occur automatically due to rotor system design. Aviators adjust AOA through normal control manipulation; even with no aviator input, however, AOA will change as an integral part of travel of the rotor blade through the rotor-disk arc. This continuous process of change accommodates rotary-wing flight. Aviators have little control over blade flapping and flexing, gusty wind, and/or turbulent air conditions. AOA is one of the primary factors determining amount of lift and drag produced by an airfoil.

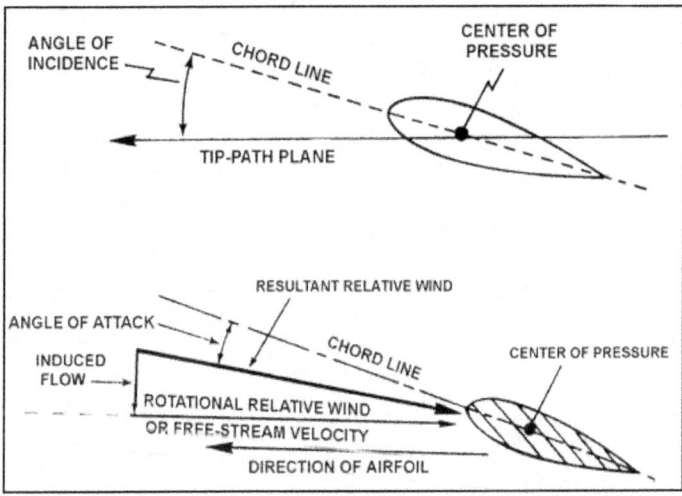

Figure 1-16. Angle of incidence and angle of attack

Effects of Airflow

1-29. As AOA is increased, there is a greater acceleration of air atop the airfoil. This results in a larger pressure differential between the top and bottom of the airfoil, producing a larger aerodynamic force. If AOA is increased beyond a critical angle, flow across the top of the airfoil will be disrupted, boundary layer separation will occur, and a stall results. When this occurs, lift rapidly decreases, drag rapidly increases, and the airfoil ceases to fly.

ROTOR BLADE ACTIONS

Rotation

1-30. Rotation of rotor blades is the most basic movement of the rotor system and produces rotational relative wind. During hovering, rotation of the rotor system produces airflow over the rotor blades. Figure 1-17, page 1-12, illustrates a typical rotor system with an arbitrary rotor diameter of 40 feet and rotor speed of 320 revolutions per minute (RPM) used to demonstrate rotational velocities. In this example, blade tip velocity is 670 feet per second, or 397 knots. At the blade root—nearer the rotor shaft or blade attachment point—blade speed is much less as the distance traveled at the smaller radius is much less. Halfway between the root and tip (point A in figure 1-17) blade speed is 198.5 knots, or one-half tip speed. Blade speed varies according to the distance or radius from the center of the main rotor shaft. While the airspeed differential between root and tip is extreme, the lift differential is more extreme because lift varies as the square of the velocity (see lift equation on page 1-26). As velocity doubles, lift increases four times. The lift at point A in figure 1-17 would be only one-fourth as much as lift at the blade tip—assuming the airfoil shape and AOA are the same at both points.

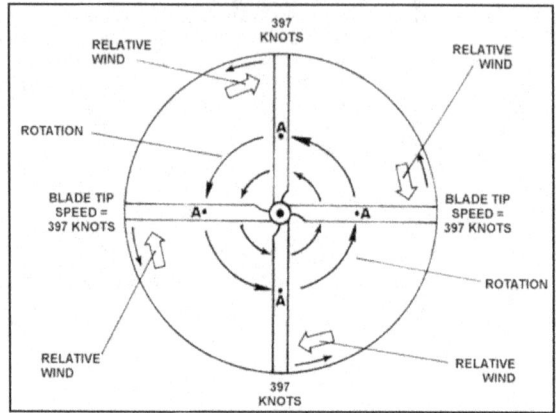

Figure 1-17. Blade rotation and blade speed

FEATHERING

1-31. Feathering is the rotation of the blade about its spanwise axis by collective/cyclic inputs causing changes in blade pitch angle (figure 1-18).

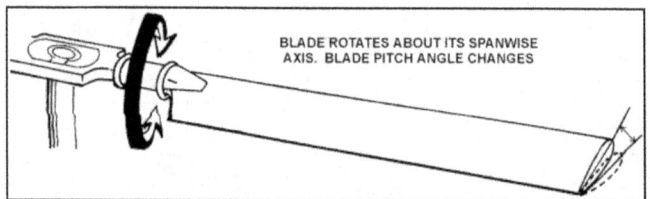

Figure 1-18. Feathering

Collective Feathering

1-32. Collective feathering changes angle of incidence equally and in the same direction on all rotor blades simultaneously. This action changes AOA, which changes coefficient of lift (C_L), and affects overall lift of the rotor system.

Cyclic Feathering

1-33. Cyclic feathering changes angle of incidence differentially around the rotor system. Cyclic feathering creates a differential lift in the rotor system by changing the AOA differentially across the rotor system. Aviators use cyclic feathering to control attitude of the rotor system. It is the means to control rearward tilt of the rotor (blowback) caused by flapping action and (along with blade flapping) counteract dissymmetry of

Aerodynamics of Flight

lift (section V). Cyclic feathering causes attitude of the rotor disk to change but does not change amount of lift the rotor system is producing.

FLAPPING

1-34. The up and down movement of rotor blades about a hinge is called flapping. It occurs in response to changes in lift due to changing velocity or cyclic feathering (figure 1-19). No flapping occurs when the tip-path plane is perpendicular to the mast. The flapping action alone, or along with cyclic feathering, controls dissymmetry of lift (section V). Flapping is the primary means of compensating for dissymmetry of lift.

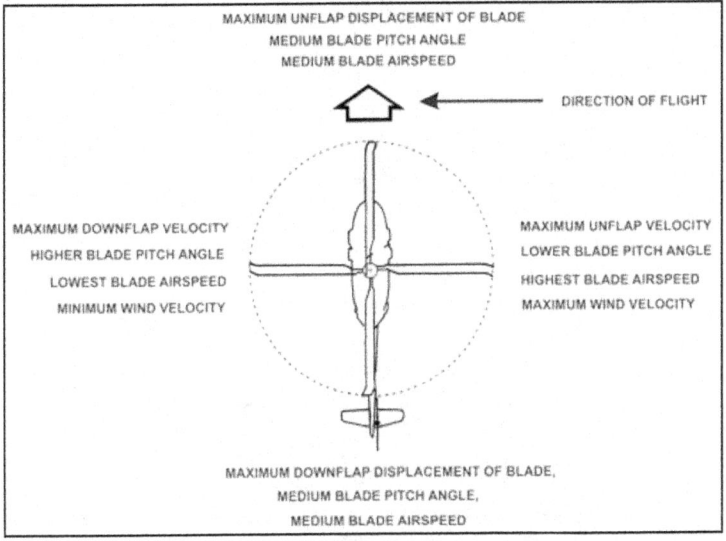

Figure 1-19. Flapping in directional flight

1-35. Flapping also allows the rotor system to tilt in the desired direction in response to cyclic input. See figure 1-20 and figures 1-21 through 1-23, page 1-14 for depictions of flapping as it occurs throughout the rotor disk.

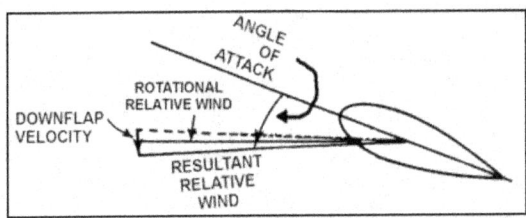

Figure 1-20. Flapping (advancing blade 3 o'clock position)

22 December 2016   TC 3-04.4   1-13

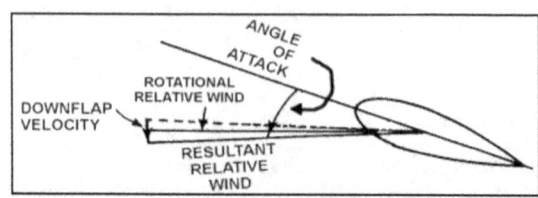

Figure 1-21. Flapping (retreating blade 9-o'clock position)

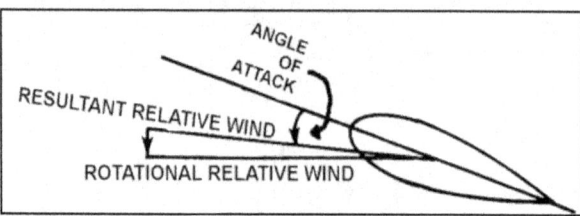

Figure 1-22. Flapping (blade over the aircraft nose)

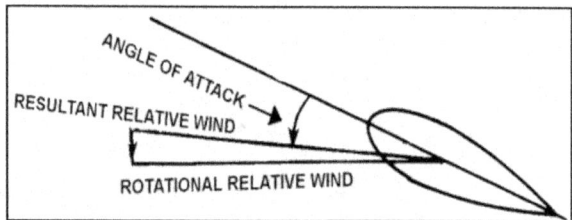

Figure 1-23. Flapping (blade over the aircraft tail)

1-36. In the semirigid rotor system, a blade is not free to flap independently of the other blades because they are affixed through the hub. The blades form one continuous unit moving together on a common teetering hinge. This hinge allows one blade to flap up as the opposite blade flaps down, although blade flex limits the amount of blade flapping. In the fully articulated rotor system, blades flap individually about a horizontal hinge pin. Therefore, each blade is free to move up and down independently from all of the other blades. Aircraft design can reduce excessive flapping in several ways; for example, a forward tilt of the transmission and mast helps minimize flapping and installation of a synchronized elevator or stabilator (UH-60/AH-64) helps maintain the desired fuselage attitude to reduce flapping.

1-37. In a rigid rotor system each blade flaps about flexible sections of the root which are rigidly attached to the rotor hub. This rotor system is mechanically simple, but structurally complex because operating loads must be absorbed in bending rather than through hinges. Rigid rotor systems tend to behave like fully articulated systems through aerodynamics, but lack flapping or lead/lag hinges.

LEAD AND LAG (HUNTING)

1-38. Lead and lag (figure 1-24, page 1-15) are fore and aft movement of the blade in the plane of rotation in response to changes in angular velocity. This rotor blade action can only occur in a fully articulated rotor system, in which the system is equipped with a vertical hinge pin (drag hinge) or elastomeric bearing providing a pivot point for each blade to move independently. In directional flight, pitch angle and the AOA of the blades are constantly changing. These changes in AOA cause changes in blade drag. To prevent undue bending stress on the blades and blade root, the blade is free to move fore and aft in the plane of rotation. The

Aerodynamics of Flight

need to lead and lag is due to the Coriolis force. It is governed by the law of conservation of angular momentum. This law states a body will continue to have the same rotational momentum unless acted on by an outside force. Two factors determine the rotational (angular) momentum—distance of the center of gravity (CG) from the center of rotation and rotational speed. If the CG moves closer to the center of rotation, the rotational speed must increase. If the CG moves farther away from the axis of rotation, rotational velocity will decrease (figure 1-24).

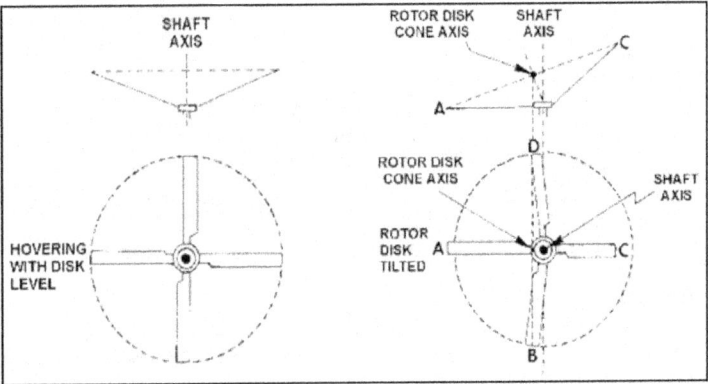

Figure 1-24. Lead and lag

Lead

1-39. As a blade flaps up, the CG of the blade (figure 1-24, point C) moves inboard toward the axis of rotation, producing a smaller radius of travel. The blade speeds up in reaction to this CG change, causing the blade to lead a few degrees ahead of its normal position in the tip-path plane (figure 1-24, point D). This motion relieves stress that would have been imposed on the blade structure.

Lag

1-40. As a blade flaps down, the CG of the blade (figure 1-24, point A) moves outboard away from the axis of rotation, producing a greater radius of travel. The blade slows down in reaction to this CG change, causing the blade to lag a few degrees behind its normal position in the tip-path plane (figure 1-24, point B). This motion relieves stress that would have been imposed on the blade structure.

SEMIRIGID ROTOR SYSTEM

1-41. Because of the design (under slung) of the semirigid rotor system, no change occurs in the travel radius of the CG of the blade associated with blade flapping (figure 1-25, page 1-16). The angular velocity of the blade does not change. Drag does impose significant stresses on the blade roots; a drag brace is normally installed at the blade root to absorb some of these bending forces.

Chapter 1

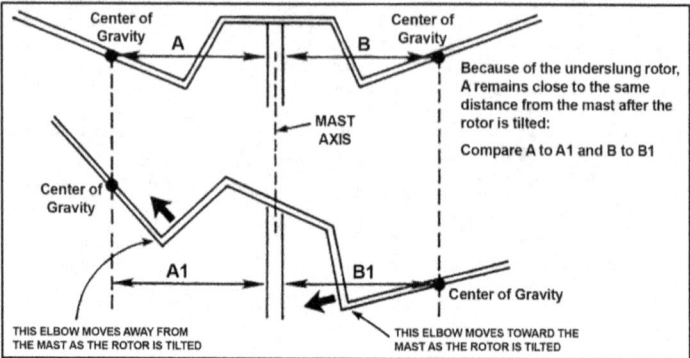

Figure 1-25. Under slung design of semirigid rotor system

RIGID ROTOR SYSTEM

1-42. The rigid rotor system behaves more like a fully articulated rotor system and its characteristics are based on how much flexibility that the blade's construction contains. The blades' flexible root provides physical behavior similar to mechanical or elastomeric hinges allowing the neutralizing forces in the same manner as Lead or Lag and Flapping do with other rotor designs."

1-43. The rigid rotor system, sometimes referred to as bearingless, is mechanically simple, but structurally complex because operating loads must be absorbed in bending rather than through hinges. In this system, the blade roots are rigidly attached to the rotor hub. Rigid rotor systems tend to behave like fully articulated systems through aerodynamics, but lack flapping or lead/lag hinges. Instead, the blades accommodate these motions by bending. They cannot flap or lead/lag, but they can be feathered. These rotor systems offer the best properties of both semirigid and fully articulated systems. The rigid rotor system is very responsive and is usually not susceptible to mast bumping like the semirigid or articulated systems because the rotor hubs are mounted solid to the main rotor mast. This allows the rotor and fuselage to move together as one entity and eliminates much of the oscillation usually present in the other rotor systems. Other advantages of the rigid rotor include a reduction in the weight and drag of the rotor hub and a larger flapping arm, which significantly reduces control inputs. Without the complex hinges, the rotor system becomes much more reliable and easier to maintain than the other rotor configurations. A disadvantage of this system is the quality of ride in turbulent or gusty air. Because there are no hinges to help absorb the larger loads, vibrations are felt in the cabin much more than with other rotor head designs.

HELICOPTER DESIGN AND CONTROL

GYROSCOPIC PRECESSION

1-44. The phenomenon of precession occurs in rotating bodies that manifest an applied force 90 degrees after application in the direction of rotation. Although precession is not a dominant force in rotary-wing aerodynamics, aviators and designers must consider it, as turning rotor systems exhibit some of the characteristics of a gyro. Figure 1-26, page 1-17, illustrates effects of precession on a typical rotor disk when force is applied at a given point. A downward force applied to the disk at point A results in a downward movement of the disk at point B.

Aerodynamics of Flight

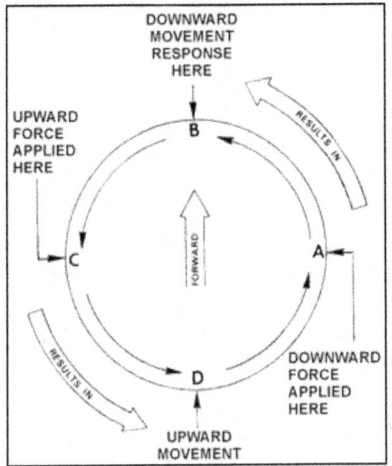

Figure 1-26. Gyroscopic precession

1-45. Table 1-2 shows reactions to forces applied to a spinning rotor disk by control input or wind gusts.

Table 1-2. Aircraft reaction to forces

Force Applied to Rotor Disk	Aircraft Reaction
Up at nose	Roll right
Up at tail	Roll left
Up on right side	Nose up
Up on left side	Nose down

1-46. This behavior explains some fundamental effects occurring during various helicopter maneuvers. For example, the helicopter behaves differently when rolling into a right turn than when rolling into a left turn. During roll into a right turn, the aviator must correct for a nose-down tendency to maintain altitude. This correction is required because precession causes a nose-down tendency. During a roll into a left turn, precession causes a nose-up tendency. Aviator input required to maintain altitude is different during a left versus right turn as gyroscopic precession acts in opposite directions.

ROTOR HEAD CONTROL

Cyclic and Collective Pitch

1-47. Aviator inputs to collective and cyclic pitch controls are transmitted to the rotor blades through a complex system. This system consists of levers, mixing units, input servos, stationary and rotating swashplates, and pitch-change arms (figure 1-27, page 1-18). In its simplest form, movement of collective pitch control causes stationary and rotating swashplates mounted centrally on the rotor shaft to rise and descend. The movement of cyclic pitch control causes the swashplates to tilt; the direction of tilt is controlled by the direction in which the aviator moves the cyclic (figure 1-28, page 1-18).

Chapter 1

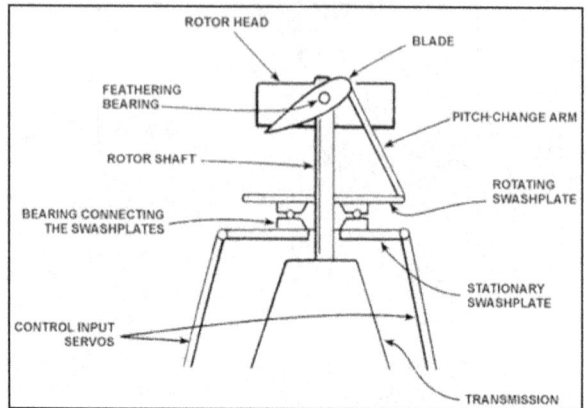

Figure 1-27. Rotor head control systems

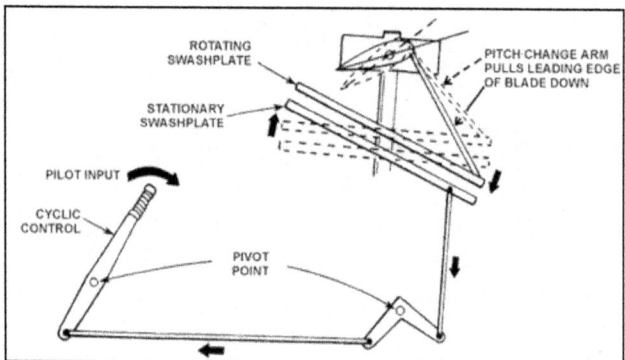

Figure 1-28. Stationary and rotating swashplates tilted by cyclic control

Tilted Swashplate Assembly

1-48. Figure 1-29, page 1-19, illustrates a swashplate tilted 2 degrees at two positions, points B and D. Points A and C form the axis about which the tilt occurs. At that axis, the swashplate remains at zero degrees. When the swashplate is moved, pitch-change arms transmit the resulting motion change to the rotor blade. As the pitch-change arms move up and down with each rotation of the swashplate, blade pitch constantly increases or decreases. If the aviator applies cyclic control to tilt the rotor, adding collective pitch does not change the tilt of the swashplate and rotor. It simply moves the swashplate upward so pitch is increased equally on all blades simultaneously, thereby increasing AOA and total lift.

Aerodynamics of Flight

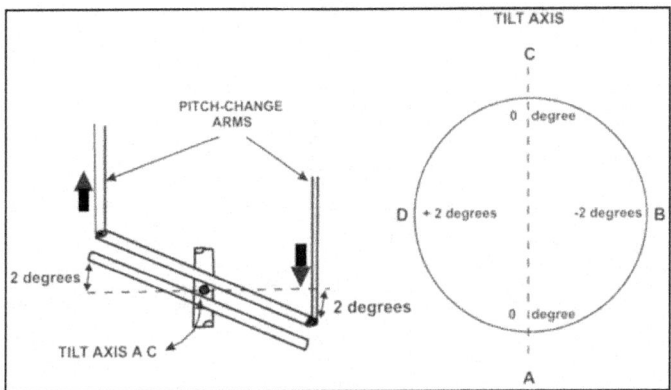

Figure 1-29. Stationary and rotating swashplates tilted in relation to mast

Pitch-Change Arms

1-49. Figure 1-30 illustrates how pitch-change arms move up and down on the tilted swashplate. The rate of vertical change throughout the rotation is not uniform. Vertical movement is larger during the 30 degrees of rotation at point A than at points B and C. This variation repeats during each 90 degrees of rotation. The rate of vertical movement is lowest at the low and high points of the swashplate and highest when the pitch-change arms pass by the tilt axis of the swashplate.

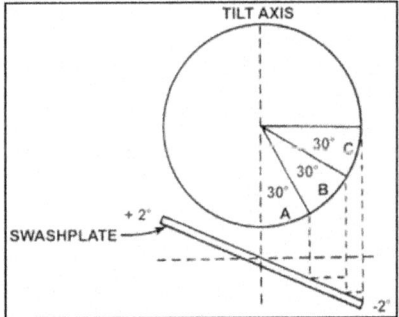

Figure 1-30. Pitch-change arm rate of movement over 90 degrees of travel

Cyclic Pitch Change

1-50. Figure 1-31, page 1-20, shows a change in cyclic pitch (cyclic feathering) causing rotor blades to climb from point A to point B then dive or descend from point B to point A. In this way, the rotor is tilted in the direction of desired flight.

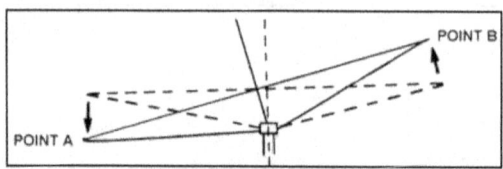

Figure 1-31. Rotor flapping in response to cyclic input

1-51. To pass through points A and B, the blades must flap up and down on a hinge or teeter on a trunnion. At the lowest flapping point (point A), the blades would appear to be at their lowest pitch angle; at the highest flapping point (point B), they would be at their highest pitch angle. If only aerodynamic considerations were involved, this might be true. However, gyroscopic precession (figure 1-26, page 1-17) causes these points to be separated by 90 degrees of rotation.

1-52. A cyclic movement decreases blade pitch at one point in the rotor disk while increasing blade pitch by the same amount 180 degrees of travel later. A decrease in lift resulting from a decrease in blade pitch angle and AOA causes the blade to flap down; the blade reaches its maximum downflapping displacement 90 degrees later in the direction of rotation. An increase in lift resulting from an increase in blade pitch angle and AOA causes the blade to flap up; the blade reaches its maximum upflapping displacement 90 degrees later in the direction of rotation. Figure 1-32, page 1-21, shows the resulting change to the rotor disk's attitude. The cyclic pitch causing blade flap must be placed on the blades 90 degrees of rotation before the lowest and highest flap are desired. This 90 degrees of phase lag due to gyroscopic precession is accounted for when rotors are designed, and it ensures when the cyclic is pushed forward, the action tilts the swashplate assembly to place the cyclic pitch accordingly. To tilt the rotor disk forward, the lowest cyclic pitch on the blade needs to be over the right side of the helicopter and the highest cyclic pitch over the left side. The rotor always tilts in the direction in which the aviator moves the cyclic.

Typical Design Features

1-53. Figure 1-33, page 1-21, illustrates a typical design feature used in most four-bladed rotor systems offsetting cyclic control input 90 degrees from where the aviator desires rotor tilt. Rotor control input locations are the left lateral servo (point A), right lateral servo (point B), and fore and aft servo (point C). Each servo is offset 45 degrees from the position corresponding to its name. The fore and aft input servo, for example, is not located at the nose or tail position but at the right front about halfway between the nose and 3 o'clock position. Similarly, the left lateral servo is located halfway between the nose and 9 o'clock position. The right lateral servo is halfway between the tail and 3 o'clock position. Locations of the input servos account for part of the offset the aviator needs to correct for gyroscopic precession. In addition, the rotor blade has a pitch-change horn extending ahead of the blade in the plane of rotation about 45 degrees. A connecting rod, called a pitch-change rod, transmits aviator control inputs from the input servos to the pitch-change horn. The design of the pitch-change horn, coupled with placement of the servo and tilt of the swashplate, provides the total offset.

Aerodynamics of Flight

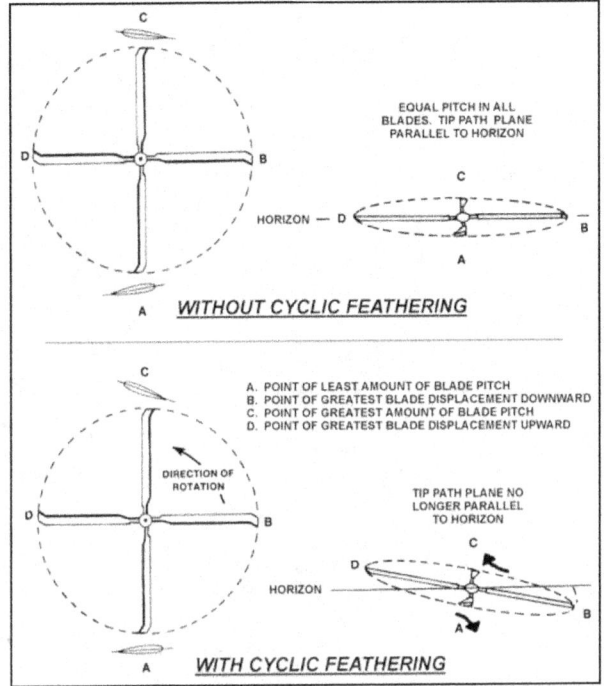

Figure 1-32. Cyclic feathering

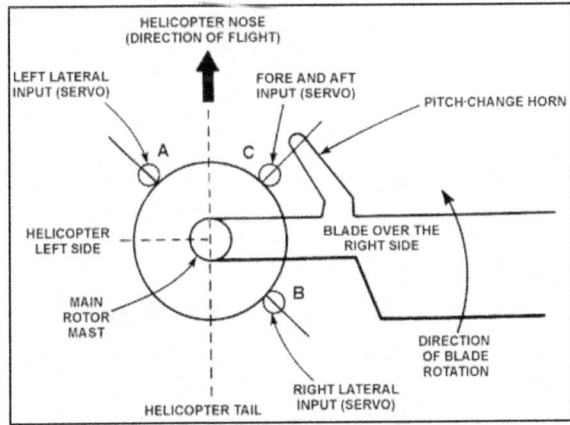

Figure 1-33. Input servo and pitch-change horn offset

Chapter 1

Cyclic Pitch Variation

1-54. Figure 1-34 illustrates typical cyclic pitch variation for a blade through one revolution with cyclic pitch control full forward. Degrees shown are for a typical aircraft rotor system; figures would vary with the type of helicopter. As described in the previous paragraph, the input servos and pitch-change horns are offset. With cyclic pitch control in the full forward position, the blade pitch angle is highest at the 9 o'clock position and lowest at the 3 o'clock position. The pitch angle begins decreasing as it passes the 9 o'clock position and continues to decrease until it reaches the 3 o'clock position. The pitch begins to increase and reaches the maximum pitch angle at the 9 o'clock position. Blade pitch angles over the nose and tail are about equal.

1-55. Figure 1-34 shows blades reach a point of lowest flapping over the nose 90 degrees in the direction of rotation from the point of lowest pitch angle. Highest flapping occurs over the tail 90 degrees in the direction of rotation from the point of the highest pitch angle. Simply stated, the force (pitch angle) causing blade flap must be applied to the blade 90 degrees of rotation before the point where the aviator desires maximum blade flap.

1-56. A pattern similar to figure 1-34 could be constructed for other cyclic positions in the circle of cyclic travel. In each case, the same principles apply. Points of highest and lowest flapping are located 90 degrees in the direction of rotation from the points of highest and lowest blade pitch.

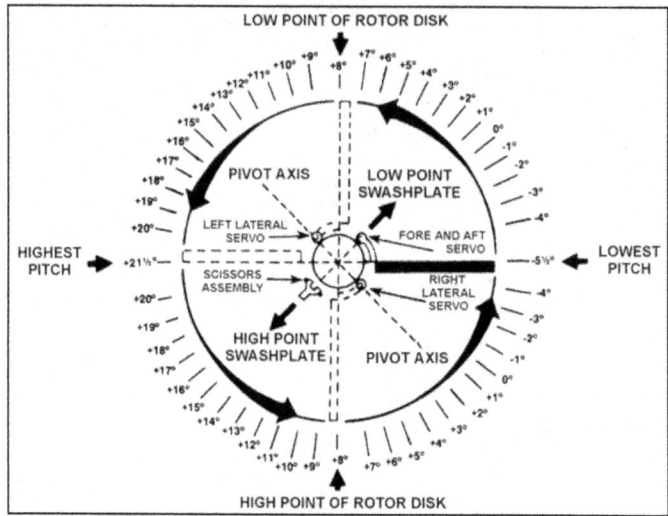

Figure 1-34. Cyclic pitch variation–full forward, low pitch

FUSELAGE HOVERING ATTITUDE

Single-Rotor Helicopter

1-57. The design of most fully articulated rotor systems includes an offset between the main rotor mast and blade attachment point. Centrifugal force acting on the offset tends to hold the mast perpendicular to the tip-path plane (figure 1-35, page 1-23). When the rotor disk is tilted left to counteract the translating tendency, the fuselage follows the main rotor mast and hangs slightly low on the left side.

1-58. A fuselage suspended under a semirigid rotor system remains level laterally unless the load is unbalanced or the tail rotor gearbox is lower than the main rotor (figure 1-36, page 1-23). The fuselage remains level because there is no offset between the rotor mast and the point where the rotor system is attached to the mast (trunnion bearings). Because trunnion bearings are centered on the mast, the mast does not tend to follow the tilt of the rotor disk during hover. In addition, the mast does not tend to remain perpendicular to the tip-path plane as it does with a fully articulated rotor system. Instead, the mast tends to hang vertically under the trunnion bearings, even when the rotor disk is tilted left to compensate for translating tendency (figure 1-36, point B). Because the mast remains vertical, the fuselage hangs level laterally unless other forces affect it.

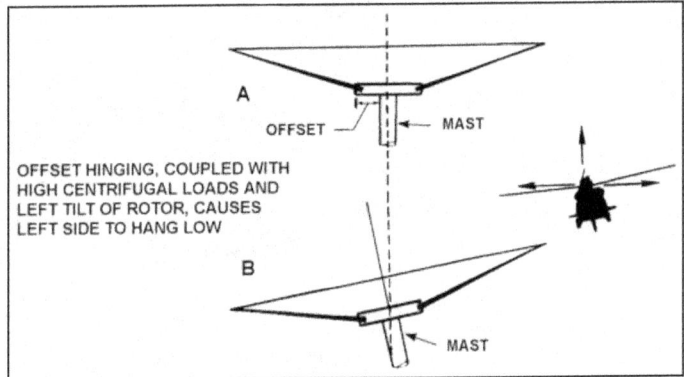

Figure 1-35. Fully articulated rotor system

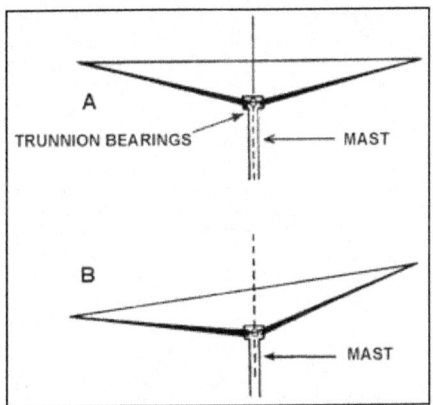

Figure 1-36. Semirigid rotor system

1-59. When there is forward tilt of the mast, the tail rotor gearbox is probably lower than the main rotor. Main rotor thrust above tail rotor thrust to the right causes the fuselage to tilt laterally left (figure 1-37, page 1-24). Although main rotor thrust to the left is equal to tail rotor thrust to the right, it acts at a greater distance from the CG, creating a greater turning moment on the fuselage. This is more pronounced in helicopters with semirigid rotor systems than those with fully articulated rotor systems. Tail rotor thrust acting at the plane of rotation of the main rotor would not change the attitude of the fuselage. The main rotor mast in semirigid and fully articulated rotor systems may be designed with a forward tilt relative to the fuselage. During forward flight, forward tilt provides a level longitudinal fuselage attitude, resulting in reduced parasite drag; during hover, it results in a tail-low fuselage attitude.

Chapter 1

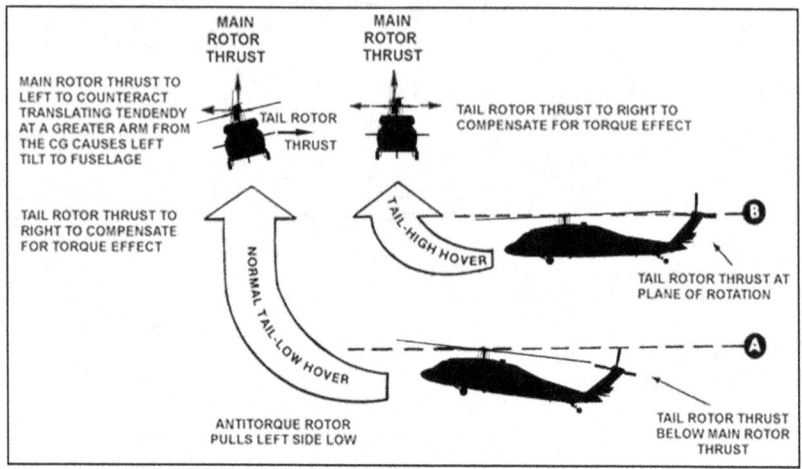

Figure 1-37. Effect of tail-low attitude on lateral hover attitude

Tandem-Rotor Helicopter

1-60. In tandem-rotor helicopters, the forward and aft rotor systems are tilted forward due to transmission mounting design. This tilt helps decrease excessive nose-low attitudes in forward flight and allows the aircraft to ground or water taxi forward. Most tandem-rotor helicopters hover at a nose-high attitude of about 5 degrees. Some models automatically compensate for this nose-high attitude through automatic programming of the rotor systems.

PENDULAR ACTION

1-61. The fuselage of the helicopter has considerable mass and is suspended from a single point (single-rotor helicopters). It is free to oscillate laterally or longitudinally like a pendulum. Normally, the fuselage follows rules governing pendulums, balance, and inertia. Rotor systems, however, follow rules governing aerodynamics, dynamics, and gyroscopes. These two unrelated systems have been designed to work well together, in spite of apparent conflict. Other factors, such as overcontrolling, cyclic-control response, and shift of attitude, affect the relationship of the rotor system and fuselage.

Overcontrolling

1-62. Overcontrolling occurs when the aviator moves the cyclic control stick, causing rotor tip-path changes not reflected in corresponding fuselage-attitude changes. Correct cyclic control movements (free of overcontrol) cause the rotor tip-path and fuselage to move in unison.

Cyclic Control Response

1-63. The rotor response to cyclic control input on a single-rotor helicopter has no lag. Rotor blades respond instantly to the slightest touch of cyclic control. The fuselage response to lateral cyclic is noticeably different from the response to fore and aft cyclic applications. Normally, considerably more fore and aft cyclic movement is required to achieve the same fuselage response as achieved from an equal amount of lateral cyclic. This is not a lag in rotor response; rather as figure 1-38, page 1-25, shows, it is due to more fuselage inertia around the lateral axis than around the longitudinal axis. For single-rotor helicopters, the normal corrective device for the lateral axis is the addition of a synchronized elevator or stabilator attached to the tail boom. This device produces lift forces keeping the fuselage of the helicopter in proper alignment with

Aerodynamics of Flight

the rotor at normal flight airspeed. This alignment helps reduce blade flapping and extends the allowable CG range of the helicopter; however, it is ineffective at slow airspeeds.

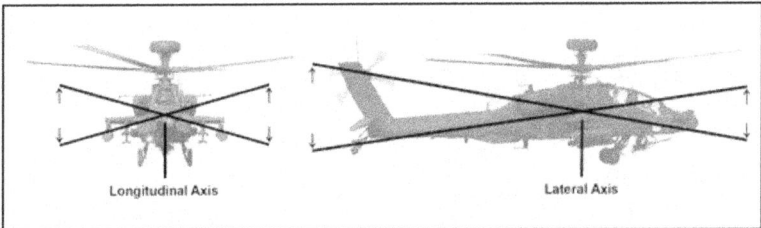

Figure 1-38. Cyclic control response around the lateral and longitudinal axes

Shift of Attitude

1-64. Fuel cells normally have a slight aft CG. As fuel is used, a slight shift to a more nose-low attitude occurs. Because of fuel expenditure and lighter fuselage, cruise attitudes tend to shift slightly lower. As fuel loads are reduced, drag affects the lighter fuselage more, resulting in a slight shift to a more nose-down attitude during flight.

SECTION III – IN-FLIGHT FORCES

TOTAL AERODYNAMIC FORCE

1-65. As air flows around an airfoil, a pressure differential develops between the upper and lower surfaces. The differential, combined with air resistance to passage of the airfoil, creates a force on the airfoil. This is known as TAF (figure 1-39). TAF acts at the center of pressure on the airfoil and is normally inclined up and rear. TAF, sometimes called resultant force, may be divided into two components, lift and drag.

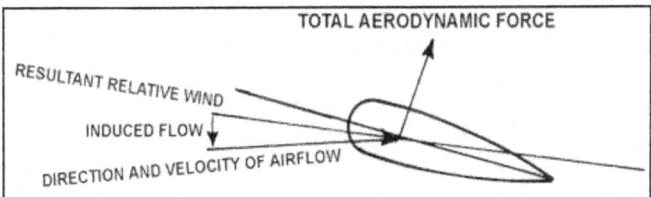

Figure 1-39. Total aerodynamic force

LIFT AND LIFT EQUATION

1-66. Lift is the component of the airfoil's TAF perpendicular to the resultant relative wind (figure 1-40, page 1-26).

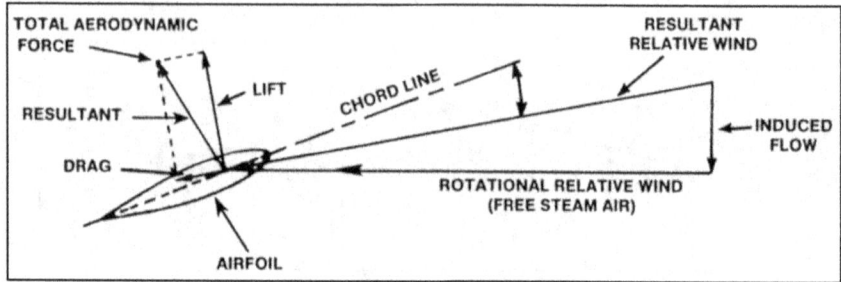

Figure 1-40. Forces acting on an airfoil

1-67. The illustration of the lift equation, accompanied by a simple explanation, helps understanding of how lift is generated. The point is to understand what an aviator can or cannot change in the equation.

Lift Equation

$L = C_L \times \rho/2 \times S \times V^2$

Where—
L = lift force
C_L = coefficient of lift
$\rho/2 = .5 \times \rho$ (rho) = density of the air (in slugs per cubic foot)
S = surface area (in square feet)
V^2 = airspeed (in feet per second)

1-68. The shape or design of the airfoil and AOA determine the coefficient of lift. Aviators have no control over airfoil design. However, they do have direct control over AOA. The aviator cannot affect ρ (rho) or S (surface area of the airfoil). With respect to V (relative wind velocity or airspeed), an increase in rotor RPM has a greater effect on lift than an increase in airspeed.

DRAG

1-69. Drag is the component of the airfoil's TAF parallel to the resultant relative wind (figure 1-40). Drag is the force opposing the motion of an airfoil through the air.

DRAG EQUATION

1-70. The illustration of the drag equation accompanied by a simple explanation (in addition to the lift equation) helps understanding of how drag is generated. The point is to understand what an aviator can and cannot change.

1-71. The shape or design of the airfoil and AOA largely determine the coefficient of drag. The aviator has no control over airfoil design but has direct control over AOA. This is one of two elements of the drag equation the aviator can change. However, an aviator cannot affect ρ (rho) which is density of the air. S represents surface area of the airfoil, a design factor also unaffected by aviator input. Finally, V represents relative wind velocity or airspeed and is the only other factor an aviator can change.

TYPES OF DRAG

1-72. Total drag acting on a helicopter is the sum of the three types of drag—parasite, profile, and induced drag. Curve D in figure 1-41, page 1-27, shows total drag and represents the sum of the other three curves.

Aerodynamics of Flight

Drag Equation

$D = C_D \times \rho/2 \times S \times V^2$

Where—
D = drag force
CD = coefficient of drag
$\rho/2 = .5 \times \rho$ (rho) = density of the air (in slugs per cubic foot)
S = surface area (in square feet)
V2 = airspeed (in feet per second)

Parasite Drag

1-73. Parasite drag is incurred from the nonlifting portions of the aircraft. It includes form drag, skin friction, and interference drag associated with the fuselage, engine cowlings, mast and hub, landing gear, wing stores, external load, and rough finish paint. Parasite drag increases with airspeed and is the dominant type at high airspeeds. Curve A in figure 1-41 shows parasite drag.

Profile Drag

1-74. Profile drag is incurred from frictional resistance of the blades passing through the air. It does not change significantly with AOA of the airfoil section but increases moderately at high airspeeds. At high airspeeds, profile drag increases rapidly with onset of blade stall or compressibility. Curve B in figure 1-41 shows profile drag.

Induced Drag

1-75. Induced drag is incurred as a result of production of lift. Higher angles of attack, which produce more lift, also generate downward velocities and vortices that increase induced drag. In rotary-wing aircraft, induced drag decreases with increased aircraft airspeed. Curve C in figure 1-41 shows induced drag.

DRAG/POWER/AIRSPEED RELATIONSHIP

1-76. Figure 1-41 illustrates the relationship between drag, power, and airspeed.

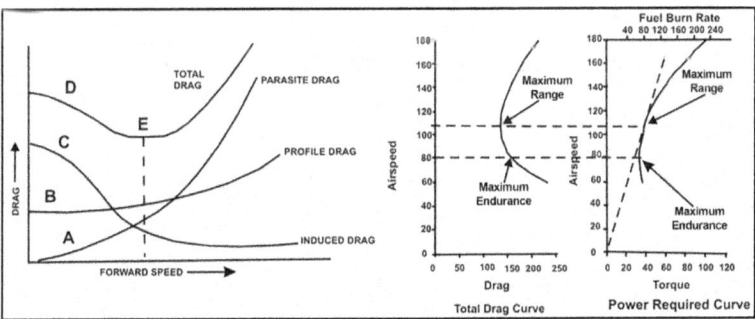

Figure 1-41. Drag and airspeed relationship

Aircraft Performance and Power Curves

1-77. Drag is a major component used in conjunction with flight testing and performance data to develop performance planning charts found in operator manuals which explains the similar appearance in these charts (figure 1-41). Performance planning charts allow aviators to compute expected performance data based on various weather conditions, loading configurations, and airspeeds. This data is required to determine predicted airspeeds, torques, and fuel flows during various mission profiles. Key information required for performance planning includes maximum range airspeed, maximum endurance airspeed, maximum rate-of-climb airspeed and the torques and fuel flows associated with those airspeeds.

1-78. Maximum range airspeed is an airspeed that should allow the helicopter to fly the furthest distance. It is determined by flying where airspeed intersects the lowest amount of total drag (point E on figure 1-41, page 1-27). However, due to flight testing and aircraft performance, cruise charts are used to determine torque and fuel flows required to maintain that airspeed. Because cruise charts are not drag charts, it can be noted the lowest point of a drag chart does not necessarily match the lowest point of the power required curve in a cruise chart.

1-79. Maximum Endurance airspeed is an airspeed that allows the helicopter to remain flying the most amount of time. It can be found on the power required curve of the cruise chart where power required is at its lowest and not necessarily where total drag is lowest on the drag chart.

1-80. Maximum rate-of-climb airspeed is maximum endurance airspeed combined with maximum torque available to achieve the fastest rate of climb.

CENTRIFUGAL FORCE AND CONING

1-81. A helicopter rotor system depends primarily on rotation to produce relative wind, which develops the aerodynamic force required for flight. This action subjects the rotor system to forces peculiar to all rotating masses. One of the forces produced is centrifugal force. The apparent force tends to make rotating bodies move away from the center of rotation. The rotating blades of a helicopter produce very high centrifugal loads on the hub and blade attachment assemblies. In rotary-wing aircraft, this is the dominant force affecting the rotor system; all other forces act to modify it. As a rotor system begins to turn, the blades begin to rise from the static position because of centrifugal force. At operating speed, the blades extend straight out although the rotor system is at flat pitch (zero degree angle of incidence) and are not producing lift. As the aircraft develops lift during takeoff and flight, the blades rise above the straight-out position and assume a coned position. The amount of coning depends on RPM, gross weight, and gravitational (G) forces experienced during flight. Figure 1-42, page 1-29, illustrates the various positions of a rotor blade in the static position, at flat pitch, and when generating lift. Excessive coning can occur if RPM is too low, gross weight is too high, an aircraft is flying in turbulent air, or the G-forces experienced are too high. This excessive coning can cause undesirable stresses on the components and a decrease in lift because of a decrease in effective disk area (figure 1-43, page 1-29).

Aerodynamics of Flight

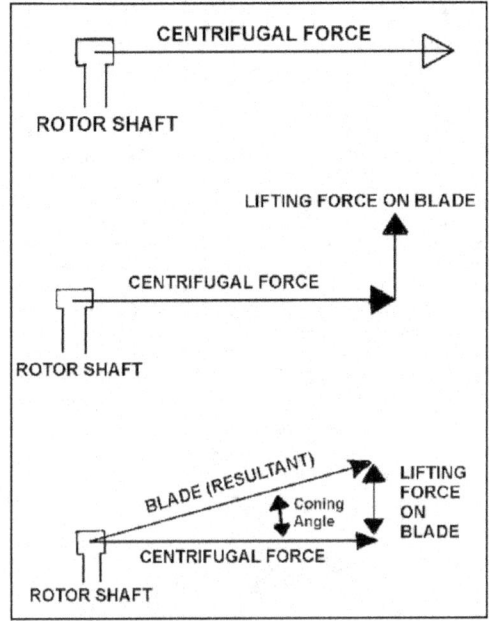

Figure 1-42. Effects of centrifugal force and lift

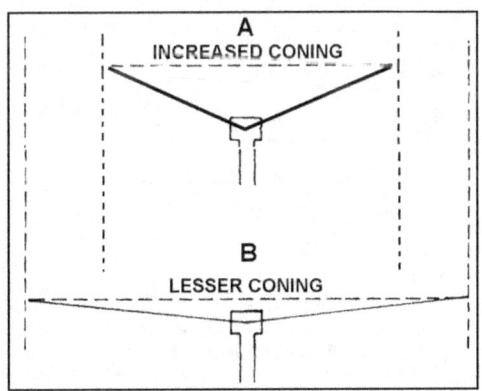

Figure 1-43. Decreased disk area (loss of lift caused by coning)

TORQUE REACTION AND ANTITORQUE ROTOR (TAIL ROTOR)

1-82. According to Newton's law of action/reaction, action created by the turning rotor system will cause the fuselage to react by turning in the opposite direction. The fuselage reaction to torque turning the main rotor is torque effect. Torque must be counteracted to maintain control of the aircraft; the antitorque rotor does this (figure 1-44). In the tandem rotor or coaxial helicopters, the two rotor systems turn in opposite directions, effectively canceling the torque effect. Most rotary-wing aircraft have a single main rotor and require a tail rotor or other means to counter the torque effect. As the initial action is generated by engine power (torque) turning the main rotor system, this torque will necessarily vary with power applied or the maneuver performed. The tail rotor is designed as a variable-pitch, antitorque rotor to accommodate the varying effects of such a system. The tail rotor is usually driven by the main transmission through a drive shaft arrangement leading to its position at the end of the tailboom. The engine power required to motor and control the tail rotor can be significant. The aviator must consider this during performance planning for varying conditions and situations. It is easy to understand why various emergency procedures have been written to compensate for problems such as loss of engine power, insufficient engine power, and tail rotor malfunction. Most American-built single-rotor helicopters turn the main rotor in a counterclockwise direction; therefore, the application of right pedal decreases pitch in the tail rotor and creates less thrust, allowing the nose of the aircraft to turn right. The opposite is true for application of left pedal.

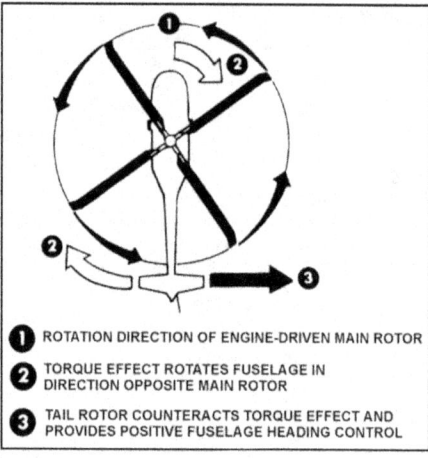

Figure 1-44. Torque reaction

HEADING CONTROL

Single-Rotor Helicopters

1-83. In addition to counteracting torque, the tail rotor and its control linkage allow the aviator to control the helicopter heading during taxi, hover, and sideslip operations on takeoffs and approaches. Applying more pedal than needed to counteract torque causes the nose of the helicopter to swing in the direction of pedal movement (left pedal to the left). Applying less pedal than needed causes the helicopter to turn in the direction of torque (nose swings to the right). Aviators must use the antitorque pedals to maintain a constant heading at a hover or during a takeoff or an approach. They apply just enough pitch on the tail rotor to neutralize torque and to hold a slip.

Aerodynamics of Flight

1-84. Heading control in forward trimmed flight is normally accomplished by cyclic control with a coordinated bank and turn to the desired heading. The antitorque pedal must be applied when power changes are made.

Tandem-Rotor Helicopters

1-85. Heading control is accomplished in tandem-rotor helicopters by differential lateral tilting of the rotor disks. When the directional pedal (right or left) is applied, the forward rotor disk tilts in the same direction and the aft rotor disk tilts in the opposite direction. The result is a hovering turn around a vertical axis, midway between the rotors.

1-86. Heading control in forward flight is accomplished by coordinated use of lateral cyclic tilt on both rotors for roll control and differential cyclic tilt on the rotors for yaw control. Only small changes in pedal trim are required for changes in longitudinal speed trim or during descents, climbs, and autorotations.

BALANCE OF FORCES

1-87. Newton's law of acceleration states the force required to produce a change in motion of a body is directly proportional to its mass and rate of change in its velocity. This means motion is started, stopped, or changed when forces acting on the body become unbalanced. Rate of change (acceleration) depends on the magnitude of the unbalanced force and on the mass of the body to which it is applied. This principle is the basis for all helicopter flight—vertical, forward, rearward, sideward, or hovering. In each case, total force generated by a rotor system is always perpendicular to the tip-path plane (figure 1-45 through figure 1-48). For this discussion, this force is divided into two components, lift and thrust. The lift component supports aircraft weight while the thrust component acts horizontally to accelerate or decelerate the helicopter in the desired direction. Aviators direct thrust in a desired direction by tilting the tip-path plane. At a hover in a no-wind condition, all opposing forces are in balance; they are equal and opposite. Therefore, lift and weight are equal, resulting in the helicopter remaining stationary (figure 1-45).

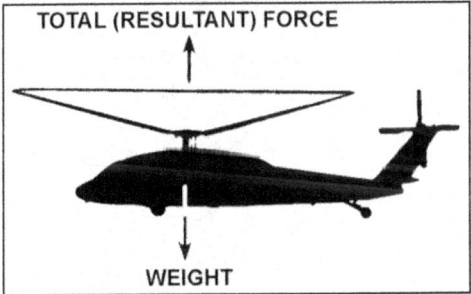

Figure 1-45. Balanced forces; hovering with no wind

1-88. To make the helicopter move in some direction, a force must be applied to cause an unbalanced condition. Figure 1-46, page 1-32, illustrates an unbalanced condition in which the aviator has changed the attitude of the rotor disk creating a lift and thrust vector, resulting in a total force forward of the vertical. No parasite drag is shown as the aircraft has not started to move forward.

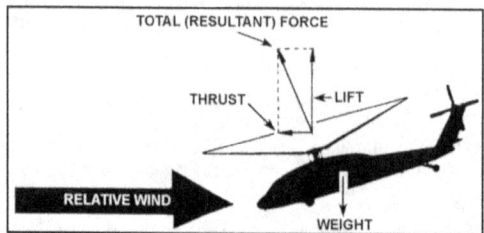

Figure 1-46. Unbalanced forces causing acceleration

1-89. As the aircraft begins to accelerate in the direction of applied thrust, parasite drag develops. When parasite drag increases to be equal to thrust, the aircraft no longer accelerates because the forces are again in balance (figure 1-47) as the aircraft has achieved steady-state (unaccelerated) flight.

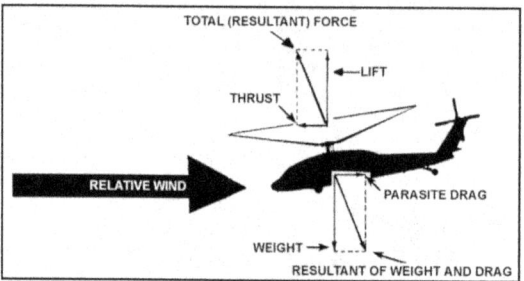

Figure 1-47. Balanced forces; steady-state flight

1-90. To return the aircraft to a hover, the aviator changes the disk attitude to unbalance the forces (figure 1-48). By tilting the rotor disk aft, the thrust force acts in the same direction as parasite drag and airspeed decreases.

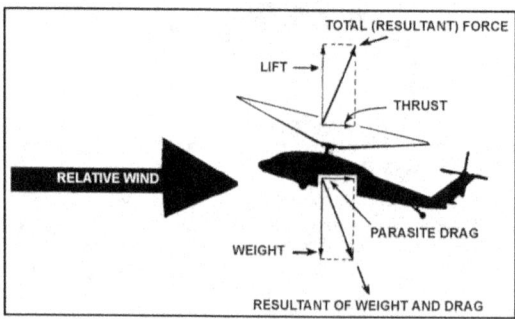

Figure 1-48. Unbalanced forces causing deceleration

Aerodynamics of Flight

SECTION IV – HOVERING

AIRFLOW IN HOVERING FLIGHT

1-91. An increase of blade pitch (through application of collective) that increases AOA, generates the additional lift necessary to hover (figure 1-49). For a helicopter to hover, lift produced by the rotor system must equal the total weight of the helicopter. In a no-wind condition, the tip-path plane remains horizontal. As forces of lift and weight are in balance during stationary hover, those forces must be altered—through application of collective—to either climb or descend vertically.

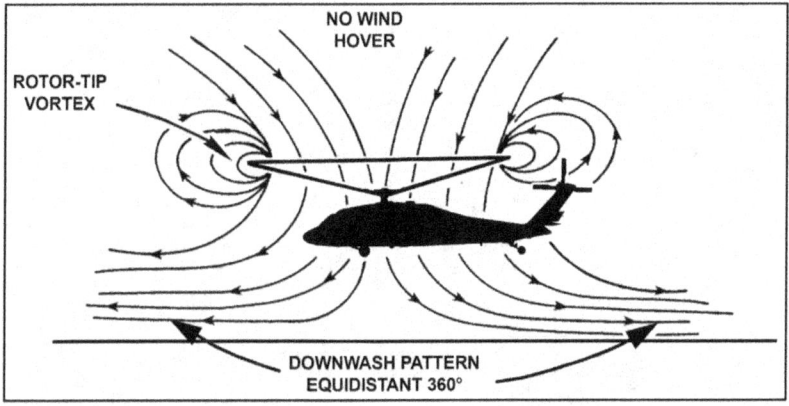

Figure 1-49. Airflow in hovering flight

1-92. At a hover, the rotor-tip vortex (air swirl at the tip of the rotor blades) reduces effectiveness of the outer blade portions. Vortices of the preceding blade affect the lift of any other blade in the rotor system. When maintaining a stationary hover, this continuous creation of vortices—combined with the ingestion of existing vortices—is the primary cause of high power requirements for hovering. Rotor-tip vortices are part of the induced flow and increase induced drag.

1-93. During hover, rotor blades move large amounts of air through the rotor system in a downward direction. This movement of air also introduces another element—induced flow—into relative wind, which alters the AOA of the airfoil. If there is no induced flow, relative wind is opposite and parallel to the flight path of the airfoil. With a downward airflow altering the relative wind, the AOA is decreased so less aerodynamic force is produced. This change requires the aviator to increase collective pitch to produce enough aerodynamic force to hover.

GROUND EFFECT

GROUND EFFECT EFFICIENCY

1-94. Ground effect is the increased efficiency of the rotor system caused by interference of the airflow when near the ground. Ground effect permits relative wind to be more horizontal, lift vector to be more vertical, and induced drag to be reduced. These allow the rotor system to be more efficient. The aviator achieves maximum ground effect when hovering over smooth hard surfaces. When the aviator hovers over such terrain as tall grass, trees, bushes, rough terrain, and water, maximum ground effect is reduced. Two reasons for this phenomenon are induced flow and vortex generation.

Chapter 1

Induced Flow

1-95. Proximity of the helicopter to the ground interrupts airflow under the helicopter by altering the velocity of induced flow. Induced flow velocity is reduced when closer to the ground, which in turn, increases AOA, reduces the amount of induced drag, allows a more vertical lift vector, and increases rotor system efficiency.

Vortex Generation

1-96. When operating close enough to a surface for ground effect to exist, the downward and outward flow of air tends to restrict vortex generation. The smaller vortexes result in the outboard portion of each blade becoming more efficient and reduce overall system turbulence caused by ingestion and recirculation of the vortex pattern.

CATEGORIES

1-97. Ground effect is categorized in two ways—in ground effect (IGE) and out of ground effect (OGE). Both are critical elements on a rotary-wing performance planning card (PPC).

In-Ground Effect

1-98. Rotor efficiency is increased by ground effect to a height of about one rotor diameter (measured from the ground to the rotor disk) for most helicopters. Figure 1-50 shows IGE hover and induced flow reduced. This increase in AOA requires a reduced blade pitch angle. This reduces the power required to hover IGE.

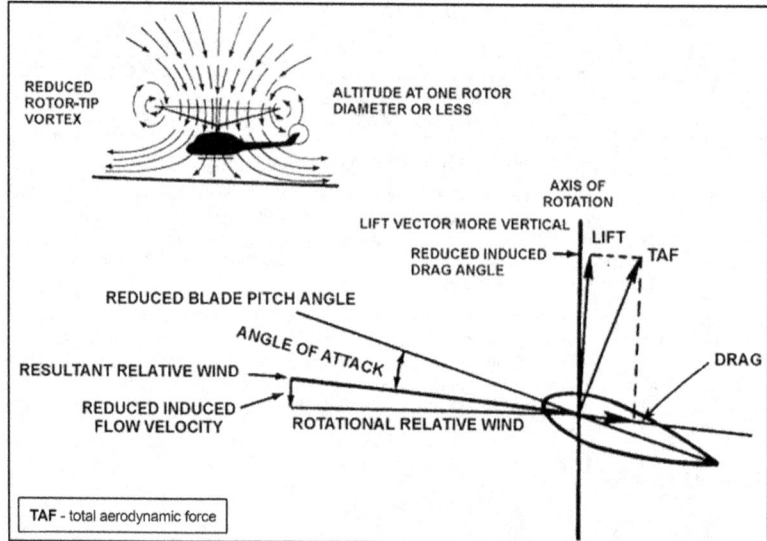

Figure 1-50. In ground effect hover

Out-of-Ground Effect

1-99. The benefit of placing the helicopter near the ground is lost above IGE altitude. Above this altitude, the power required to hover remains nearly constant, given similar conditions (such as wind). Figure 1-51, page 1-35, shows OGE hover. Induced flow velocity is increased causing a decrease in AOA. A higher blade pitch angle is required to maintain the same AOA as in IGE hover. The increased pitch angle also creates more drag. More power to hover OGE than IGE is required by this increased pitch angle and drag.

Aerodynamics of Flight

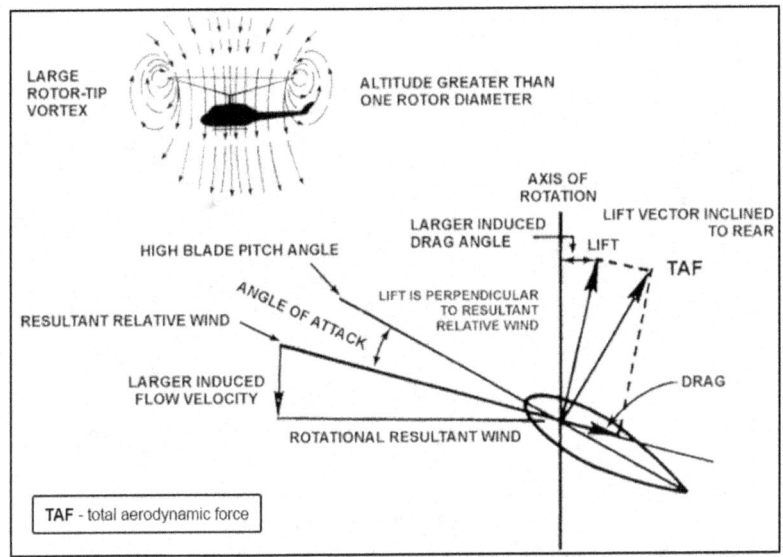

Figure 1-51. Out of ground effect hover

TRANSLATING TENDENCY

1-100. During hovering flight, the counterclockwise rotating, single-rotor helicopter has a tendency to drift laterally to the right. The translating tendency (figure 1-52, page 1-36) results from right lateral tail-rotor thrust exerted to compensate for main rotor torque (main rotor turning in a counterclockwise direction). The aviator must compensate for this right translating tendency of the helicopter by tilting the main rotor disk to the left. This lateral tilt creates a main rotor force to the left compensating for the tail-rotor thrust to the right. Helicopter design usually includes one or more of the following features, which help the aviator compensate for translating tendency:

- Flight control rigging may be designed so the rotor disk is tilted slightly left when the cyclic control is centered.
- Transmission may be mounted so the mast is tilted slightly left when the helicopter fuselage is laterally level.
- The collective pitch control system may be designed so the rotor disk tilts slightly left as collective pitch is increased.
- Programmed mechanical inputs/automatic flight-control systems/stabilization augmentation systems.

Chapter 1

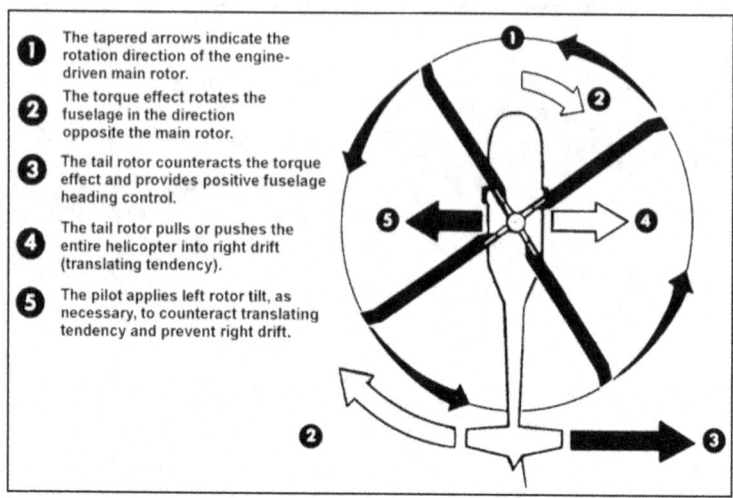

① The tapered arrows indicate the rotation direction of the engine-driven main rotor.

② The torque effect rotates the fuselage in the direction opposite the main rotor.

③ The tail rotor counteracts the torque effect and provides positive fuselage heading control.

④ The tail rotor pulls or pushes the entire helicopter into right drift (translating tendency).

⑤ The pilot applies left rotor tilt, as necessary, to counteract translating tendency and prevent right drift.

Figure 1-52. Translating tendency

SECTION V – ROTOR IN TRANSLATION

AIRFLOW IN FORWARD FLIGHT

1-101. Airflow across the rotor system in forward flight varies from airflow at a hover. In forward flight, air flows opposite the aircraft's flight path. The velocity of this air flow equals the helicopter's forward speed. Because the blades turn in a circular pattern, the velocity of airflow across a blade depends on the position of the blade in the plane of rotation at a given instant, its rotational velocity, and airspeed of the helicopter. Therefore, the airflow meeting each blade varies continuously as the blade rotates. The highest velocity of airflow occurs over the right side (3 o'clock position) of the helicopter (advancing blade in a rotor system that turns counterclockwise) and decreases to rotational velocity over the nose. It continues to decrease until the lowest velocity of airflow occurs over the left side (9-o'clock position) of the helicopter (retreating blade). As the blade continues to rotate, velocity of the airflow then increases to rotational velocity over the tail. It continues to increase until the blade is back at the 3 o'clock position.

1-102. The advancing blade (figure 1-53, blade A, page 1-37) moves in the same direction as the helicopter. The velocity of the air meeting this blade equals rotational velocity of the blade plus wind velocity resulting from forward airspeed. The retreating blade (blade C) moves in a flow of air moving in the opposite direction of the helicopter. The velocity of airflow meeting this blade equals rotational velocity of the blade minus wind velocity resulting from forward airspeed. The blades (B and D) over the nose and tail move essentially at right angles to the airflow created by forward airspeed; the velocity of airflow meeting these blades equals the rotational velocity. This results in a change to velocity of airflow all across the rotor disk and a change to the lift pattern of the rotor system. Figure 1-54, page 1-38, depicts force vectors acting on various blade areas in forward flight.

Aerodynamics of Flight

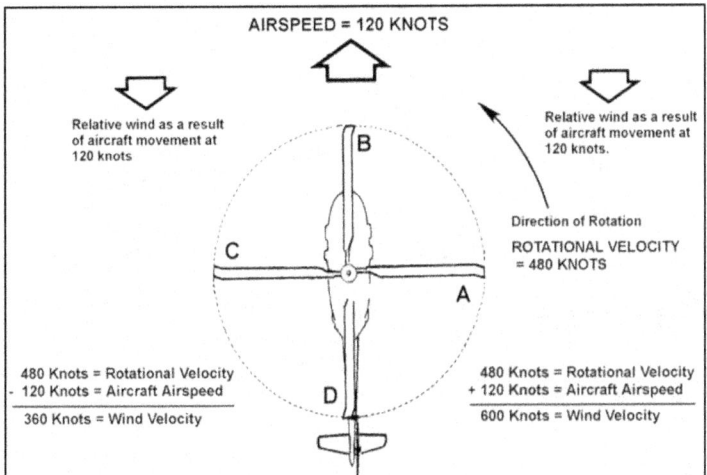

Figure 1-53. Differential velocities on the rotor system caused by forward airspeed

NO-LIFT AREAS

1-103. The no-lift areas are reverse flow, negative stall, and negative lift.

Reverse Flow

1-104. Part A of figure 1-54, page 1-38, shows reverse flow. At the root of the retreating blade is an area where the air flows backward from the trailing to the leading edge of the blade. This is due to wind created by forward airspeed being greater than rotational velocity at this point on the blade.

Negative Stall

1-105. Part B of figure 1-54 shows negative stall. In the negative stall area, rotational velocity exceeds forward flight velocity, causing resultant relative wind to move toward the leading edge. The resultant relative wind is so far above the chord line, a negative AOA above the critical AOA results. The blade stalls with a negative AOA.

Negative Lift

1-106. Part C of figure 1-54 shows negative lift. In the negative lift area, rotational velocity, induced flow, and blade flapping combine to reduce the AOA from a negative stall to an AOA that causes the blade to produce negative lift.

POSITIVE LIFT AND POSITIVE STALL

1-107. Figure 1-54, parts D and E, show positive lift and positive stall. That portion of the blade outboard of the no-lift areas produces positive lift. In the positive lift area, the resultant relative wind produces a positive AOA. Under certain conditions, it is possible to have a positive stall area near the blade tip. Section VIII covers retreating blade stall.

Chapter 1

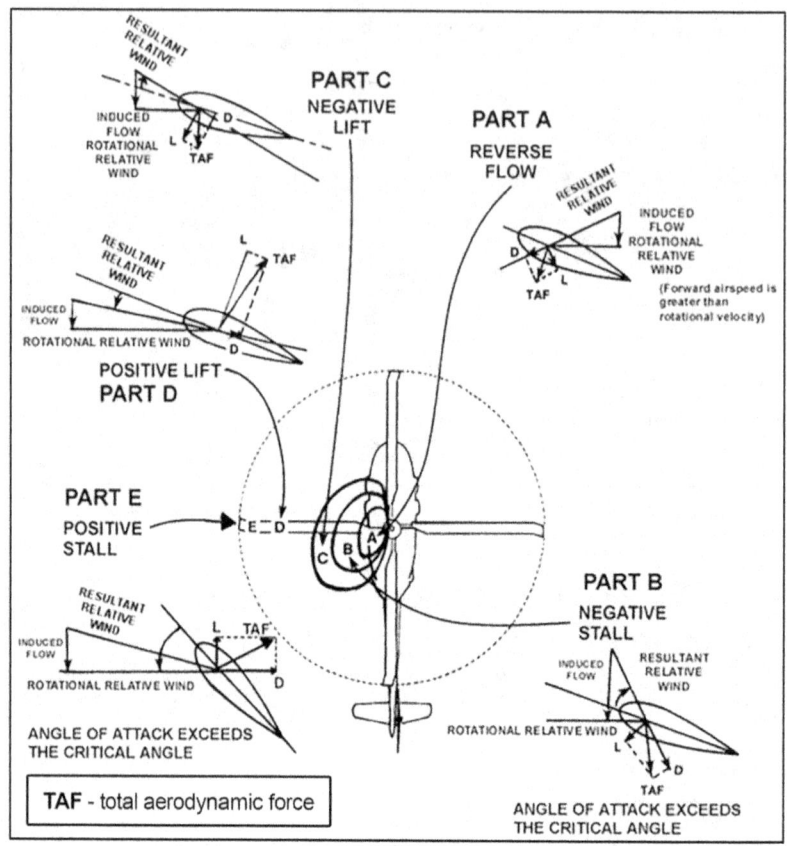

Figure 1-54. Blade areas in forward flight

DISSYMMETRY OF LIFT

1-108. Dissymmetry of lift is the differential (unequal) lift between advancing and retreating halves of the rotor disk caused by the different wind flow velocity across each half. This difference in lift would cause the helicopter to be uncontrollable in any situation other than hovering in a calm wind. There must be a means of compensating, correcting, or eliminating this unequal lift to attain symmetry of lift.

1-109. In forward flight, two factors in the lift equation, blade area and air density, are the same for the advancing and retreating blades. Airfoil shape is fixed for a given blade, and air density cannot be affected; the only remaining variables are blade speed and AOA. Rotor RPM controls blade speed. Because rotor RPM must remain relatively constant, blade speed also remains relatively constant. This leaves AOA as the one variable remaining that can compensate for dissymmetry of lift. This is accomplished through blade flapping and/or cyclic feathering.

Aerodynamics of Flight

Blade Flapping

1-110. When blade flapping compensates for dissymmetry of lift, the upward and downward flapping motion changes induced flow velocity. This changes AOA on the advancing and retreating blades.

Advancing Blade

1-111. As the relative wind speed of the advancing blade increases, the blade gains lift and begins to flap up (figure 1-55). It reaches its maximum upflap velocity at the 3-o'clock position, where the wind velocity is the greatest. This upflap creates a downward flow of air and has the same effect as increasing the induced flow velocity by imposing a downward vertical velocity vector to the relative wind. This decreases the AOA.

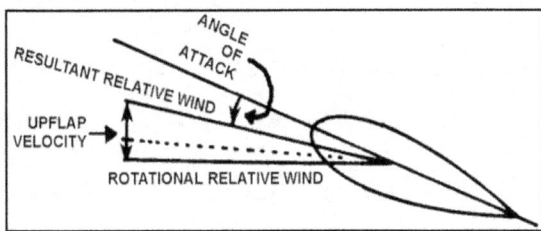

Figure 1-55. Flapping (advancing blade, 3-o'clock position)

Retreating Blade

1-112. As relative wind speed of the retreating blade decreases, the blade loses lift and begins to flap down (figure 1-56). It reaches its maximum downflap velocity at the 9 o'clock position, where wind velocity is the least. This downflap creates an upward flow of air and has the same effect as decreasing the induced flow velocity by imposing an upward velocity vertical vector to the relative wind. This increases AOA.

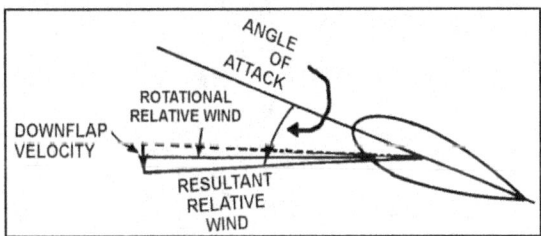

Figure 1-56. Flapping (retreating blade, 9-o'clock position)

Over the Aircraft Nose and Tail

1-113. Blade flapping over the nose and tail of the helicopter are essentially equal. The net result is an equalization, or symmetry, of lift across the rotor system. Up flapping and down flapping do not change the total amount of lift produced by the rotor blades. When blade flapping has compensated for dissymmetry of lift, the rotor disk is tilted to the rear, called blowback. The maximum upflap occurring over the nose and the maximum downflap occurring over the tail cause blowback. This would cause airspeed to decrease. The aviator uses cyclic feathering to compensate for dissymmetry of lift allowing him or her to control the attitude of the rotor disk.

Cyclic Feathering

1-114. Cyclic feathering compensates for dissymmetry of lift (changes the AOA) in the following way. At a hover, equal lift is produced around the rotor system with equal pitch and AOA on all the blades and at all points in the rotor system (disregarding compensation for translating tendency). The rotor disk is parallel to the horizon. To develop a thrust force, the rotor system must be tilted in the desired direction of movement. Cyclic feathering changes the angle of incidence differentially around the rotor system. Forward cyclic movements decrease the angle of incidence at one part on the rotor system while increasing the angle in another part. Maximum down flapping of the blade over the nose and maximum up flapping over the tail tilt the rotor disk and thrust vector forward. To prevent blowback from occurring, the aviator must continually move the cyclic forward as velocity of the helicopter increases. Figure 1-57 illustrates the changes in pitch angle as the cyclic is moved forward at increased airspeeds. At a hover, the cyclic is centered and the pitch angle on the advancing and retreating blades is the same. At low forward speeds, moving the cyclic forward reduces pitch angle on the advancing blade and increases pitch angle on the retreating blade. This causes a slight rotor tilt. At higher forward speeds, the aviator must continue to move the cyclic forward. This further reduces pitch angle on the advancing blade and further increases pitch angle on the retreating blade. As a result, there is even more tilt to the rotor than at lower speeds.

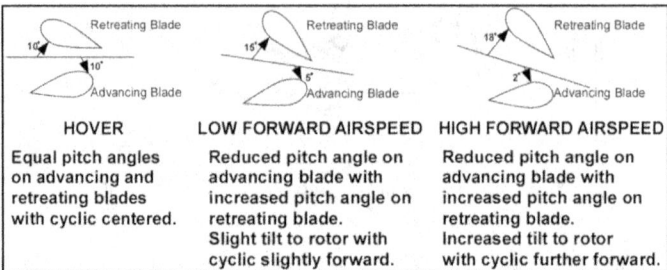

Figure 1-57. Blade pitch angles

1-115. This horizontal lift component (thrust) generates higher helicopter airspeed. The higher airspeed induces blade flapping to maintain symmetry of lift. The combination of flapping and cyclic feathering maintains symmetry of lift and desired attitude on the rotor system and helicopter.

Tandem-Rotor Helicopter Dissymmetry of Lift

1-116. The biggest difference between single-rotor and tandem-rotor helicopters is the aviator does not manually compensate for dissymmetry of lift when applying forward cyclic. Automatic cyclic-feathering systems are installed on tandem-rotor helicopters. These systems are activated through computer-generated commands at specified airspeeds, usually starting around 70 knots. At low airspeeds, blade flapping compensates for dissymmetry of lift. As airspeed increases, these systems program allowing a more level fuselage attitude and reduce stresses on the rotor driving mechanisms. If the cyclic-feathering system fails to properly feather the rotor system at higher airspeeds, greater blade-flapping angles and nose-low flight attitudes occur and induce increased stresses on the rotor-driving mechanisms.

TRANSLATIONAL LIFT

1-117. Improved rotor efficiency resulting from directional flight is translational lift. The efficiency of the hovering rotor system is improved with each knot of incoming wind gained by horizontal movement or surface wind. As the incoming wind enters the rotor system, turbulence and vortexes are left behind and the flow of air becomes more horizontal. In addition, the tail rotor becomes more aerodynamically efficient during the transition from hover to forward flight. As the tail rotor works in progressively less turbulent air, this improved efficiency produces more thrust, causing the nose of the aircraft to yaw left (with a main rotor turning counterclockwise) and forces the aviator to apply right pedal (decreasing the AOA in the tail rotor blades) in response.

1-118. Figure 1-58 shows the airflow pattern for 1 to 5 knots of forward airspeed. Note how the downwind vortex is beginning to dissipate and induced flow down through the rear of the rotor system is more horizontal.

Aerodynamics of Flight

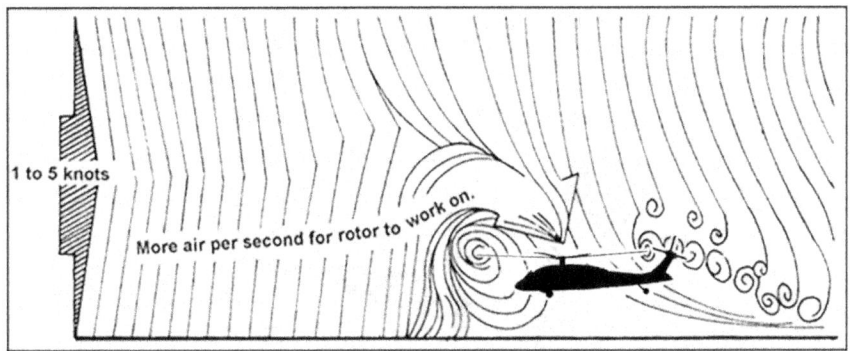

Figure 1-58. Translational lift (1 to 5 knots)

1-119. Figure 1-59 shows the airflow pattern at a speed of 10 to 15 knots. At this increased airspeed, the airflow continues to become more horizontal. The leading edge of the downwash pattern is being overrun and is well back under the nose of the helicopter.

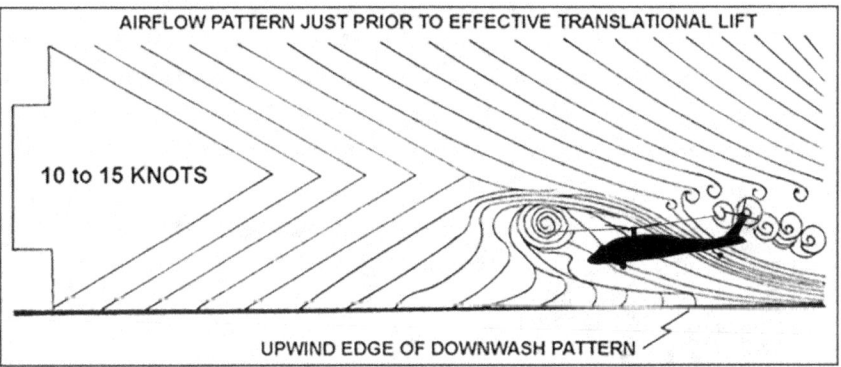

Figure 1-59. Translational lift (10 to 15 knots)

TRANSVERSE FLOW EFFECT

1-120. In forward flight, air passing through the rear portion of the rotor disk has a greater downwash angle than air passing through the forward portion. This is due to the fact the greater the distance air flows over the rotor disk, the longer the disk has to work on it and the greater the deflection on the aft portion. Downward flow at the rear of the rotor disk causes a reduced AOA, resulting in less lift. The front portion of the disk produces an increased AOA and more lift because airflow is more horizontal. These differences in lift between the fore and aft portions of the rotor disk are called transverse flow effect (figure 1-60, page 1-42). This effect causes unequal drag in the fore and aft portions of the rotor disk and results in vibration easily recognizable by the aviator. It occurs between 10 and 20 knots. Transverse flow effect is most noticeable during takeoff and, to a lesser degree, during deceleration for landing. Gyroscopic precession causes the effects to be manifested 90 degrees in the direction of rotation, resulting in a right rolling motion.

Chapter 1

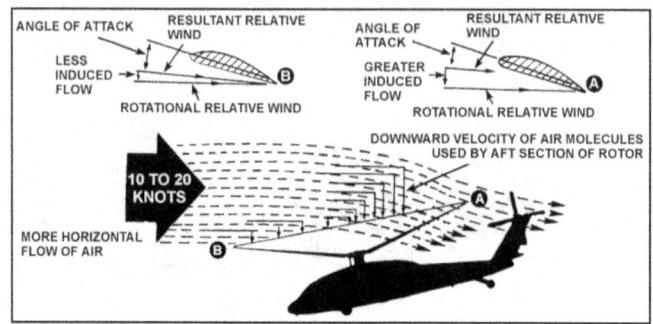

Figure 1-60. Transverse flow effect

EFFECTIVE TRANSLATIONAL LIFT

1-121. Effective translational lift (ETL) (figure 1-61) occurs with the helicopter at about 16 to 24 knots, when the rotor—depending on size, blade area, and RPM of the rotor system—completely outruns the recirculation of old vortexes and begins to work in relatively undisturbed air. The rotor no longer pumps the air in a circular pattern but continually flies into undisturbed air. The flow of air through the rotor system is more horizontal, therefore induced flow and induced drag are reduced. The AOA is subsequently increased, which makes the rotor system operate more efficiently. This increased efficiency continues with increased airspeed until the best climb airspeed is reached, when total drag is at its' lowest point. Greater airspeeds result in lower efficiency due to increased parasite drag.

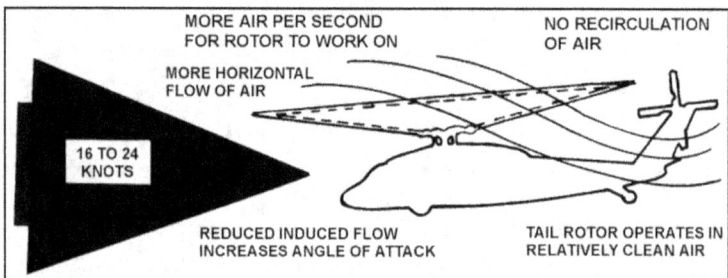

Figure 1-61. Effective translational lift

1-122. As single-rotor aircraft speed increases, translational lift becomes more effective, nose rises or pitches up, and aircraft rolls to the right. The combined effects of dissymmetry of lift, gyroscopic precession, and transverse flow effect cause this tendency. Aviators must correct with additional forward and left lateral cyclic input to maintain a constant rotor-disk attitude.

AUTOROTATION

AERODYNAMICS OF VERTICAL AUTOROTATION

1-123. During powered flight, rotor drag is overcome with engine power. When the engine fails or is deliberately disengaged from the rotor system, some other force must sustain rotor RPM so controlled flight can be continued to the ground. Adjusting the collective pitch to allow a controlled descent generates this

Aerodynamics of Flight

force. Airflow during helicopter descent provides energy to overcome blade drag and turn the rotor. When the helicopter descends in this manner, it is in a state of autorotation. In effect, the aviator exchanges altitude at a controlled rate in return for energy to turn the rotor at a RPM that provides aircraft control and a safe landing. Helicopters have potential energy based on their altitude above the ground. As this altitude decreases, potential energy is converted into kinetic energy used in turning the rotor. Aviators use this kinetic energy to slow the rate of descent to a controlled rate and affect a smooth touchdown.

1-124. Most autorotations are performed with forward airspeed. For simplicity, the following aerodynamic explanation is based on a vertical autorotative descent (no forward airspeed) in still air. Under these conditions, forces that cause the blades to turn are similar for all blades, regardless of their position in the plane of rotation. Therefore, dissymmetry of lift resulting from helicopter airspeed is not a factor. During autorotation, the rotor disk is divided into three regions—driven, driving, and stall (figure 1-62).

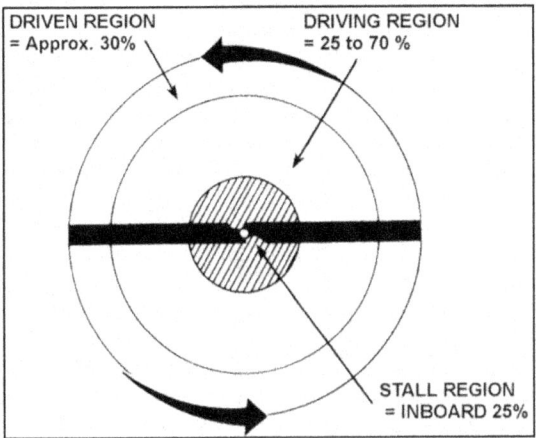

Figure 1-62. Blade regions in vertical autorotation descent

Driven Region

1-125. This region is also called the propeller region and nearest the blade tip. It normally consists of about 30 percent of the disk radius. In the driven region, the TAF acts above the blade and behind the axis of rotation. This region creates lift, which slows the rate of descent and drag, which slows rotation of the blade. Region size varies with the blade pitch setting, rate of descent, and rotor RPM. Any change of these factors also changes the size of the regions along the blade span.

Driving Region

1-126. This region extends from about the 25 to 70 percent radius of the blade. It lies between the driven and stall regions. It can also be identified as the area of autorotative force because it is the region of the blade that produces the force necessary to turn the blades during autorotation. TAF in the driving region is inclined slightly forward of the axis of rotation and produces a continual acceleration force. This direction of force supplies thrust, which tends to accelerate the rotation of the blade. The size of the region varies with the blade pitch setting, rate of descent, and rotor RPM. Any change of these factors also changes the size of the regions along the blade span.

Stall Region

1-127. This region includes the inboard 25 percent of the blade radius. It operates above the stall AOA and causes drag, which tends to slow the rotation of the blade.

Blade Region Relationships

1-128. Figure 1-63, page 1-45, illustrates the three regions. Additional information in the figure pertains to force vectors on those regions and two additional equilibrium points. This figure serves to locate those regions/points on the blade span and depict the interplay of force vectors. Force vectors are different in each region because rotational relative wind is slower near the blade root and increases continually toward the blade tip. In addition, blade twist gives a more positive AOA in the driving region than in the driven region. The combination of inflow up through the rotor with rotational relative wind produces different combinations of aerodynamic force at every point along the blade.

1-129. There are two points of equilibrium on the blade (figure 1-63, page 1-45)—point B, between the driven and driving regions, and point D, between the driving and stall regions. At this point, TAF is aligned with the axis of rotation. Lift and drag are produced, but overall, there is neither acceleration nor deceleration force developed.

1-130. The aviator manipulates these regions to control all aspects of the autorotative descent. For example, if the collective pitch is increased, the pitch angle increases in all regions. This causes point of equilibrium B to move inboard and point of equilibrium D to move outboard along the blade span, thus increasing the size of the driven and stall regions while reducing the driving region. The stall region also becomes larger while the driving region is reduced in size. Reducing the size of the driving region decreases acceleration force and rotor RPM. An aviator can achieve a constant rotor RPM by adjusting the collective pitch so blade acceleration forces from the driving region are balanced with deceleration forces from the driven and stall regions.

AERODYNAMICS OF AUTOROTATION IN FORWARD FLIGHT

1-131. Aerodynamic forces in forward flight (figure 1-64, page 1-46) are produced in exactly the same manner as in vertical autorotation. However, because forward speed changes the inflow of air up through the rotor disk, this changes the location and size of the regions on the retreating and advancing sides of the rotor disk. Because the retreating side experiences an increased AOA, all three regions move outboard along the blade span with the stall region growing larger and an area nearest the hub experiencing a reversed flow. Because the advancing side experiences a decreased AOA, the driven region takes up more of that blade span.

Aerodynamics of Flight

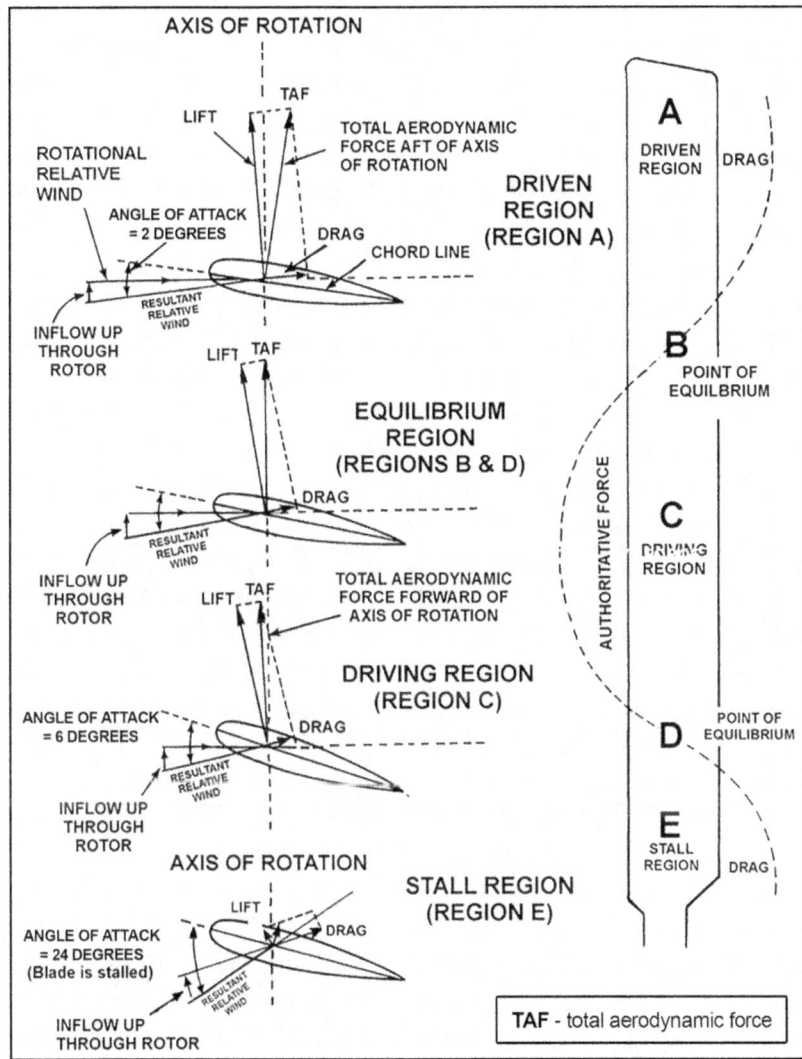

Figure 1-63. Force vectors in vertical autorotative descent

Chapter 1

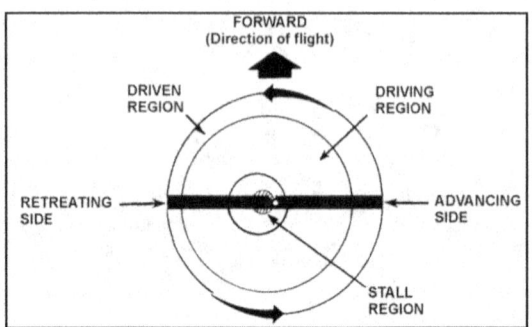

Figure 1-64. Autorotative regions in forward flight

AUTOROTATIVE PHASES

1-132. Autorotations may be divided into three distinct phases—entry, steady-state descent, and deceleration and touchdown. Each phase is aerodynamically different from the others.

Entry

1-133. This phase is entered after loss of engine power. The loss of engine power and rotor RPM is more pronounced when the helicopter is at high gross weight, high forward speed, or in high-density altitude conditions. Any of these conditions demand increased power (high collective position) and a more abrupt reaction to loss of that power. In most helicopters, it takes only seconds for RPM decay to fall into a minimum safe range requiring a quick collective response from the aviator. Entry is a combination of figures 1-65 and 1-66.

Level-Powered Flight at High Speed

1-134. Figure 1-65 shows the airflow and force vectors for a blade in this configuration. Lift and drag vectors are large, and the TAF is inclined well to the rear of the axis of rotation. An engine failure in this mode will cause rapid rotor RPM decay. To prevent this, an aviator must lower the collective quickly, reducing drag and inclining the TAF vector forward, nearer the axis of rotation.

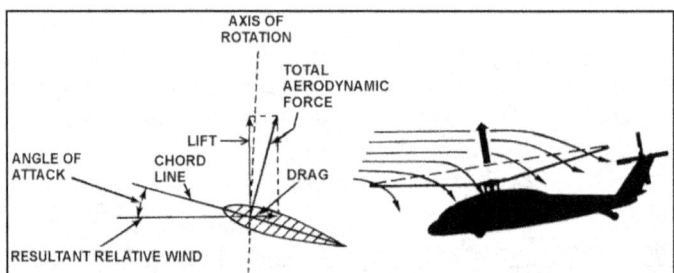

Figure 1-65. Force vectors in level-powered flight at high speed

Collective Pitch Reduction

1-135. Figure 1-66, page 1-47, shows airflow and force vectors for a blade immediately after power loss and subsequent collective reduction, yet before the aircraft has begun to descend. Lift and drag are reduced, with the TAF vector inclined further forward than it is in powered flight. As the helicopter begins to descend, the airflow begins to flow upward and under the rotor system. This causes the TAF to incline further forward until it reaches an equilibrium that maintains a safe operating RPM.

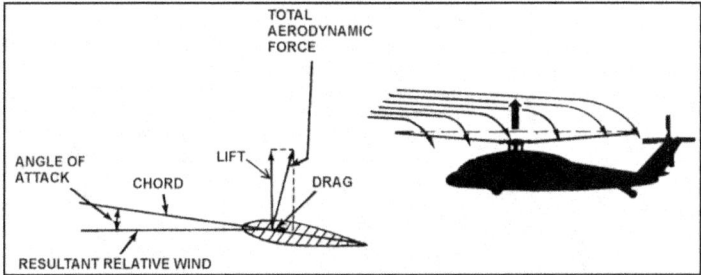

Figure 1-66. Force vectors after power loss–reduced collective

Steady-State Descent

1-136. Figure 1-67 shows airflow and force vectors for a blade in steady-state autorotative descent. Airflow is now upward through the rotor disk because of the descent. This inflow of air creates a larger AOA although blade pitch angle has not changed since the descent began. TAF on the blade is increased and inclined further forward until equilibrium is established, rate of descent and rotor RPM are stabilized, and the helicopter is descending at a constant angle. Angle of descent is normally 17 to 20 degrees, depending on airspeed, density altitude, wind, and type of helicopter.

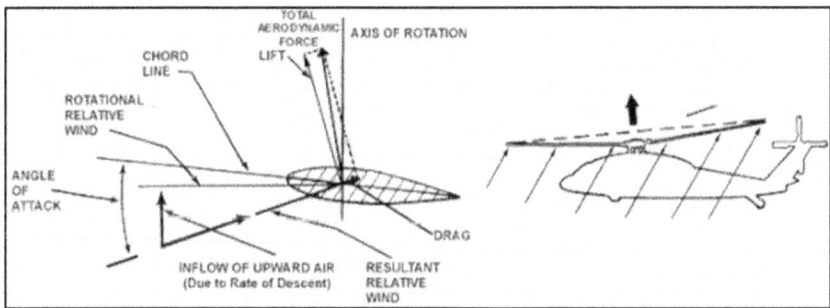

Figure 1-67. Force vectors in autorotative steady-state descent

Deceleration and Touchdown

1-137. Figure 1-68, page 1-48, shows airflow and force vectors for a blade in autorotative deceleration. To make an autorotative landing, aviators reduce airspeed and rate of descent just before touchdown. They can partially accomplish both actions by applying aft cyclic, which changes the attitude of the rotor disk in relation to the relative wind. This attitude change inclines the resultant lift of the rotor system to the rear, slowing forward speed. It also increases AOA on all blades by changing direction of airflow through the rotor system, thereby increasing rotor RPM. The lifting force of the rotor system is increased and rate of descent is reduced. After an aviator reduces forward speed to a safe landing speed, the helicopter is placed in a landing attitude while applying collective pitch to cushion the touchdown.

Chapter 1

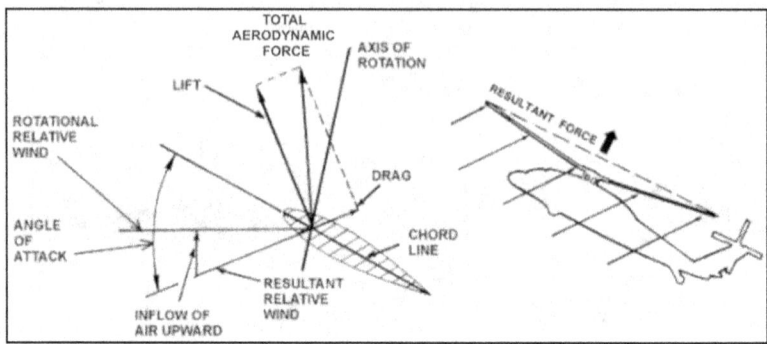

Figure 1-68. Autorotative deceleration

GLIDE AND RATE OF DESCENT IN AUTOROTATION

1-138. Helicopter airspeed and drag are significant factors affecting rate of descent in autorotation. The rate of descent is high at very low airspeeds, decreases to a minimum at some intermediate speed and increases again at faster speeds. Airspeeds for minimum rate of descent and maximum glide distance vary by helicopter type and can be found in individual operator manuals (figure 1-69, page 1-49).

Circle of Action

1-139. The circle of action is a point on the ground that has no apparent movement in the pilot's field of view (FOV) during a steady-state autorotation. The circle of action would be the point of impact if the pilot applied no deceleration, initial pitch, or cushioning pitch during the last 100 feet of autorotation. Depending on the amount of wind present and the rate and amount of deceleration and collective application, the circle of action is usually two or three helicopter lengths short of the touchdown point.

Last 50 to 100 Feet

1-140. It can be assumed autorotation ends at 50 to 100 feet and landing procedures then begin. To execute a power-off landing for rotary-wing aircraft, an aviator exchanges airspeed for lift by decelerating the aircraft during the last 100 feet. When executed correctly, deceleration is applied and timed so rate of descent and forward airspeed are minimized just before touchdown. At about 10 to 15 feet, this energy exchange is essentially complete. Initial pitch application occurs at 10 to 15 feet. This is used to trade some of the rotor energy to slow the rate of descent prior to cushioning. The primary remaining control input is application of collective pitch to cushion touchdown. Because all helicopter types are slightly different, aviator experience in that particular aircraft is the most useful tool for predicting useful energy exchange available at 100 feet and the appropriate amount of deceleration and collective pitch needed to execute the exchange safely and land successfully.

Aerodynamics of Flight

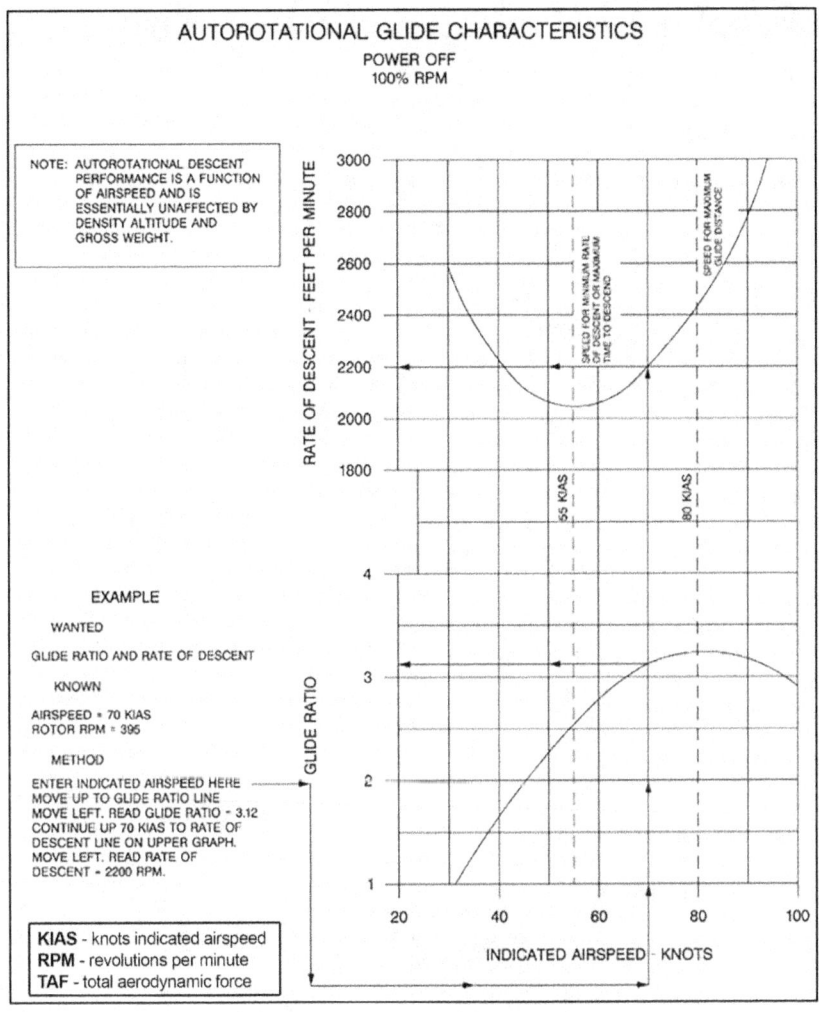

Figure 1-69. Drag and airspeed relationship

SECTION VI – MANEUVERING FLIGHT

AERODYNAMICS

1-141. There are several characteristics aviators must be aware of to successfully perform combat maneuvers.

BEST RATE-OF-CLIMB/MAXIMUM ENDURANCE AIRSPEED

1-142. This airspeed has the following characteristics—
- Total drag at the minimum.
- Largest amount of excess power available.
- Lowest fuel flow during powered flight.
- Maximum single engine gross weight that can be carried (for dual engine aircraft).

1-143. Aviators should always be aware of their best rate-of-climb airspeed as it is where the aircraft will turn and climb the best, maximize available power margin, and get the lowest fuel flow.

BUCKET SPEED

1-144. Bucket speed is the airspeed range providing the best power margin for maneuvering flight. Using the cruise chart for current conditions, enter at 50 percent of maximum torque available, go up to gross weight, over to the lowest and highest airspeed intersecting the aircraft gross weight, and note speeds between which there is the greatest power margin for maneuvering flight. The most critical is lower speed since at higher speeds airspeed energy may be traded to maintain altitude while maneuvering. When below minimum bucket speed reduce bank angle. Otherwise, altitude loss may become unavoidable.

TRANSIENT TORQUE

1-145. Transient torque is a phenomenon occurring in single-rotor helicopters when lateral cyclic is applied and is caused by aerodynamic forces. For conventional American helicopters where the main rotor turns counterclockwise, (figure 1-70) a left cyclic input causes a temporary rise in torque and a right cyclic input causes a temporary drop in torque.

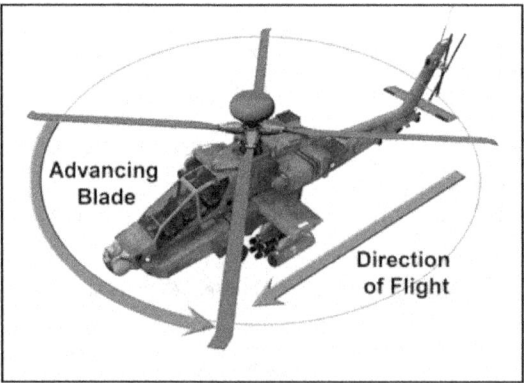

Figure 1-70. Counterclockwise blade rotation

1-146. At the rear half of the rotor disk, downwash is greater than seen at the forward half of the rotor disk. This effect is more pronounced for heavier aircraft which exhibit greater coning due to their weight, causing

Aerodynamics of Flight

even greater downwash at the rear of the rotor disk. If a left cyclic input is made by the pilot, the following events occur leading to a temporary increase in torque:
- The swashplate commands an increased blade AOA as each blade passes over the tail.
- The increase in blade AOA causes the rotor disk to tilt left, which is felt as a left roll on the aircraft.
- With increased lift on the rotor blades passing over the tail, there is also increased drag (induced drag).
- The increased rotor drag due to the left turn will initially try to slow the rotor, but is sensed by the applicable engine computer. The engine responds by delivering more torque to the rotor system to maintain rotor speed.

1-147. The opposite holds true for right cyclic turns, but is less pronounced. Unlike the left hand turn, in right turns blade pitch is being changed at the front of the rotor disk where induced downwash is lower, so the drag penalty is lower. Transient torque is not as prevalent at slower airspeeds because the induced downwash distribution is nearly uniform across the rotor disk.

1-148. Five factors affect how much torque change occurs during transient torque—
- Torque transients are proportional with the amount of power applied. The higher the torque setting when lateral cyclic inputs are made, the higher or lower the transient.
- Rate of movement of the cyclic. The faster the rate of movement the higher resultant torque spike.
- Magnitude of cyclic displacement directly affects the torque transient. An example of worst-case scenario occurs when a pilot initiates a rapid right roll, then due to an unexpected event breaks left. The transition from right cyclic applied to left cyclic applied results in a large amount of pitch change in the advancing blade, resulting in large torque transients.
- Drag is increased or decreased by the factor of velocity squared. Thus, the higher the forward airspeed, the higher the torque transient results.
- High aircraft weight increases coning, which will make transient torque more pronounced.

1-149. Extreme caution must be used when maneuvering at near maximum torque available especially at high airspeeds. It is not uncommon to experience as much as 50 percent torque changes in uncompensated maneuvers with high power settings at high forward airspeeds. In these situations, the pilot must ensure collective is reduced as left lateral cyclic is applied and increased for right cyclic inputs. When recovering from these inputs, opposite collective inputs must be made so aircraft limitations are not exceeded.

1-150. As a good basic technique, imagine a piece of string tied between the cyclic and collective (right cyclic-collective increase/left cyclic-collective decrease). Also, inputs must be made to keep the aircraft from descending due to torque reductions (when recovering from left cyclic inputs with collective reduced).

Note. 701C/D/DD/E equipped helicopters employ maximum torque rate attenuator (MTRA) which attempts to prevent transient torque related over-torques but may produce a rotor droop and loss of roll rate. Once the pilot has gained confidence in the ability of the MTRA to prevent over-torques resulting from transient torque, he or she can aggressively maneuver the aircraft without closely monitoring engine torque.

MUSHING

1-151. Mushing is a temporary stall condition occurring in helicopters when rapid aft cyclic is applied at high forward airspeeds. Normally associated with dive recoveries, which result in a significant loss of altitude, this phenomenon can also occur in a steep turn resulting in an increased turn radius. Mushing results during high G-maneuvers when at high forward airspeeds aft cyclic is abruptly applied. This results in a change in the airflow pattern on the rotor exacerbated by total lift area reduction as a result of rotor disc coning. Instead of an induced flow down through the rotor system, an upflow is introduced which results in a stall condition on portions of the entire rotor system. While this is a temporary condition (because in due time the upflow will dissipate and the stall will abate), the situation may become critical during low altitude recoveries or when maneuvering engagements require precise, tight turning radii. High aircraft gross weight and high density altitude are conditions conducive to and can aggravate mushing.

1-152. Mushing can be recognized by the aircraft failing to respond immediately but continuing on the same flight path as before the application of aft cyclic. Slight feedback and mushiness may be felt in the controls. When mushing occurs, the tendency is to pull more aft cyclic which prolongs stall and increases recovery times. Make a forward cyclic adjustment to recover from the mushing condition. This reduces the induced flow, improves the resultant AOA, and reduces rotor disc coning which increases the total lift area of the disc. The pilot will immediately feel a change in direction of the aircraft and increased forward momentum as the cyclic is moved forward. To avoid mushing, the pilot must use smooth and progressive application of the aft cyclic during high G-maneuvers such as dive recoveries and tight turns.

CONSERVATION OF ANGULAR MOMENTUM

1-153. The law of conservation of angular momentum states the value of angular momentum of a rotating body will not change unless external torques are applied. In other words, a rotating body will continue to rotate with the same rotational velocity until some external force is applied to change the speed of rotation. Angular momentum can be expressed as—

> **Law of Conservation of Angular Momentum**
> Mass x Angular Velocity x Radius Squared

1-154. Changes in angular velocity, known as angular acceleration or deceleration, take place if the mass of a rotating body is moved closer to or further from the axis of rotation. The speed of the rotating mass will increase or decrease in proportion to the square of the radius.

1-155. An excellent example for this principle is when watching a figure skater on ice skates. The skater begins a rotation on one foot, with the other leg and both arms extended. The rotation of the skater's body is relatively slow. When a skater draws both arms and one leg inward, the moment of inertia (mass times radius squared) becomes much smaller and the body is rotating almost faster than the eye can follow. Because the angular momentum must, by law of nature, remain the same (no external force applied), the angular velocity must increase.

1-156. The mathematician, Coriolis, was concerned with forces generated by such radial movements of mass on a rotating disc or plane. These forces cause acceleration and deceleration. It may be stated as a mass moving radically—
- Outward on a rotating disk will exert a force on its surroundings opposite to rotation.
- Inward on a rotating disk will exert a force on its surroundings in the direction of rotation.

1-157. The major rotating elements in the system are the rotor blades. As the rotor begins to cone due to G-loading maneuvers, the diameter of the disc shrinks. Due to conservation of angular momentum, the blades continue to travel the same speed even though the blade tips have a shorter distance to travel due to reduced disc diameter. This action results in an increase in rotor RPMs. Most pilots arrest this increase with an increase in collective pitch.

1-158. Conversely, as G-loading subsides and the rotor disc flattens out from the loss of G-load induced coning, the blade tips now have a longer distance to travel at the same tip speed. This action results in a reduction of rotor RPMs. However, if this droop in rotor continues to the point it attempts to decrease below normal operating RPM, the engine control system adds more fuel/power to maintain the specified engine RPM. If the pilot does not reduce collective pitch as the disc unloads, the combination of the engines compensating for the RPM slow down and the additional pitch added as G-loading increased may result in exceeding the torque limitations or power the engines can produce. This problem is exacerbated by effects of the TAF encountered during maneuvering flight.

HIGH BANK ANGLE TURNS

1-159. As the angle of bank increases, the amount of lift opposite the vertical weight decreases (figure 1-71, page 1-53). If adequate excess engine power is available, increasing collective pitch enables continued flight while maintaining airspeed and altitude. If sufficient excess power is not available, the result is altitude loss unless airspeed is traded (aft cyclic) to maintain altitude or altitude is traded to maintain airspeed.

Aerodynamics of Flight

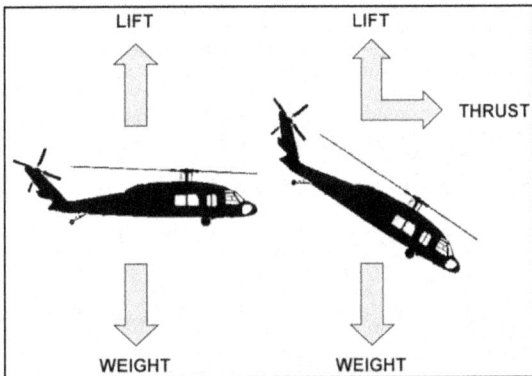

Figure 1-71. Lift to weight

1-160. At some point (airspeed/angle of bank) sufficient excess power will not be available and the aviator must apply aft cyclic to maintain altitude (table 1-3). The percentages shown are not a direct torque percentage, but percentage of torque increase required based on aircraft torque to maintain straight and level flight. If indicated cruise torque is 48 percent and a turn to 60 degrees is initiated, a torque increase of 48 percent (96 percent torque indicated) is required to maintain airspeed and altitude.

Table 1-3. Bank angle versus torque

Bank Angle - Degree	Increase in TR - Percent
0	---
15	3.6
30	15.4
45	41.4
60	100.0
TR=torque	

1-161. Additionally, rotor system capability may limit the maneuver as opposed to insufficient excess power. In high energy maneuvering, the rotor is normally a limiting factor. It is not unusual for a reduction in collective to be required to achieve maximum performance when maneuvering at increased G-loads, altitudes, or high weights.

1-162. Aviators must be familiar with this characteristic, anticipate cyclic input results, and apply appropriate control inputs to conduct combat maneuvers successfully. Aviators unfamiliar with this characteristic may be surprised at the rapid build of sink rates when turning the aircraft to bank angles approaching 60 degrees. When flying heavy aircraft in a high hot environment, sufficient time and altitude may not be available to arrest the resultant descent.

MANEUVERING FLIGHT AND TOTAL AERODYNAMIC FORCE

1-163. The cyclic inputs and associated rotor disc pitch changes required to accomplish successful combat maneuvers have a substantial effect on TAF. Large aft cyclic inputs increase the inflow through the rotor system. Since lift is perpendicular to the resultant relative wind, the TAF of each rotor blade may move to a point aligned with or forward of the axis of rotation (much like the driving and driven region of a blade during autorotational flight). While the engine control system reduces fuel flow to reduced load, the rotor system may still climb to transient ranges or attempt to overspeed.

1-164. Conversely, when the cyclic is rapidly repositioned to a more forward position, the inflow through the rotor is rapidly reduced resulting in the blade TAF moving aft of the axis of rotation and a slowing of rotor RPM (figure 1-72). The engine control systems sense this and increase fuel flow to the engines to maintain rotor RPM causing torque to increase. As a general rule, when traveling at airspeeds above bucket speed, aft cyclic results in a reduction in torque and an increase in rotor RPMs. Recovery from an aft cyclic input (pushover or high G-turn recovery) results in torque increase as the engines compensate for the rotor system slow down. In aggressive maneuvers, this may result in an overtorque or overspeed if appropriate collective input is not made to keep torque and rotor consistent.

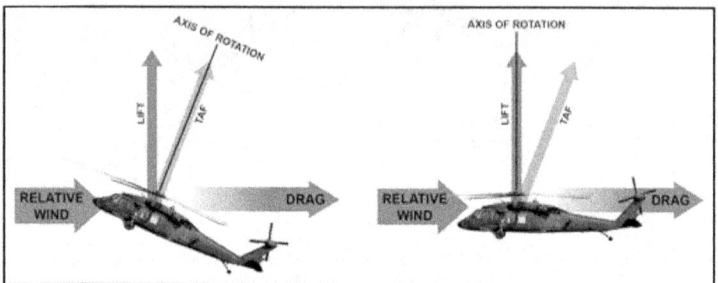

Figure 1-72. Aft cyclic results

1-165. This phenomenon is exacerbated by high gross weight and also affected by ambient temperature and density altitude. Typically, cold dry air results in more rapid rotor RPM increase during aft cyclic input and a corresponding higher torque increase with a forward cyclic input. Hot temperatures and higher density altitudes result in more collective input required to arrest a climbing rotor.

ANGULAR MOMENTUM AND TOTAL AERODYNAMIC FORCE COMBINED EFFECTS

1-166. Angular momentum and TAF combine during cyclic pitch changes. During aft cyclic or G-loading, the rotor increases and torque goes down. During G-load recovery, torque increases as the engine control systems work to maintain a rotor RPM attempting to decrease. Aviators must be able to apply appropriate and timely collective inputs to maintain consistent torque and keep rotor RPM within limits.

DIG-IN

1-167. While making large aft cyclic movements, the pilot must be aware of the helicopter's tendency to rapidly and unpredictably build G-forces. As the cyclic is moved aft, the rotor disk responds by tilting aft, which tilts the thrust vector aft and ultimately causes the aircraft to pitch nose-up. This rapid pitch-up also increases the length of the aircraft thrust vector, which will in turn increase the pitch-up rate. The rapid onset of the pitch-up motion due to the tilting and then lengthening of the thrust vector is considered destabilizing and countered by the helicopter's horizontal tail or stabilizer, which will try to drive the nose back down. For large pitch-up rates, the tendency of the main rotor to continue pitching-up will overpower the horizontal tail/stabilizer and the aircraft will dig-in and slow down rapidly. Dig-in is usually accompanied by airframe vibration and sometimes controls feedback.

1-168. Aft cyclic movements give predictable increases in G-load up to the dig-in point; however, the dig-in occurs at different G-levels for each model of helicopter. The point at which dig-in occurs depends on a number of factors, but most important is the size of the horizontal tail/stabilizer and amount of rotor offset. For most helicopters, this point is between 1.5 and 2.0 Gs. Pilots should be prepared for dig-in during aggressive aft cyclic inputs, especially during break turns.

GUIDELINES

1-169. Below are good practices to follow during maneuvering flight:
- Never move the cyclic faster than trim, torque, and rotor can be maintained. When entering a maneuver and the trim, rotor, or torque reacts quicker than anticipated, pilot limitations have been exceeded. If continued, an aircraft limitation will be exceeded. Perform the maneuver with less intensity until all aspects of the machine can be controlled.
- Anticipate changes in aircraft performance due to loading or environmental condition. The normal collective increase to check rotor speed at sea level standard (SLS) may not be sufficient at 4,000 feet pressure altitude (PA) and 95 degrees F (4K95).
- Anticipate the following characteristics during aggressive maneuvering flight and adjust or lead with collective as necessary to maintain trim and torque:
    - Left turns, torque increases.
    - Right turns, torque decreases.
    - Application of aft cyclic, torque decreases and rotor climbs.
    - Application of forward cyclic (especially when immediately following aft cyclic application), torque increases and rotor speed decreases.
- Always leave a way out.
- Know where the winds are.
- Most engine malfunctions occur during power changes.
- If combat maneuvers have not been performed in a while, start slowly to develop proficiency.
- Crew coordination is critical. Everyone needs to be fully aware of what is going on and each crewmember has a specific duty.
- In steep turns the nose will drop. In most cases, energy (airspeed) must be traded to maintain altitude as the required excess engine power may not be available (to maintain airspeed in a 2g/60 degree turn rotor thrust/engine power has to increase by 100 percent). Failure to anticipate this at low altitude endangers the crew and passengers. The rate of pitch change is proportional to gross weight and density altitude.
- Many maneuvering flight over-torques occur as the aircraft unloads Gs. This is due to insufficient collective reduction following the increase to maintain consistent torque and rotor as G-loading increased (dive recovery or recovery from high G-turn to the right).

SECTION VII – FACTORS AFFECTING PERFORMANCE

1-170. A helicopter's performance is dependent upon the power output of the engine and lift production of the rotors. Any factor affecting engine and rotor efficiency affects performance. The three major factors affecting performance are density altitude, weight, and wind.

DENSITY ALTITUDE

1-171. As air density increases, engine power output, rotor efficiency, and aerodynamic lift also increase. Density altitude is the altitude above mean sea level (MSL) at which a given atmospheric density occurs in the standard atmosphere. It can also be interpreted as PA corrected for nonstandard temperature differences.

1-172. PA is displayed as the height above a standard datum plane, which in this case, is a theoretical plane where air pressure is equal to 29.92 inches mercury (Hg). PA is the indicated height value when the altimeter setting is adjusted to 29.92 inches Hg. PA, as opposed to true altitude, is an important value for calculating performance as it more accurately represents the air content at a particular level. The difference between true altitude and PA must be clearly understood. True altitude means the vertical height above MSL and is displayed on the altimeter when the altimeter is correctly adjusted to the local setting.

1-173. For example, if the local altimeter setting is 30.12 inches Hg and adjusted to this value, it indicates exact height above sea level. However, this does not reflect conditions found at this height under standard conditions. Since the altimeter setting is more than 29.92 inches Hg, the air in this example has a higher

Chapter 1

pressure and is more compressed, indicative of air found at a lower altitude. Therefore, the PA is lower than the actual height above MSL. To calculate PA without use of an altimeter, remember pressure decreases approximately 1 inch of mercury for every 1,000-foot increase in altitude. For example, if the current local altimeter setting at a 4,000 foot elevation is 30.42, the PA would be 3,500 feet (30.42 – 29.92 = .50 inches Hg/.50 x 1,000 feet = 500 feet; subtracting 500 feet from 4,000 equals 3,500 feet.). Four factors affecting density altitude most are atmospheric pressure, altitude, temperature, and moisture content of the air.

ATMOSPHERIC PRESSURE

1-174. Due to changing weather conditions, atmospheric pressure at a given location changes from day to day. If the pressure is lower, the air is less dense. This means a higher density altitude and less helicopter performance.

ALTITUDE

1-175. As altitude increases, air becomes thinner. This is because the atmospheric pressure acting on a given volume of air is less, allowing air molecules to move further apart. Dense air contains air molecules spaced closely together, while thin air contains air molecules spaced further apart. As altitude increases, density altitude increases.

TEMPERATURE

1-176. As warm air expands the air molecules move further apart, creating less dense air. Since cool air contracts, air molecules move closer together creating denser air. High temperatures cause even low elevations to have high density altitudes.

MOISTURE (HUMIDITY)

1-177. The water content of air also changes air density as water vapor weighs less than dry air. Therefore, as the water content of the air increases, air becomes less dense, increasing density altitude and decreasing performance.

1-178. Humidity, also called relative humidity, refers to the amount of water vapor contained in the atmosphere and is expressed as a percentage of the maximum amount of water vapor air can hold. This amount varies with temperature; warm air can hold more water vapor, while colder air holds less. Perfectly dry air that contains no water vapor has a relative humidity of 0 percent, while saturated air that cannot hold any more water vapor has a relative humidity of 100 percent.

1-179. Humidity alone is usually not considered an important factor in calculating density altitude and helicopter performance however, it does contribute. There are no rules-of-thumb or charts used to compute the effects of humidity on density altitude. Aviators should expect a decrease in hovering and takeoff performance in high humidity conditions.

HIGH AND LOW DENSITY ALTITUDE CONDITIONS

1-180. A thorough understanding of the terms high density altitude and low density altitude are required. In general, high density altitude refers to thin air, while low density altitude refers to dense air. Those conditions resulting in a high density altitude (thin air) are high elevations, low atmospheric pressure, high temperatures, high humidity, or some combination thereof. Lower elevations, high atmospheric pressure, low temperatures, and low humidity are more indicative of low density altitude (dense air). However, high density altitude s may be present at lower elevations on hot days, so it is important to calculate density altitude and determine performance before a flight.

1-181. One of the ways density altitude can be determined (CPU-26A/P is another) is through use of charts designed for that purpose (figure 1-73, page 1-57). The graph is used to find density altitude either on the ground or aloft. Set altimeter at 29.92 inches to indicate PA. Read outside air temperature (OAT). Enter the graph at that PA and move horizontally to the temperature. Read density altitude from the sloping lines.

- **Example 1.** Find density altitude in flight. PA is 9,500 feet and temperature is 18 degrees F. Find 9,500 feet on the left of the graph and move across to 18 degrees F. density altitude is 9,000 feet (marked 1 on the graph).
- **Example 2.** Find density altitude for takeoff. PA is 4,950 feet and temperature is 97 degrees F. Enter the graph at 4,950 feet and move across to 97 degrees F. density altitude is 8,200 feet (marked 2 on graph).

Note. In warm air, density altitude is considerably higher than PA.

1-182. Most performance charts do not require computation of density altitude. Instead, the computation is built into the performance chart. All that remains is to enter the correct PA and temperature.

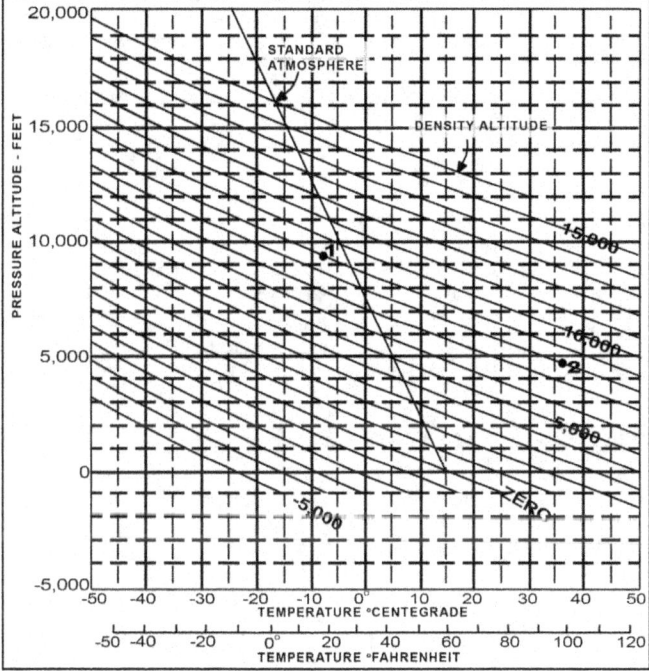

Figure 1-73. Density altitude computation

WEIGHT

1-183. Weight is the force opposing lift. As weight increases, power required to produce lift needed to compensate for the added weight must also increase. Most performance charts include weight as one of the variables. By reducing weight, the helicopter is able to safely takeoff or land at a location otherwise impossible. However, if in doubt, delay takeoff until more favorable density altitude conditions exists. If airborne, try to land at a location that has more favorable conditions, or one where a landing can be made that does not require a hover.

1-184. In addition, at higher gross weights the increased power required to hover produces more torque, which means more antitorque thrust is required. In some helicopters, during high altitude operations, the maximum antitorque produced by the tail rotor during a hover may not be sufficient to overcome torque even if the gross weight is within limits.

WINDS

1-185. Wind direction and velocity also affect hovering, takeoff, and climb performance. Translational lift occurs anytime there is relative airflow over the rotor disc. This occurs whether the relative airflow is caused by helicopter movement or wind. As wind speed increases, translational lift increases, resulting in less power required to hover.

1-186. Wind direction is also an important consideration. Headwinds are desirable as they contribute to the most increase in performance. Strong crosswinds and tailwinds may require use of more tail rotor thrust to maintain directional control. This increased tail rotor thrust absorbs power from the engine, which means less power is available to the main rotor for production of lift. Some helicopters even have a critical wind azimuth or maximum safe relative wind chart. Operating the helicopter beyond these limits could cause loss of tail rotor effectiveness.

1-187. Takeoff and climb performance is greatly affected by wind. When taking off into a headwind ETL is achieved earlier, resulting in more lift and a steeper climb angle. When taking off with a tailwind more distance is required to accelerate through translation lift.

PERFORMANCE CHARTS

1-188. In developing performance charts, aircraft manufacturers make certain assumptions about the condition of the helicopter and ability of the pilot. It is assumed the helicopter is in good operating condition and the engine is developing its rated power. The pilot is assumed to be following normal operating procedures and to have average flying abilities. Average means a pilot capable of doing each of the required tasks correctly and at appropriate times.

1-189. Using these assumptions, the manufacturer develops performance data for the helicopter based on actual flight tests. However, they do not test the helicopter under each and every condition shown on a performance chart. Instead, they evaluate specific data and mathematically derive the remaining data.

HOVERING PERFORMANCE

1-190. Helicopter performance revolves around whether or not hover is possible. More power is required during hover than in any other flight regime. Obstructions aside, if hover can be maintained, takeoff can be made, especially with the additional benefit of translational lift. Charts are provided for IGE and OGE under various conditions of gross weight, altitude, temperature, and power. The IGE hover ceiling is higher than OGE hover ceiling due to the added lift benefit produced by ground effect.

1-191. As density altitude increases more power is required to hover. At some point, the power required is equal to the power available. This establishes the hovering ceiling under existing conditions. Any adjustment to gross weight by varying fuel, payload, or both, affects the hovering ceiling. The heavier the gross weight, the lower the hovering ceiling. As gross weight is decreased, the hover ceiling increases.

1-192. Being able to hover at the takeoff location with a certain gross weight does not ensure the same performance at the landing point. If the destination point is at a higher density altitude because of higher elevation, temperature, and/or relative humidity, more power is required to hover. You should be able to predict whether hovering power will be available at the destination by knowing the temperature and wind conditions, using performance charts in the helicopter flight manual, and making certain power checks during hover and in flight prior to commencing the approach and landing.

CLIMB PERFORMANCE

1-193. Most factors affecting hover and takeoff performance also affect climb performance. In addition, turbulent air, pilot techniques, and overall condition of the helicopter can cause climb performance to vary.

Aerodynamics of Flight

1-194. A helicopter flown at the best rate-of-climb speed obtains the greatest gain in altitude over a given period of time. This speed is normally used during the climb after all obstacles have been cleared and is usually maintained until reaching cruise altitude. Rate of climb must not be confused with angle of climb. Angle of climb is a function of altitude gained over a given distance. The best rate-of-climb speed results in the highest climb rate, but not the steepest climb angle and may not be sufficient to clear obstructions. The best angle-of-climb speed depends upon power available. If there is a surplus of power available the helicopter can climb vertically, so the best angle-of-climb speed is zero.

1-195. Wind direction and speed have an effect on climb performance, but it is often misunderstood. Airspeed is the speed at which the helicopter is moving through the atmosphere and is unaffected by wind. Atmospheric wind affects only the ground speed and ground track.

SECTION VIII – EMERGENCIES

SETTLING WITH POWER

1-196. Settling with power (figures 1-74 through 1-76) is a condition of powered flight in which the helicopter settles in its own downwash. This condition may also be referred to as vortex ring state. Under certain conditions the helicopter may descend at a high rate which exceeds the normal downward induced flow rate of the inner blade sections (inner section of the rotor disk). Therefore, the airflow of the inner blade sections is upward relative to the disk. This produces a secondary vortex ring in addition to the normal tip vortex system. The secondary vortex ring is generated about the point on the blade where airflow changes from up to down. The result is an unsteady turbulent flow over a large area of the disk which causes loss of rotor efficiency although engine power is still supplied to the rotor system.

1-197. Figure 1-74 shows normal induced flow velocities along the blade span during hovering flight. Downward velocity is highest at the blade tip where blade speed is highest. As blade speed decreases nearer the center of the disk, downward velocity is less.

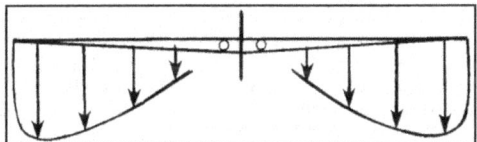

Figure 1-74. Induced flow velocity during hovering flight

1-198. Figure 1-75 shows the induced airflow velocity pattern along the blade span during a descent conducive to settling with power. The descent is so rapid, induced flow at the inner portion of the blades is upward rather than downward. The upflow caused by the descent has overcome the downflow produced by blade rotation and pitch angle.

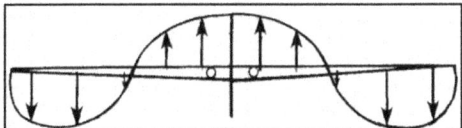

Figure 1-75. Induced flow velocity before vortex ring state

1-199. If this rate of descent exists with insufficient power to slow or stop the descent, it will enter the vortex ring state (figure 1-76, page 1-60). During this vortex ring state, roughness and loss of control occur due to turbulent rotational flow on the blades and unsteady shifting of the flow along the blade span.

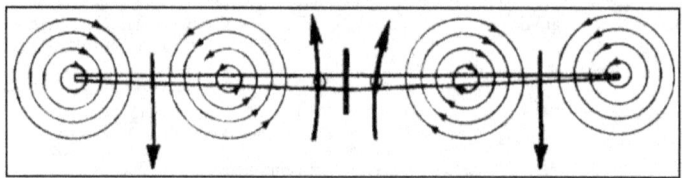

Figure 1-76. Vortex ring state

1-200. The following conditions must exist simultaneously for settling with power to occur:
- A vertical or near-vertical descent of at least 300 feet per minute (FPM). Actual critical rate depends on gross weight, rotor RPM, density altitude, and other pertinent factors.
- Slow forward airspeed (less than ETL).
- Rotor system must be using 20 to 100 percent of the available engine power with insufficient power remaining to arrest the descent. Low rotor RPM could aggravate this.

1-201. The following flight conditions are conducive to settling with power:
- Steep approach at a high rate of descent.
- Downwind approach.
- Formation flight approach (where settling with power could be caused by turbulence of preceding aircraft).
- Hovering above the maximum hover ceiling.
- Not maintaining constant altitude control during an OGE hover.
- During masking/unmasking.

1-202. Recovery from settling with power may be affected by one, or a combination, of the following ways:
- During the initial stage (when a large amount of excess power is available), a large application of collective pitch may arrest rapid descent. If done carelessly or too late, collective increase can aggravate the situation resulting in more turbulence and an increased rate of descent.
- In single-rotor helicopters, aviators can accomplish recovery by applying cyclic to gain airspeed and arrest upward induced flow of air and/or by lowering the collective (altitude permitting). Normally, gaining airspeed is the preferred method as less altitude is lost.
- In tandem-rotor helicopters, fore and aft cyclic inputs aggravate the situation. By lowering thrust (altitude permitting) and applying lateral cyclic input or pedal input to arrest this upward induced flow of air, the aviator can accomplish recovery.

1-203. Several conclusions can be drawn from figure 1-77, page 1-61—
- The vortex ring state can be completely avoided by descending on flight paths shallower than about 30 degrees (at any speed).
- For steeper approaches, the vortex ring state can be avoided by using rates of descent versus horizontal velocity either faster or slower than those passing through the area of severe turbulence and thrust variation.
- At very shallow angles of descent, the vortex ring wake is dispersed behind the helicopter. Forward airspeed coupled with induced-flow velocity prevents the upflow from materializing on the rotor system.
- At steep angles, the vortex ring wake is below the helicopter at slow rates of descent and above the helicopter at high rates of descent. Low rates of descent prevent the upflow from exceeding the induced flow velocities. High rates of descent result in autorotation or the windmill brake state.

Aerodynamics of Flight

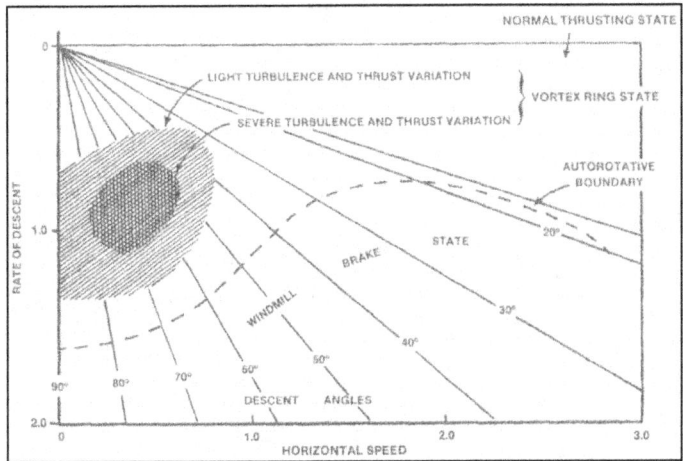

Figure 1-77. Settling with power region

DYNAMIC ROLLOVER

1-204. A helicopter is susceptible to a lateral-rolling tendency called dynamic rollover. Dynamic rollover can occur on level ground as well as during a slope or crosswind landing and takeoff. Three conditions are required for dynamic rollover—pivot point, rolling motion, and exceed critical angle.

Pivot Point

1-205. Dynamic rollover begins when the helicopter starts to pivot around its skid, wheel, or any portion of the aircraft in contact with the ground. When this happens, lateral cyclic control response is more sluggish and less effective than for a free hovering helicopter. This can occur for a variety of reasons including failure to remove a tiedown or skid securing device, the skid or wheel contacts a fixed object while hovering sideward, or the gear is stuck in ice, soft asphalt, or mud. Dynamic rollover may also occur if proper landing or takeoff technique is not used or while performing slope operations. If the gear or skid becomes a pivot point, dynamic rollover is possible if proper corrective techniques are not used.

Rolling Motion

1-206. The rate of rolling motion is vital. As the roll rate increases, the critical angle is reduced. In a fully articulated rotor system, all three control inputs (collective, cyclic, and pedals) can contribute to the rolling motion.

Exceed Critical Angle

1-207. To understand critical angle we must first discuss static rollover angle. Each helicopter has a static rollover angle that, if exceeded, will cause the aircraft to rollover. The static angle is based on CG and pivot point. This angle is described as being the point where the aircraft CG is located over the pivot point.

1-208. When a rolling motion is present the dynamic rollover angle is introduced and is called the critical angle. The dynamic angle varies based on the rate of the rolling motion of the helicopter. The greater the rolling motion the earlier (less bank angle) the critical angle will be exceeded. If the dynamic rollover angle is exceeded, momentum will carry the helicopter through the static rollover angle, regardless of corrections by the aviator.

TYPES

1-209. Certain factors influence dynamic rollover including right skid down, left pedal inputs (single-rotor aircraft), the effects of pilot input (lateral motion) in a rigid rotor aircraft, lateral loading (asymmetrical loading), crosswind, and high roll rates. Smooth and moderate collective inputs are most effective in preventing dynamic rollover as it reduces the rate at which lift/thrust is applied. A smooth and moderate collective reduction is recommended if the onset of dynamic rollover is encountered. There are three main rollover types normally encountered—rolling over on level ground (takeoff), rolling downslope (takeoff or landing) and rolling upslope (takeoff).

Rolling Over on Level Ground

1-210. A rollover condition can occur during takeoff from level ground if one skid or wheel is stuck on the ground. As collective pitch is increased, the stuck skid or wheel becomes the pivot point which sets dynamic rollover into motion. A smooth and moderate collective reduction is recommended lowering the aircraft back to the ground until the stuck skid or wheel is free. Then the aircraft may be picked up normally.

Rolling Downslope

1-211. A downslope rollover during landing (figure 1-78) occurs when the steepness of the slope causes the helicopter to tilt beyond the lateral cyclic control limits. If the steepness of the slope, a crosswind component, or CG conditions exceeds lateral cyclic control limits, the mast forces the rotor to tilt downslope. The resultant rotor vector has a downslope component even with full upslope cyclic applied. To prevent downslope rollover during landing, the aviator slowly descends vertically until ground contact with the upslope skid/wheel occurs. At this point, aircrew members can better assess slope conditions. After stabilizing the helicopter in this position, the aviator smoothly reduces collective until the downslope skid/wheel contacts the ground or cyclic nears lateral limits. If the cyclic is near the lateral limit, the aviator must carefully evaluate remaining distance to ensure enough cyclic travel remains to land without exceeding aircraft limits. If not enough travel remains the aviator should abort the landing, return the aircraft to a hover, and select an area of lesser slope.

1-212. A downslope rollover during takeoff (figure 1-78) can occur when the aviator lands the helicopter on too steep a slope, then attempts takeoff. If the upslope skid/wheel begins to rise first, the aviator should lower the collective to prevent a downslope rollover condition. If, with full cyclic applied, the resultant lift of the main rotor is not vertical or directed upslope enough to raise the downslope gear first, and then further takeoff attempts result in the mast causing resultant rotor lift to move further downslope and cause dynamic rollover. The aviator should consider some adjustments before making additional takeoff attempts. These adjustments include awaiting different wind conditions, changing the CG of the helicopter by moving/removing some of the internal load, or contacting a recovery crew.

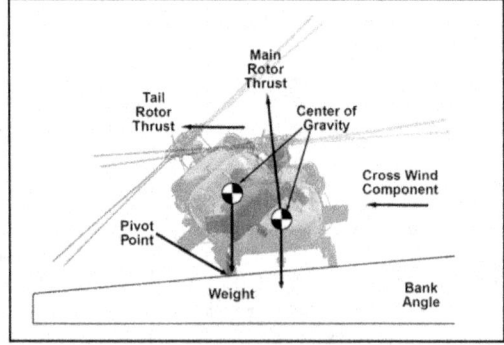

Figure 1-78. Downslope rolling motion

Aerodynamics of Flight

Rolling Upslope

1-213. An upslope rollover during takeoff (figure 1-79) occurs when the aviator applies too much cyclic into the slope to hold the skid/wheel firmly on the slope. If the aviator improperly applies collective, the helicopter then rapidly pivots upslope around the upslope skid/wheel. To prevent this, the aviator needs to cautiously apply collective while neutralizing the cyclic. When the cyclic is neutral and upslope skid/wheel has no side pressure applied, the aviator performs a vertical lift-off to a hover, then a normal takeoff.

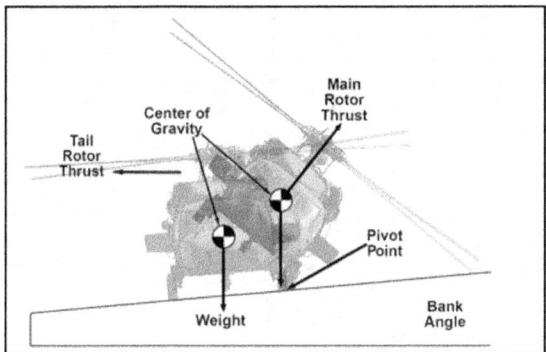

Figure 1-79. Upslope rolling motion

PREVENTION

1-214. Dynamic rollover usually occurs due to a combination of physical and human factors. Physical factors considered in the prevention of dynamic rollover include main rotor thrust, CG, tail-rotor thrust, crosswind component, ground surface, sloped landing area, and in some aircraft, presence of a low fuel condition which might cause the CG to move upward. The aviator can prevent dynamic rollover by avoiding the physical factors causing it; however, human factors can interfere in the avoidance process. Human factors considered in the prevention of dynamic rollover include—

- **Inattention.** Dynamic rollover is more likely if the aviator at the controls is inattentive to aircraft position and attitude when lifting off or touching down to the ground, effectively losing situational awareness (SA).
- **Inexperience.** Most dynamic rollover accidents occur while inexperienced aviators are at the controls. The pilot in command (PC) must remain vigilant.
- **Failure to take timely corrective action.** Timely action must be exercised before a roll rate develops.
- **Inappropriate control input.** Applying inappropriate or incorrect control input is the root cause of nearly all dynamic rollovers. If the aviator applies appropriate control input smoothly and carefully, dynamic rollover is avoidable.
- **Loss of visual reference.** Loss of visual reference may allow the aircraft to drift unnoticed by the crew. If the aircraft contacts the ground while drifting sideward, rollover can occur. Therefore, if visual reference is lost while the aircraft nears the ground, the aviator should execute a takeoff or go-around using instrument techniques if necessary.

COMMON ERRORS

1-215. The following are examples of common errors:

- Aviator fails to detect the aircraft's lateral motion across the ground before landing.
- Aviator makes abrupt cyclic displacements (with or without thrust) in fully articulated rotor systems.

Chapter 1

- Aviator makes large and/or uncoordinated antitorque pedal inputs.
- Aviator performs slope landing/takeoff maneuvers while using rapidly increasing or decreasing collective control applications.

RETREATING BLADE STALL

1-216. The retreating blade of a helicopter will eventually stall in forward flight (figures 1-80 through 1-82). As the stall of an airplane wing limits the low speed of a FW aircraft, the stall of a rotor blade limits the high speed of a rotary-wing aircraft. In forward flight, decreasing velocity of airflow on the retreating blade demands a higher AOA to generate the same lift as the advancing blade. Figure 1-80 illustrates the lift pattern at a normal hover with distribution/production of lift evenly spread throughout the rotor disk.

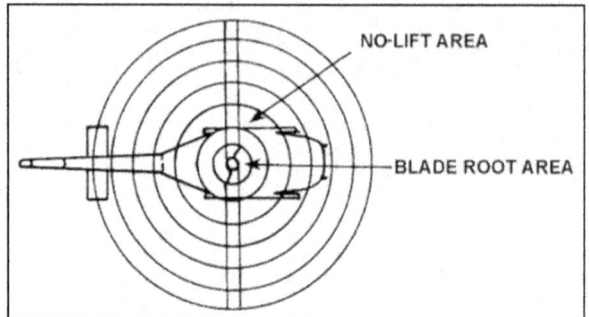

Figure 1-80. Retreating blade stall (normal hovering lift pattern)

1-217. Figure 1-81 illustrates the normal cruise lift pattern where the smaller area of the retreating blade, with its high angles of attack, must still produce an amount of lift equal to the larger area of the advancing blade with its lower angles of attack. This figure shows the advancing blade producing lift throughout its span while the retreating blade is producing lift in only part of its span due to effects of forward airspeed. When forward speed increases, the no-lift areas of the retreating blade grow larger, placing an even greater demand for production of lift on a progressively smaller section of the retreating blade. This smaller section of blade demands a higher AOA until the tip of the blade (area of the highest AOA) stalls.

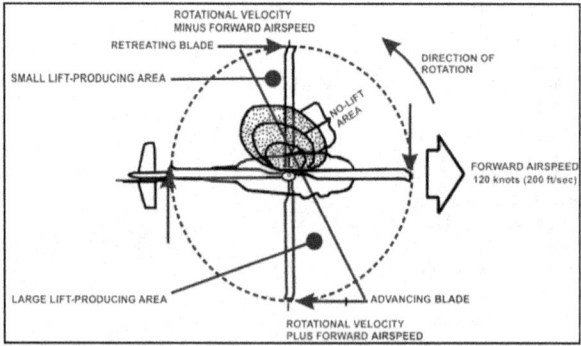

Figure 1-81. Retreating blade stall (normal cruise lift pattern)

1-218. Figure 1-82, page 1-65 illustrates the same disk at a critical airspeed with the retreating blade producing less than sufficient lift due to the no-lift area growing larger and effects of tip stall. Tip stall causes vibration and buffeting which spread inboard and aggravate the situation while the aircraft may roll left and nose pitches up. While this may be subtle, it will worsen if aft cyclic is not applied or collective is reduced (altitude permitting). The effects of retreating blade stall in a tandem-rotor helicopter create a different response. With the forward and aft rotor systems turning in opposite directions, effects of retreating blade stall on the separate rotors tend to counteract themselves. The pitch-up of the nose will be insignificant. Blade stall will probably occur on the aft system first as it operates in the turbulent wake of the forward rotor system. The most likely effect will be an increasing vibration which is easily reduced by slowing down and reducing collective pitch (thrust).

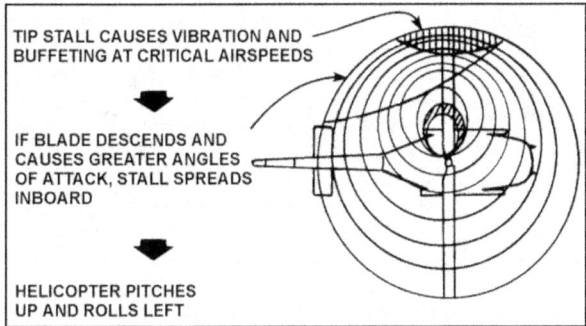

Figure 1-82. Retreating blade stall (lift pattern at critical airspeed–retreating blade stall)

Conditions Producing Blade Stall

1-219. In operations at high forward speeds, the following conditions are most likely to produce blade stall in either single- or tandem-rotor helicopters—
- High blade loading (high gross weight).
- Low rotor RPM.
- High- density altitude.
- High G-maneuvers.
- Turbulent air.

Recovering From Blade Stall

1-220. The following steps enable the aviator to recover from retreating blade stall—
- Reduce collective.
- Reduce airspeed.
- Descend to a lower altitude (if possible).
- Increase rotor RPM to normal limits.
- Reduce the severity of the maneuver.

GROUND RESONANCE

1-221. Ground resonance may develop in helicopters having fully articulated rotor systems when a series of shocks causes the rotor blades in the system to become positioned in unbalanced displacement. If this oscillating condition progresses, it can be self-energizing and extremely dangerous. It can easily cause structural failure. It is most common to three-blade helicopters with landing wheels. The rotor blades in a three-blade system are equally spaced (120 degrees), but are constructed to allow some horizontal lead and lag action. Ground resonance occurs when the helicopter contacts the ground during landing or takeoff (figure 1-83, page 1-66). If one wheel of the helicopter strikes the ground ahead of the others, a shock is transmitted through the fuselage to the rotor. Another shock is transmitted when the next wheel hits. The first shock causes the blades straddling the contact point to jolt out of angular balance. If repeated by the next contact, resonance is established setting up a self-energizing oscillation of the fuselage. Severity of the oscillation increases rapidly. The helicopter can quickly disintegrate without immediate corrective action. Corrective action may consist of an immediate takeoff to a hover or a change in rotor RPM to alleviate the condition and disrupt the pattern of oscillation. In the event takeoff is not an option, all personnel should remain in the aircraft until main rotors have stopped. Ground resonance usually occurs when the aircraft is nearly airborne (80 to 90 percent hover power applied).

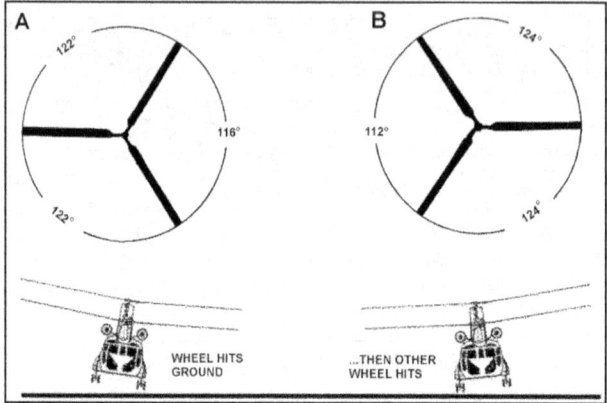

Figure 1-83. Ground resonance

1-222. The following conditions can cause ground resonance—
- Defective drag dampers allowing excessive lead and lag and creating angular unbalance.
- Improperly serviced or defective landing-gear struts.
- Hard landings on one skid or wheel.
- Ground taxiing over rough terrain.
- Hesitant or bouncing landings.

COMPRESSIBILITY EFFECTS

COMPRESSIBLE AND INCOMPRESSIBLE FLOW

1-223. At low airspeeds, air is incompressible. Incompressible airflow is similar to the flow of water, hydraulic fluid, or any other incompressible fluid. At low speeds, air experiences relatively small changes in pressure with little change in density. However, at high speeds greater pressure changes occur causing compression of air which results in significant changes to air density. This compressible flow occurs when there is a transonic or supersonic flow of air across the airfoil. Because helicopters are being flown at increasingly higher speeds, aviators must learn more about coping with effects of compressible flow.

1-224. The major factor in high-speed airflow is the speed of sound. Speed of sound is the rate at which small pressure disturbances move through the air. This propagation speed is solely a function of air temperature. Table 1-4, page 1-67, shows the variation of speed of sound with temperature at various altitudes in the standard atmosphere.

Aerodynamics of Flight

Table 1-4. Speed of sound variation with temperature and altitude

Altitude	Temperature		Speed of Sound
Feet	°F	°C	Knots
Sea Level	59.0	15.0	661.7
5,000	41.2	5.1	650.3
10,000	23.3	-4.8	638.6
15,000	5.5	-14.7	626.7
20,000	-12.3	-24.6	614.6
25,000	-30.2	-34.5	602.2
30,000	-48.0	-44.4	589.6
35,000	-65.8	-54.3	576.6
40,000	-69.7	-56.5	573.8
50,000	-69.7	-56.5	573.8
60,000	-69.7	-56.5	573.8
C-Celcius F-Fahrenheit			

1-225. Compressibility effects are not limited to blade speeds at and above the speed of sound. The aerodynamic shape of an airfoil causes local flow velocities greater than blade speed. Thus a blade can experience compressibility effects at speeds well below the speed of sound because both subsonic and supersonic flows can exist on a blade.

1-226. Differences between subsonic and supersonic flow are due to compressibility of supersonic flow. Figure 1-84, page 1-68, compares incompressible and compressible flow through a closed tube. In this example, the mass flow along the tube is constant.

Subsonic Incompressible Flow

1-227. The example of subsonic incompressible flow is simplified because density of flow is constant throughout the tube. As the flow approaches a constriction and streamlines converge, velocity increases as static pressure decreases. A convergence of the tube requires an increasing velocity to accommodate the continuity of flow. Also, as the subsonic incompressible flow enters a diverging section of the tube, velocity decreases and static pressure increases; density remains unchanged.

Supersonic Compressible Flow

1-228. The example of supersonic compressible flow is complicated because variations of flow density are related to changes in velocity and static pressure. The behavior of supersonic compressible flow is a convergence causing compression; a divergence causes expansion. Therefore, as the supersonic compressible flow approaches a constriction and streamlines converge, velocity decreases and static pressure increases. Continuity of mass flow is maintained by the increase in flow density accompanying the decrease in velocity. As the supersonic compressible flow enters a diverging section of the tube, velocity increases and static pressure decreases; density decreases to accommodate the condition of continuity.

Chapter 1

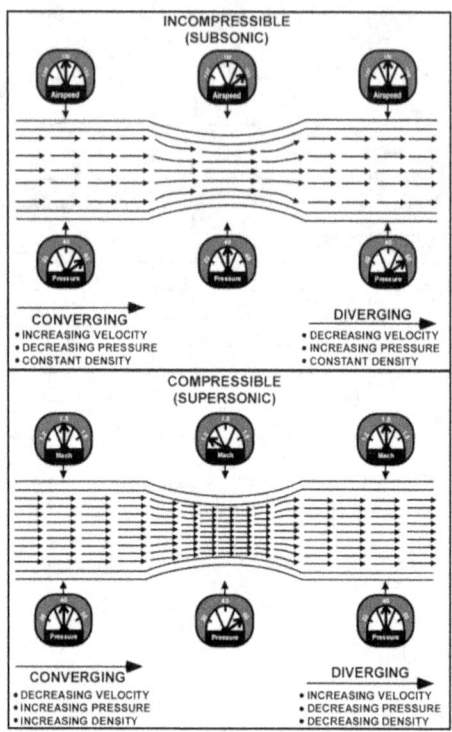

Figure 1-84. Compressible and incompressible flow comparison

TRANSONIC FLOW PATTERNS

1-229. In subsonic flight, an airfoil producing lift has local velocities on the surface greater than the free stream velocity. Compressibility effects can then be expected to occur at flight speeds less than the speed of sound. Mixed subsonic and supersonic flow may be encountered in the transonic regime of flight. The first significant effects of compressibility occur in this regime. Compressibility effects on the helicopter increase the power required to maintain rotor RPM and cause rotor roughness, vibration, cyclic shake, and an undesirable structural twisting of the blade.

1-230. Critical Mach number is the highest blade speed without supersonic airflow. As the critical Mach number is exceeded, an area of supersonic airflow is created. A normal shock wave then forms the boundary between supersonic and subsonic flow on the aft portion of the airfoil surface. The acceleration of airflow from subsonic to supersonic is smooth and without shock waves if the surface is smooth and transition gradual. However, transition of airflow from supersonic to subsonic is always accompanied by a shock wave. When airflow direction does not change, the wave formed is a normal shock wave.

1-231. The normal shock wave is detached from the leading edge of the airfoil and perpendicular to the upstream flow. The flow immediately behind the wave is subsonic. Figure 1-85 illustrates how an airfoil at high subsonic speeds has local supersonic flow velocities. As the local supersonic flow moves aft, a normal shock wave forms slowing the flow to subsonic. As supersonic air passes through shock wave, air density

Aerodynamics of Flight

increases, heat is created, velocity of the air decreases, static pressure increases, and boundary layer separation may occur.

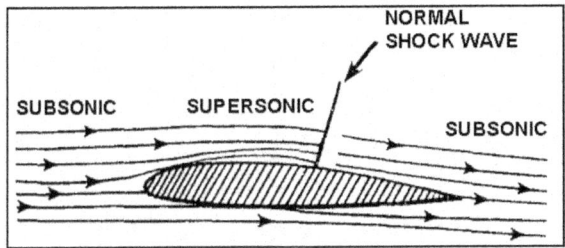

Figure 1-85. Normal shock wave formation

1-232. As the shock waves move toward the trailing edge of the airfoil, the aerodynamic center begins to move away from its normal location of 25 percent chord. By the time the shock wave has reached the trailing edge of the airfoil, the aerodynamic center has retreated to the 50 percent chord. This causes the leading edge of the airfoil to be deflected down, which may result in structural failure of the blade (skin deformation or separation).

1-233. Because speed of the helicopter is added to the speed of rotation of the advancing blade, the highest relative velocities occur at the tip of the advancing blade. When the Mach number of the advancing blade tip section exceeds the critical Mach number for the rotor blade section, compressibility effects result. The critical Mach number is the free stream Mach number producing the first evidence of local sonic flow. The principal effects of compressibility are large increase in drag and rearward shift of the airfoil aerodynamic center.

ADVERSE COMPRESSIBILITY CONDITIONS

1-234. The following operating conditions represent the most adverse compressibility conditions:
- High airspeed.
- High rotor RPM.
- High gross weight
- High- density altitude.
- High G-maneuvers.
- Low temperature. Speed of sound is proportional to the square root of the absolute temperature; therefore, the aviator more easily obtains sonic velocity at low temperatures.
- Turbulent air. Sharp gusts momentarily increase the blade AOA and thus, lower the critical Mach number to the point where compressibility effects may be encountered on the blade.

CORRECTIVE ACTIONS

1-235. Corrective actions are any actions decreasing AOA or velocity of airflow that help the situation. There are similarities in the critical conditions for compressibility and retreating blade stall, with notable exceptions—compressibility occurs at high rotor RPM, and retreating blade stall occurs at low rotor RPM. With the exception of RPM control, the recovery technique is identical for both. Such techniques include decreasing—
- Blade pitch by lowering collective, if possible.
- Rotor RPM.
- Severity of maneuver.
- Airspeed.

This page intentionally left blank.

Chapter 2

Weight, Balance, and Loads

Flight performance of aircraft are directly dependent upon ambient conditions and aircraft weight and balance. Gross weight and center of gravity (CG) have bearing on performance, stability, and control of the aircraft. Hazardous flight conditions and accidents can be prevented by adherence to the principles of weight and balance set forth in this chapter.

SECTION I – WEIGHT

2-1. Weight is one of the most important factors considered from the time the aircraft is designed until it is removed from service. It is of prime importance to the manufacturer through all phases of production and must remain foremost in the pilot's mind when planning and performing missions. Changes in basic aircraft design weight, either in initial production or subsequent modifications by maintenance activities, have direct bearing on aircraft performance. Cargo/troop loading and aircraft gross weight must be examined closely by the pilot as these factors may determine the safety and success of a mission. Gross weight limitations have been established and are in applicable operator's manuals.

WEIGHT DEFINITIONS

EMPTY WEIGHT

2-2. Empty weight is used for design purposes and usually does not affect service activities. Empty weight includes aircraft structure weight plus communications, control, electrical, hydraulic, instrument, and power plant systems; furnishings; anti-icing equipment; auxiliary power plant; flotation landing gear; and armament, anchor, and towing provisions.

BASIC WEIGHT

2-3. Basic weight of an aircraft is weight including all hydraulic and oil systems full, trapped and unusable fuel, and all fixed equipment. From basic weight total, it is only necessary to add crew, fuel, cargo, and ammunition (if carried) when determining the aircraft's gross weight. Basic weight varies with structural modifications and changes of fixed aircraft equipment.

OPERATING WEIGHT

2-4. Operating weight includes basic weight plus aircrew, aircrew's baggage, emergency gear, and other equipment required. Operating weight does not include weight of fuel, ammunition, bombs, cargo, or external auxiliary fuel tanks if such tanks are to be disposed of during flight.

GROSS WEIGHT

2-5. Gross weight is total weight of an aircraft and its contents.

TAKEOFF GROSS WEIGHT

2-6. Takeoff gross weight includes operating weight plus fuel, cargo, ammunition, bombs, auxiliary fuel tanks, and other material carried.

Landing Gross Weight

2-7. Landing gross weight is takeoff gross weight minus items expended during flight.

Useful Load

2-8. Useful load is the difference between empty and gross weight and includes fuel, oil, crew, passengers, cargo, and other material carried.

Total Aircraft Weight

2-9. Total aircraft weight includes the sum of operating weight and weight of takeoff fuel.

WEIGHT VERSUS AIRCRAFT PERFORMANCE

2-10. Specific weight limitations of an aircraft cannot be exceeded without compromising safety. Overloading an aircraft may cause structural failure or result in reduced engine and airframe life. An increase in gross weight affects the aircraft's performance as follows:

- Increases takeoff distance.
- Reduces hover performance.
- Reduces rate of climb.
- Reduces cruising speed.
- Increases stalling speed (FW).
- Decreases retreating blade stall speed (rotary-wing).
- Reduces maneuverability.
- Reduces ceiling.
- Reduces range.
- Increases landing distances.
- Promotes instability.

SECTION II – BALANCE

2-11. Balance is of primary importance to aircraft stability. The CG is the point about which an aircraft would balance if it were possible to support the aircraft at that point. An aircraft should never be flown if the pilot is not satisfied with its loading and balance condition.

CENTER OF GRAVITY

2-12. The CG is defined as the theoretical point where all the aircraft's weight is considered to be concentrated. If an aircraft is suspended by a cable attached to the CG point, it balances like a teeter-totter. For aircraft with a single main rotor, the CG is usually close to the main rotor mast. The CG is not necessarily a fixed point; its location depends on distribution of items loaded in the aircraft. As variable load items are shifted or expended, there is a resultant shift in CG location. If mass center of an aircraft is displaced too far forward on the longitudinal axis, a nose heavy condition results. Conversely, if mass center is displaced too far aft on the longitudinal axis, a tail heavy condition results. An unfavorable location of the CG could possibly produce such an unstable condition that the pilot could lose control of the aircraft.

LATERAL BALANCE

2-13. Location of the CG with reference to the lateral axis is important. The design of an aircraft is such that symmetry is assumed to exist about a vertical plane through the longitudinal axis. This means for each item of weight existing to the left of the fuselage centerline there is generally an equal weight existing at a corresponding location on the right. Lateral mass symmetry, however, may be easily upset due to unbalanced lateral loading. Location of the lateral CG is not only important from the aspect of loading rotary-wing aircraft, but is also extremely important when considering FW exterior drop loads. The position of the lateral

CG is not computed, but the crew must be aware adverse effects will arise as a result of a laterally unbalanced condition.

BALANCE DEFINITIONS

2-14. Definitions of the more important terms pertaining to balance and its relationship to aircraft weight distribution are as follows.

ARM

2-15. The arm is the horizontal distance from the datum to any component of the aircraft or any object located within the aircraft. Another term used interchangeably with arm is station. If the component or object is located rear of the datum, it is measured as a positive number and usually referred to as inches aft the datum. Conversely, if the component or object is located forward of the datum, it is indicated as a negative number and usually referred to as inches forward the datum.

MOMENT

2-16. If the weight of an object is multiplied by its arm the result is known as its moment. Moment is a force resulting from an object's weight acting at a distance. Moment is also referred to as tendency of an object to rotate or pivot about a point. The farther an object is from a pivotal point, the greater its force.

REFERENCE DATUM

2-17. Reference datum is an imaginary plane perpendicular to the longitudinal axis of the aircraft and is usually located at or near the nose of the aircraft to eliminate arms with a minus value. If a negative arm is encountered, the corresponding moment will also be negative. Simplified moment is one which has been reduced in magnitude through division by a constant. For example, 3,201 inch pounds/1,000 is the simplified expression of 3,200,893 divided by 1,000 and rounded to the nearest whole number.

2-18. The advantage of simplification is seen in application when a column of moments is added. Inaccuracies resulting from rounding figures to the nearest whole number tend to cancel.

AIRCRAFT STATION

2-19. An aircraft station is a position defined by a plane perpendicular to the longitudinal aircraft axis. The number designation of this station signifies its distance from the reference datum. A station forward of the reference datum is negative, while a station aft of the reference datum is positive. Figure 2-1, page 2-4, illustrates location of stations.

Chapter 2

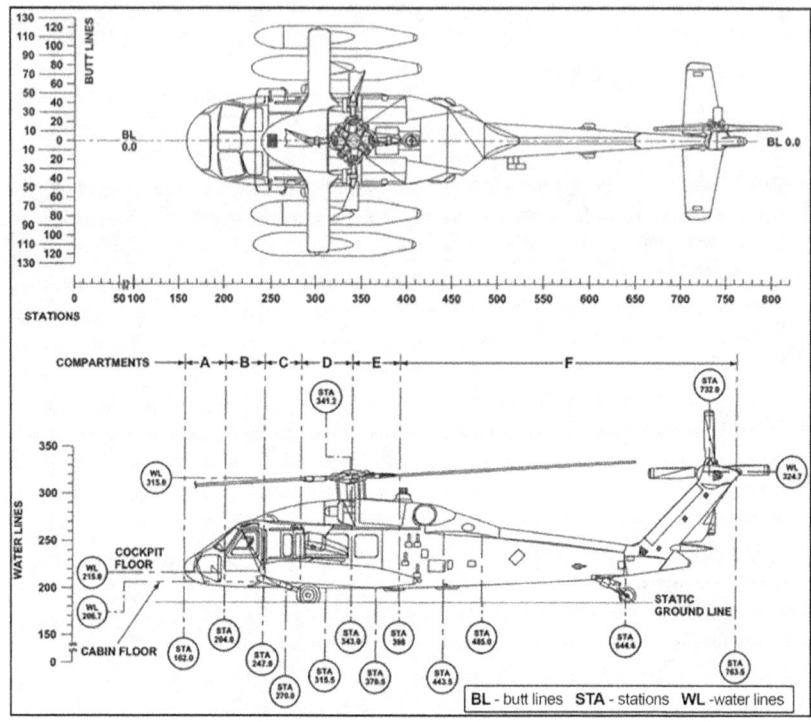

Figure 2-1. Helicopter station diagram

GROSS WEIGHT MOMENT

2-20. Gross weight moment is the sum of moments of all items making up the aircraft in the gross weight condition. Gross weight moment is the product of gross weight multiplied by gross weight arm.

BASIC ARM

2-21. Basic arm is the distance from the reference datum to the aircraft's CG in basic condition. It is obtained by dividing basic moment by basic weight.

GROSS WEIGHT ARM

2-22. Gross weight arm is the distance from reference datum to the aircraft CG in gross weight condition.

Weight, Balance, and Loads

> **Gross Weight Army**
> Gross weight arm (in) = Gross weight moments (in lb) divided by gross weight (lb)

PRINCIPLE OF MOMENTS

2-23. To understand balance, a working knowledge of the principle of moments is necessary. To calculate a moment, force (or weight) and distance must be known. The distance is measured from some desired known point (reference point or reference datum) to the point through which the force acts. A moment is meaningless unless the reference point about which the moment was calculated is specified.

2-24. For the purpose of illustration, an aircraft may be compared to a seesaw with the sum of the moments on each side of the balance point or fulcrum equal in magnitude (figure 2-2). The moment produced about the fulcrum by the 200-pound weight is 200 pounds x -50 inches = -10,000 inch pounds counterclockwise. The moment produced about the same reference point by the 100-pound weight is 100 pounds x 100 inches = 10,000 inch pounds clockwise. In this case, the clockwise moment counterbalances the counterclockwise moment and the system is in equilibrium. This example illustrates the principle of moments for a system to be in static equilibrium, the sum of the moments about any point must equal zero.

2-25. The clockwise moment is arbitrarily given a positive sign while the counterclockwise moment is given a negative sign. In determining balance of an aircraft, the fulcrum or CG is the unknown and must be determined.

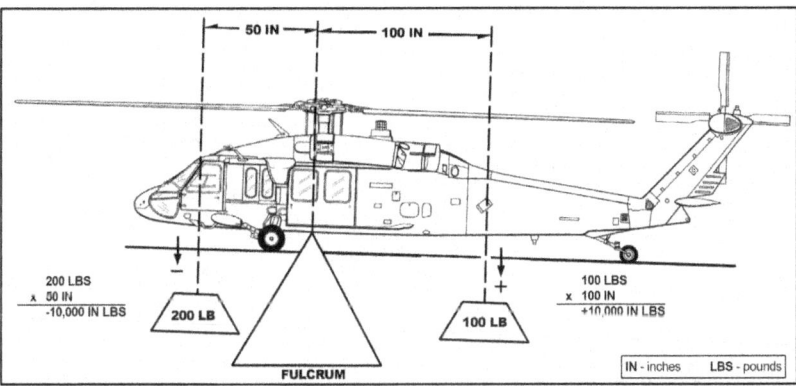

Figure 2-2. Aircraft balance point

SECTION III – WEIGHT AND BALANCE CALCULATIONS

2-26. When determining whether an aircraft is properly loaded, crews must answer two questions—
- Is gross weight less than or equal to maximum allowable gross weight?
- Is the CG within the allowable range and will it stay within that range as fuel is burned off?

2-27. To answer the first question, add the weight of the items comprising the useful load (pilot, passengers, fuel, oil [if applicable] cargo, and baggage) to the aircraft's basic empty weight to determine total weight does not exceed maximum allowable gross weight.

2-28. To answer the second question, use CG or moment information from loading charts, tables, or graphs in the operator's manual. Then, using one of the methods described below, calculate loaded moment and/or loaded CG and verify it falls within allowable CG range.

Chapter 2

2-29. By totaling the weights and moments of all components and objects carried, crews can determine the point where a loaded aircraft will balance. This point is known as the CG.

WEIGHT AND BALANCE METHODS

2-30. Since weight and balance are critical to safe operation of an aircraft, it is important to know how to check this condition for each loading arrangement. Most aircraft manufacturers use one of two methods, or a combination of these methods, to check weight and balance conditions.

COMPUTATIONAL METHOD

2-31. To determine CG (figure 2-3) location of a loaded aircraft—
- Obtain the aircraft's basic weight and moment from Department of Defense (DD) Form 365-3 (Weight and Balance Record, Chart C- Basic) and DD Form 365-4 (Weight and Balance Clearance Form F-Transport/Tactical).
- Obtain gross weight by adding the weight of the items being loaded to the aircraft's basic weight.
- Compute the moment of each load item by multiplying its weight by its arm.
- Find gross weight moment by adding the basic aircraft moment and moments of the load items.
- Determine the CG location by dividing gross weight moment by gross weight.

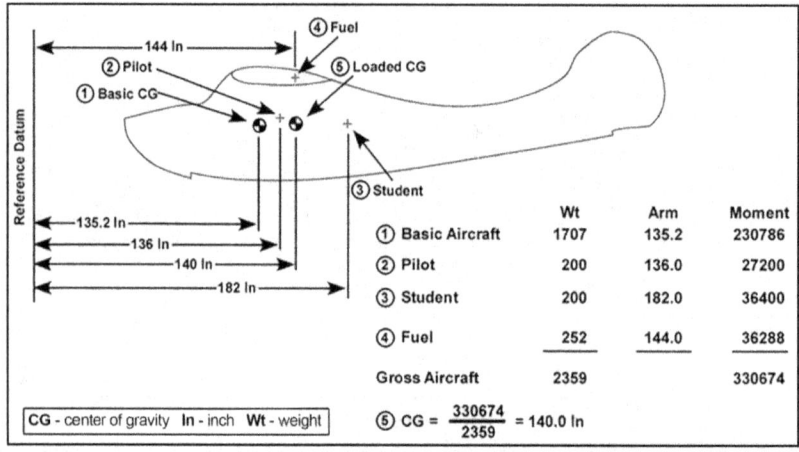

Figure 2-3. Locating aircraft center of gravity

LOADING CHART METHOD

2-32. This method can use line tracing or table format to obtain moments. Figure 2-4, page 2-7, illustrates the chart method of obtaining moments for calculation of CG.

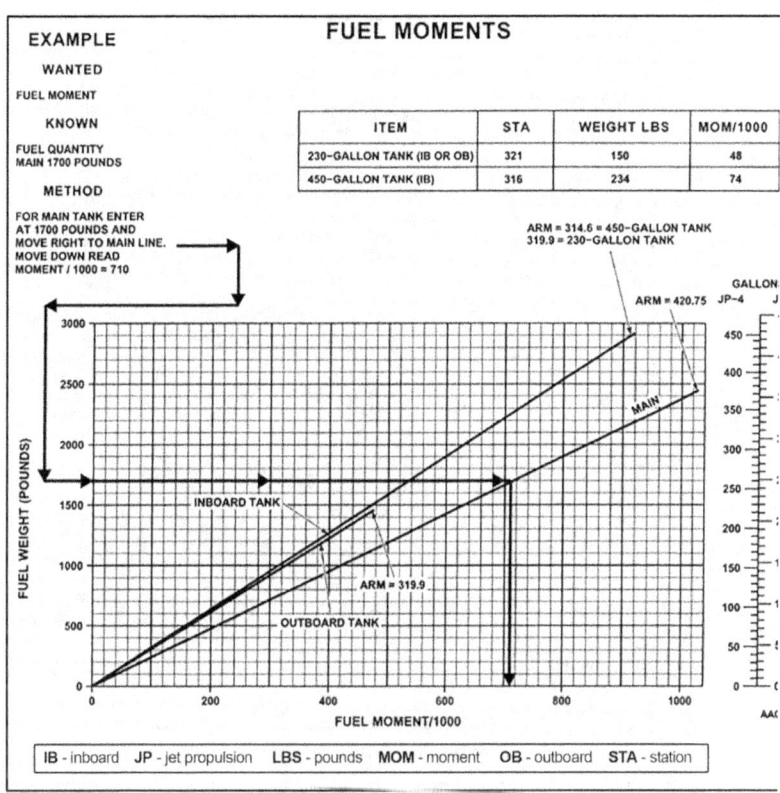

Figure 2-4. Fuel moments

CENTER OF GRAVITY LIMITS

2-33. The CG limit chart (figure 2-5, page 2-8) allows the CG (inches) to be determined when total and total moment are known. Individuals can also use this chart to determine allowable CG range by the arm at the intersection of gross weight and forward/aft limits line.

Chapter 2

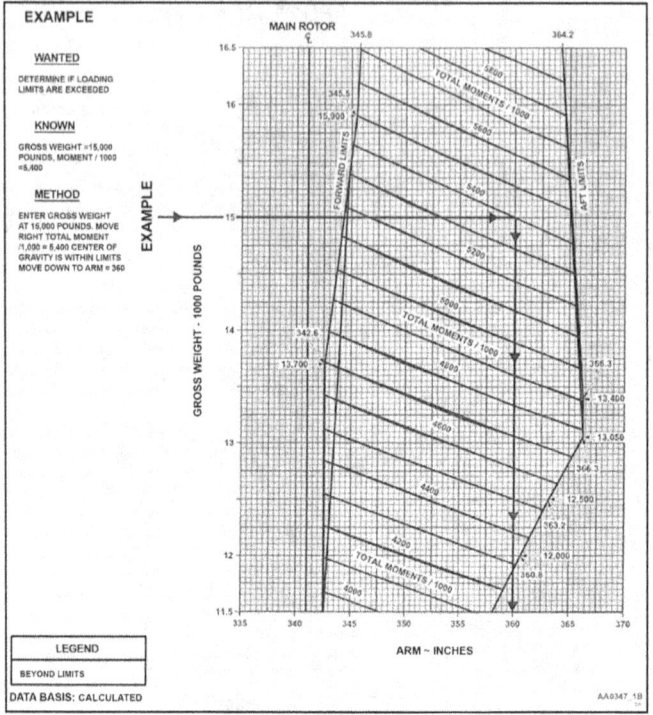

Figure 2-5. Center of gravity limits chart

SECTION IV – LOADS

PLANNING

SUPPORTED UNIT

2-34. The supported unit establishes liaison with the aviation unit to coordinate transport requirements. In particular, the supported unit is responsible for the following—
- Establishing priority for transport of cargo.
- Providing trained personnel, materiel, or handling equipment required to accomplish cargo preparation, rigging, hook-up release, and derigging. This should include all equipment required to contain or rig an external load enabling it to be attached to the helicopter hook (vehicles, containers, pallets, slings, straps, and clevises).
- Preparing internal cargo by aircraft load including shoring if required.
- Preparing external cargo by aircraft loads. Crews should prepare and rig external loads minimizing load oscillation during flight. Loads must not exceed allowable cargo weight established by the helicopter unit.
- Preparing dangerous cargo in accordance with appropriate regulations.

- Providing the helicopter unit with information on cargo weight, CG, load density, dimensions, axle weights of vehicles, and descriptions and quantities of all cargo. Whenever possible crews will mark weight and load density on each cargo element and complete cargo load. When weight and density of a load/element is not known, the supported unit provides the helicopter pilot with an estimated weight and density.
- Providing any static-electricity discharge probes or protective equipment and clothing required for ground hook-up personnel during external-load operations.
- Selecting and preparing pickup and release sites with technical advice provided as required by the supporting helicopter unit.

Supporting Aviation Unit

2-35. The helicopter unit is responsible for—
- Providing liaison with the supported unit to coordinate planning. The helicopter unit provides information and advice on aircraft availability, allowable cargo load (ACL), and special loading instructions such as selection of internal or external load transport methods. It also provides guidance on selection and preparation of pickup and release sites, safety and security instructions, and procedures ensuring maximum recovery of all rigging equipment. In addition, it ensures internal and external cargo is properly secured or rigged.
- Supplying special equipment for internal and external loads not available to the supported unit, such as lashings, tie-downs, and equipment organic to the helicopter unit required exclusively for cargo transport and helicopter operations.
- Supplying technical supervision to supported unit during loading, tie-down, and off-loading of cargo.

2-36. Table 2-1 discusses remaining responsibilities.

Table 2-1. Responsibilities

	Loading
Supported unit	Normally loads and restrains internal cargo under supervision of a helicopter crew member.
Supporting avn unit	Large or heavy loads are rigged so the helicopter cargo hook position is as close to the load's center of gravity (CG) as possible. The pilot in command (PC) has final responsibility for accepting a load to include distribution and restraint of internal cargo.
	Unloading
Supported unit	Normally responsible for unloading internal cargo and recovering slings, nets, and other equipment.
Supporting aviation unit	May assist in recovery of slings, nets, and other equipment by arranging for backloading in helicopters returning empty to the supported unit. The PC has final responsibility for safe unloading or release of cargo.
	Marshalling
Supported unit	Provides specially trained personnel to marshal (ground guide) helicopters to landing points for pickup and release of external loads. Units equip these personnel with distinctively colored clothing such as fluorescent international orange or yellow wherever practicable.
	Restricts personnel in the danger area around helicopter to those directly involved in marshalling, loading, hookup, release, or unloading of cargo.
	Provides ground personnel with lighting devices when performing night operations. Light intensity varies, depending on whether the aircrew is using unaided vision or night vision devices (NVDs).
	Provides additional reference lighting if requested by the helicopter unit to aid the pilot in hookup of loads at night.
Supporting avn unit	If necessary, the helicopter unit may issue special instructions on hook-up procedures.

Chapter 2

INTERNAL LOADS

ADVANTAGES AND DISADVANTAGES

2-37. Helicopters are ideally suited for moving troops, supplies, weapons, ammunition, and equipment rapidly across the battlefield. Internal and external loading are the two methods used to transport cargo. Table 2-2 provides internal loading considerations.

Table 2-2. Internal loading considerations

Advantages
Flight can be conducted at nap-of-the-earth (NOE) altitudes.
Flight can be conducted at higher airspeeds.
Cargo is protected from weather.
Fragile equipment is afforded better protection.
Less power is required and aircraft endurance is increased.
Disadvantages
The helicopter must land to load and unload.
Pickup zones (PZs) and landing zones (LZs) may require some preparation.
Loading and unloading are time consuming.
Planning
Cargo to be transported must fit inside the aircraft.
Tie-down equipment must be available to properly secure cargo.
Care must be exercised to avoid damaging the aircraft while loading.
Personnel must be cautious of turning rotors.
Cargo must not block aircraft exits or access to emergency equipment.
Floor contact pressure of the cargo must not exceed floor loading limitations.
Items to be unloaded first must be loaded last.
Heavy or bulky items should be loaded on the floor, with lighter or more fragile items on top to prevent damage.
Loading instructions provided on equipment, shipping containers (this side up), or in equipment item technical manuals (TMs) should be observed.
Prior to loading, cargo center of gravity (CG) should be determined.

CARGO FLOOR CONTACT PRESSURE

2-38. Aircraft cargo floors are structural components of the aircraft, crews must place particular emphasis on proper distribution of cargo weight as damage to them may weaken the airframe. Aircraft operator's manuals provide either floor-loading limits or a plan view of the cargo floor, showing differences in floor strength and weight concentration for various compartments. Exercise care during loading and unloading ensuring the cargo floor is not damaged.

Shoring

2-39. Shoring is lumber, planking, or similar material used to spread highly concentrated loads over a greater cargo floor area than occupied by the cargo alone and protects the floor from damage. In general, shoring lumber should be 1 to 2 inches thick, 10 or 12 inches wide, and should not exceed 12 feet in length. Plywood sheets of various thicknesses may also be used. Defects in shoring reduce its strength. Split lumber will not transfer weight horizontally past a split. When used, shoring should extend at least a distance equal to the thickness of the shoring beyond the base of the item being supported.

Weight-Spreading Effect of Shoring

2-40. In figure 2-6, cargo weight resting on shoring does not extend over the entire shoring area in contact with the cargo floor. In general, shoring only increases the area a load is distributed over to the area developed. This area can be determined by extending a line drawn downward and outward from the outside edge of the cargo's base at a 45-degree angle until it meets the surface on which the shoring rests. When shoring is used, the area the load is distributed is enlarged by a border equal to the thickness of the shoring all around the cargo's base.

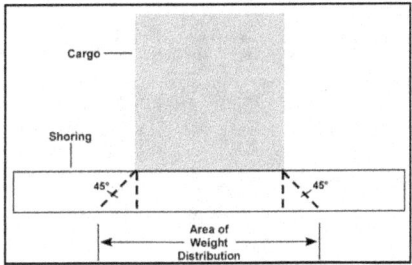

Figure 2-6. Weight-spreading effect of shoring

Load Contact Pressure

2-41. To determine the contact pressure of a load, divide its total weight by the area of contact to include the extended weight distribution area gained by using shoring (figure 2-7).

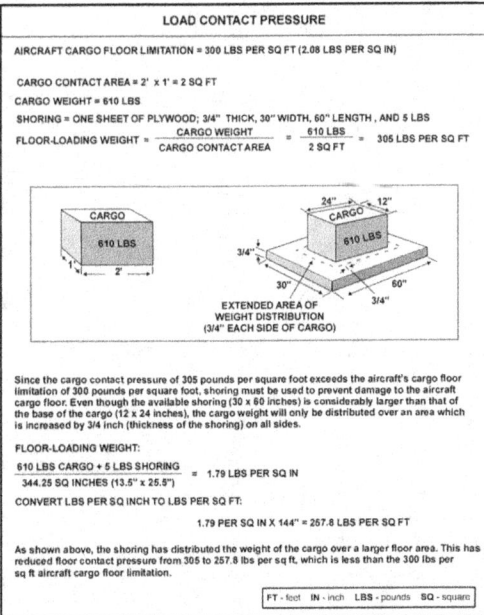

Figure 2-7. Load contact pressure

Chapter 2

Load Pressure Formulas

2-42. As stated earlier, surface contact pressure of an item is determined by dividing weight of the item by the area in contact with the aircraft cargo floor. Figure 2-8 provides sample formulas used in load-pressure calculations.

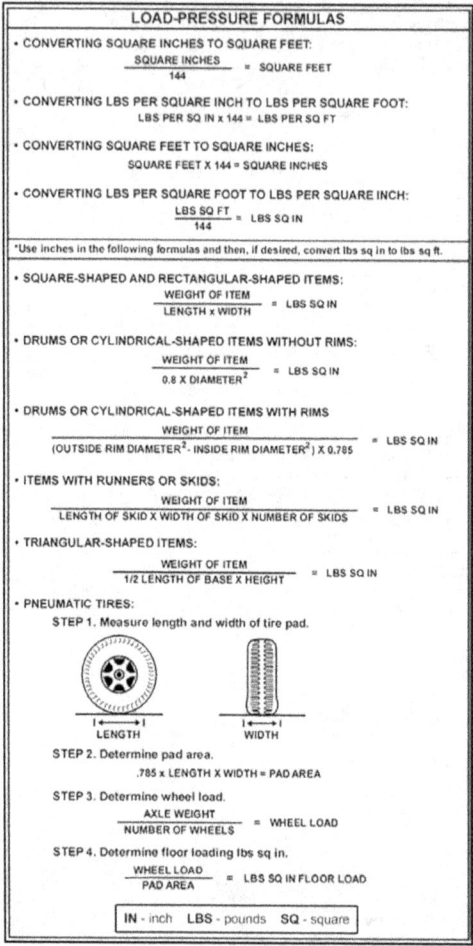

Figure 2-8. Formulas for load pressure calculations

DETERMINATION OF CARGO CENTER OF GRAVITY

2-43. Computing and marking the CG on cargo enables load crews to properly position cargo within an aircraft and accurately compute the weight and balance condition of a loaded aircraft. Procedures for determining the CG of general cargo and vehicles are provided below.

General Cargo

2-44. The CG of general cargo may be determined by balancing the item on a roller (figure 2-9) and then marking the balance point.

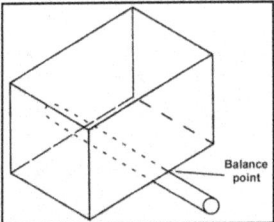

Figure 2-9. Determining general cargo center of gravity

Wheeled Vehicles

Individuals can determine the CG of a wheeled vehicle by finding the weight on each axle. Vehicle data plates or applicable operator's manuals provide axle weights for empty vehicles, while axle weights of loaded vehicles can be determined by running the wheels on a suitable scale (figure 2-10). The CG is then determined using the following formula:

> **Wheeled Vehicle Center of Gravity Formula**
> (Rear Axle Load x Wheelbase) ÷ Vehicle Gross Weight = CG Location Aft of Front Axle

Placement of Cargo

2-45. For weight and balance purposes, weight of an item is concentrated at the item's CG. CG markings on cargo enable load crews to place cargo at precise locations or fuselage stations within the aircraft aiding in accurately computing weight and balance of a loaded aircraft.

Cargo Load Center of Gravity

2-46. Compartment and station methods are used to compute the CG of a cargo load.

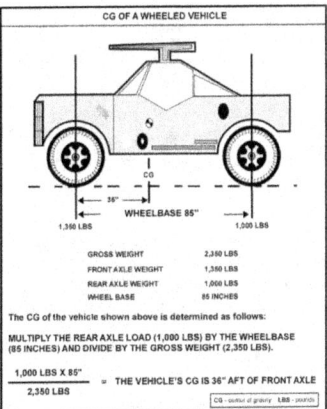

Figure 2-10. Determining center of gravity of wheeled vehicle

Chapter 2

Compartment Method

2-47. For cargo helicopters, loading by compartments provides a rapid means of computing the CG of a load. This method can be used whenever a load consists of a number of items.

2-48. The CH-47 cargo area is divided into three compartments—C, D, and E. The centroid (also known as center of gravity or CG) of each compartment is at stations 181, 303, and 425, respectively. When using the compartment method, it is assumed the weight of all cargo in the compartment is concentrated at the compartment's CG. If an item extends into two or three compartments, the weight of the item should be proportionately distributed in each compartment (figure 2-11).

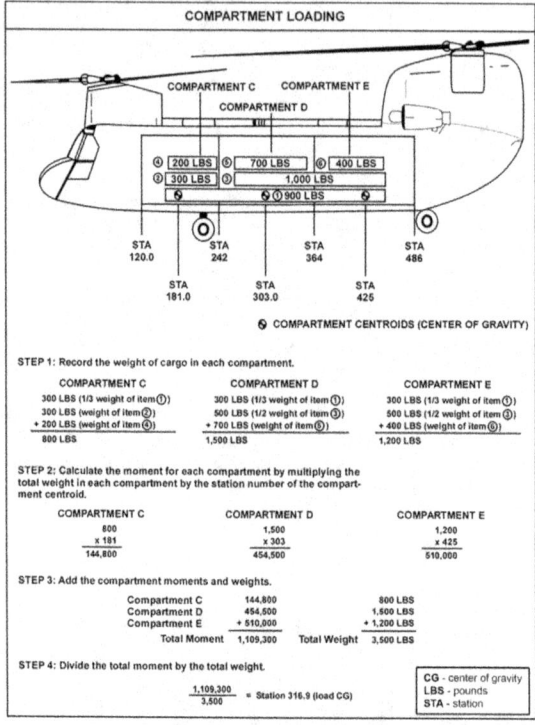

Figure 2-11. Compartment method steps

Station Method

2-49. The station method is a more precise method of computing the CG of a load and should be used whenever possible. To use this method, it is necessary to know the CG of each item of cargo. Station loading requires the CG of each item of cargo be placed precisely on a specific fuselage station number. Figure 2-12, page 2-15, provides a sample application of the station method for a UH-60 helicopter.

Weight, Balance, and Loads

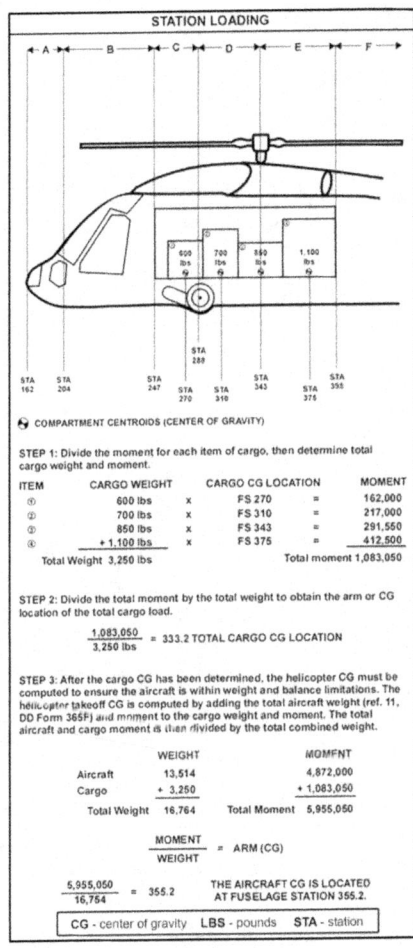

Figure 2-12. Station method steps

Cargo Restraints

Restraint Criteria

2-50. Aircraft are subjected to G-forces resulting from air turbulence, acceleration, rough or crash landings, and aerial maneuvers. Since the cargo is moving at the same rate of speed as the aircraft, forward movement is the strongest force likely to act on cargo if the aircraft is suddenly slowed or stopped. Other forces which tend to shift cargo aft, laterally, or vertically will be less severe. Restraining or tie-down devices prevent cargo movement that could result in injury to occupants, damage to the aircraft or cargo, or cause the aircraft CG to move out of limits. The amount of restraint required to keep cargo from moving in any direction is called restraint criteria and is expressed in Gs. The maximum force exerted by an item of cargo is equal to

Chapter 2

its normal weight times the number of Gs specified in restraint criteria. Restraint criteria are normally different for each type of aircraft and provided in the operator's manual. To prevent cargo movement, the amount of restraint applied should equal or exceed the amount of restraint required. Restraint is referred to by the direction in which it keeps cargo from moving. For example, forward restraint keeps cargo from moving forward and aft restraint keeps cargo from moving aft.

Cargo Classification

2-51. Cargo is generally classified as either prepared or miscellaneous. Prepared cargo is carried in containers equipped with tie-down devices, or equipment with attached tie-down points. Miscellaneous cargo is all other cargo, or cargo without tie-down provisions.

Restraint Devices

2-52. Restraint equipment includes cargo nets, chains, webbed-nylon straps, and various types of attaching hooks and tightening devices.

APPLICATION OF TIE-DOWN DEVICES

2-53. Most aircraft operator's manuals provide specific instructions for use of tie-down devices. A tie-down device will withstand a force equal to its rated strength only when the force is exerted parallel to the length of the device. It is seldom possible to fasten a device in this manner. Instead, it is usually necessary to fasten the device to the cargo at some point above the floor, resulting in a partial loss of restraint strengths. The strength of restraint is reduced in ratio to the angles formed by the device with the floor and the axis of the aircraft. Based on calculations, a 30-degree angle of attachment in the intended restraint direction causes a restraint loss of 25 percent in that direction and is the most desirable angle. While causing a loss of restraint in one direction, angled tie-down devices furnish restraint in two other directions so one device provides restraint in three directions simultaneously. The effective holding strength of devices applicable at a 30-degree and 45-degree angle is illustrated in figure 2-13, page 2-17.

2-54. General rules for the application of tie-down devices are—
- Fasten devices so they form, as nearly as possible, 30-degree angles with the cargo floor and longitudinal axis of the aircraft.
- Consider the strength of the aircraft tie-down fittings and the points of attachment on the load. A tie-down device is no stronger than its weakest component. A 10,000-pound device attached to a 5,000-pound rated fitting will only provide 5,000 pounds of restraint. Axles, tow hooks, bumper supports, and vehicle frames are good points of attachment for securing most vehicles. Since general cargo items may not have points of attachment, the devices should be applied over or across the cargo items. Additionally, cargo nets will aid in restraining items of miscellaneous cargo.
- For prepared cargo, it is desirable to use an even number of tie-downs of the same length and attach them symmetrically in pairs.
- When tie-down devices providing forward and aft restraint are crisscrossed over the cargo, adequate restraint is automatically provided in the lateral and vertical directions. If devices providing forward and aft restraint are applied across the front and rear of the cargo, lateral and vertical restraint will have to be provided. Vehicles will have sufficient lateral and vertical restraint if forward and aft restraints are applied properly.

Weight, Balance, and Loads

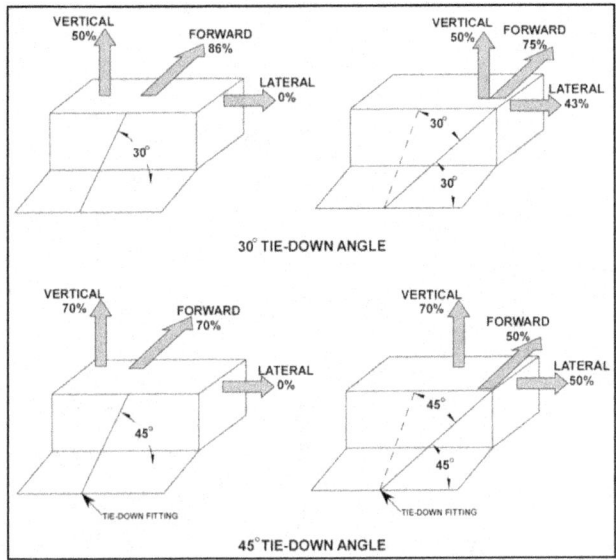

Figure 2-13. Effectiveness of tie-down devices

CALCULATION OF TIE-DOWN DEVICE REQUIREMENTS

2-55. To calculate the number of tie-down devices required to restrain a load in any given direction, these factors must be known—
- Weight of cargo.
- Restraint criteria. This data is normally found in aircraft operator's manuals.
- Angle of tie-down and percent effectiveness of a tie-down device. The effectiveness of a tie-down device is determined from the percentage restraint chart (table 2-3, page 2-18).
- Rated strength of weakest link or component of a tie-down.

Chapter 2

Table 2-3. Percentage restraint chart

		5°	10°	15°	20°	25°	30°	35°	40°	45°	50°	55°	60°	65°	70°	75°	80°
	VERTICAL	8.7	17.4	25.9	34.2	42.3	50.0	57.4	64.3	70.7	76.6	81.9	86.6	90.6	93.9	96.6	98.5
5°	LONG	99.2	98.1	96.2	93.6	90.2	86.3	81.6	76.3	70.4	64.0	57.2	49.8	42.1	34.1	25.8	17.3
	LAT	8.7	8.6	8.4	8.2	7.9	7.5	7.1	6.7	6.2	5.8	4.9	4.4	3.7	2.9	2.3	1.5
10°	LONG	98.1	97.0	95.2	92.6	89.2	85.3	80.7	75.5	69.9	63.3	56.5	49.3	41.7	33.7	25.5	17.1
	LAT	17.3	17.1	16.8	16.6	15.8	15.1	14.3	13.3	12.3	11.2	9.9	8.7	7.4	5.9	4.5	3.0
15°	LONG	96.2	95.2	93.3	90.8	87.5	83.7	79.1	73.9	68.3	62.1	55.4	48.3	40.9	33.0	25.0	16.8
	LAT	25.8	25.5	25.0	24.3	23.5	22.4	21.2	19.8	18.3	16.7	14.9	12.9	10.9	8.9	6.7	4.5
20°	LONG	93.6	92.6	90.8	88.4	85.2	81.4	76.9	72.0	66.5	60.4	53.9	47.0	39.8	32.1	24.3	16.6
	LAT	34.1	33.7	33.0	32.1	30.9	29.6	28.0	26.2	24.2	21.9	19.6	17.1	14.5	11.7	8.9	5.9
25°	LONG	90.2	89.2	87.5	85.2	82.1	78.5	74.2	69.4	64.1	58.3	52.0	45.3	38.3	30.9	23.5	15.8
	LAT	42.1	41.7	40.9	39.8	38.3	36.6	34.6	32.4	29.9	27.2	24.3	21.2	17.9	14.5	10.9	7.4
30°	LONG	86.3	85.3	83.7	81.4	78.5	74.9	70.9	66.3	61.2	55.7	49.7	43.3	36.6	29.6	22.4	15.1
	LAT	49.8	49.3	48.3	47.0	45.3	43.3	40.9	38.3	35.4	32.2	28.7	25.0	21.2	17.1	12.9	8.7
35°	LONG	81.6	80.7	79.1	76.9	74.2	70.9	67.1	62.7	57.9	52.7	47.0	40.9	34.6	28.0	21.2	14.3
	LAT	57.2	56.5	55.4	53.9	52.0	49.7	47.0	43.9	40.6	36.9	32.9	28.7	24.3	19.6	14.9	9.9
40°	LONG	76.3	75.5	73.9	72.0	69.4	66.3	62.7	58.7	54.2	49.3	43.9	38.3	32.4	26.2	19.8	13.3
	LAT	64.0	63.3	62.1	60.4	58.3	55.7	52.7	49.3	45.5	41.3	36.9	32.2	27.2	21.9	16.7	11.2
45°	LONG	70.4	69.6	68.3	66.5	64.1	61.2	57.9	54.2	49.9	45.5	40.6	35.4	29.9	24.2	18.3	12.3
	LAT	70.4	69.6	68.3	66.5	64.1	61.2	57.9	54.2	49.9	45.5	40.6	35.4	29.9	24.2	18.3	12.3
50°	LONG	64.0	63.3	62.1	60.4	58.3	55.7	52.7	49.3	45.5	41.3	36.9	32.2	27.2	21.9	16.7	11.2
	LAT	76.3	75.5	73.9	72.0	69.4	66.3	62.7	58.7	54.2	49.3	43.9	38.3	32.4	26.2	19.8	13.3
55°	LONG	57.2	56.5	55.4	53.9	52.0	49.7	47.0	43.9	40.6	36.9	32.9	28.7	24.3	19.6	14.9	9.9
	LAT	81.6	80.7	79.1	76.9	74.2	70.9	67.1	62.7	57.9	52.7	47.0	40.9	34.6	28.0	21.2	14.3
60°	LONG	49.8	49.3	48.3	47.0	45.3	43.3	40.9	38.3	35.4	32.2	28.7	25.0	21.2	17.1	12.9	8.7
	LAT	86.3	85.3	83.7	81.4	78.5	74.9	70.9	66.3	61.2	55.7	49.7	43.3	36.6	29.6	22.4	15.1
65°	LONG	42.1	41.7	40.9	39.8	38.3	36.6	34.6	32.4	29.9	27.2	24.3	21.2	17.9	14.5	10.9	7.4
	LAT	90.2	89.2	87.5	85.2	82.1	78.5	74.2	69.4	64.1	58.3	52.0	45.3	38.3	30.9	23.5	15.8
70°	LONG	34.1	33.7	33.0	32.1	30.9	29.6	28.0	26.2	24.2	21.9	19.6	17.1	14.5	11.7	8.9	5.9
	LAT	93.6	92.6	90.8	88.4	85.2	81.4	76.9	72.0	66.5	60.4	53.9	47.0	39.8	32.1	24.3	16.6
75°	LONG	25.8	25.5	25.0	24.3	23.5	22.4	21.2	19.8	18.3	16.7	14.9	12.9	10.9	8.9	6.7	4.5
	LAT	96.2	95.2	93.3	90.8	87.5	83.7	79.1	73.9	68.3	62.1	55.4	48.3	40.9	33.0	25.0	16.8
80°	LONG	17.3	17.1	16.8	16.6	15.8	15.1	14.3	13.3	12.3	11.2	9.9	8.7	7.4	5.9	4.5	3.0
	LAT	98.1	97.0	95.2	92.6	89.2	85.3	80.7	75.5	69.6	63.3	56.5	49.3	41.7	33.7	25.5	17.1

(1) Angles across the top are those formed between the tie-down device and the cabin floor.
(2) Angles down the side are those formed between the tie-down device and the longitudinal axis of the aircraft.
(3) Vertical restraint is related only to the angle between the tie-down device and the cabin floor. The lateral angle has no bearing on it.
(4) The shaded area indicates the "best compromise" position.
LAT-latitude
LONG=longitude

2-56. Figure 2-14, page 2-19, provides application for determining number of tie-down devices required to restrain an item of cargo. The following formula is used to calculate the number of tie-down devices required to restrain a load from moving in any direction:

$$\frac{\text{Weight of Load} \times \text{Restraint Criteria}}{\text{Weakest Link of Tie-Down} \times \text{Percent Effectiveness}} = \text{Required Tie-Downs}$$

Note: If the formula yields fractional results, for miscellaneous cargo, the number will be rounded up to the next whole number. For prepared cargo, the number is rounded up to the next whole even number providing an even number of tie-down devices.

Weight, Balance, and Loads

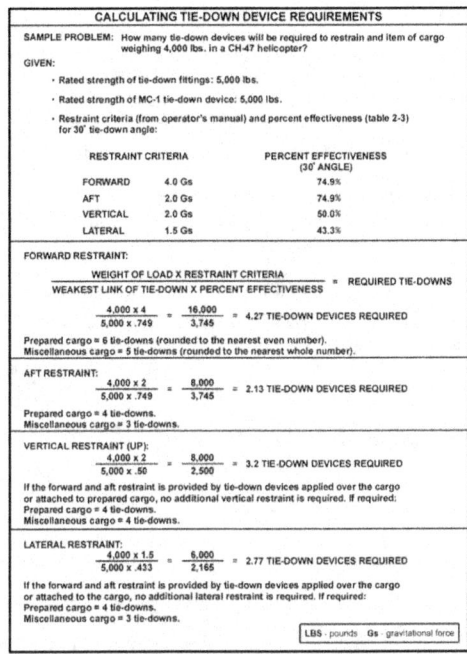

Figure 2-14. Calculating tie-down requirements

EXTERNAL LOADS

PLANNING CONSIDERATIONS

Application

2-57. Helicopters are frequently used to move cargo externally (sling loads) when heavy, outsized, or needed-now items are required to be rapidly transported over untenable terrain. The following situations favor the use of external loads:

- Cargo compartment of the aircraft is too small.
- Aircraft CG would be exceeded, due to the characteristics of the load, if loaded internally.
- Loading and/or unloading must be accomplished in the shortest possible time.
- Pickup zone (PZ)/landing zone (LZ) conditions prevent aircraft from touching down.
- Nature of cargo is such that rapid cargo-jettison capability is desirable.

Load Categories

2-58. All external loads are divided into three basic categories—high-density, low-density, and aerodynamic. Each exhibits different characteristics in flight. High-density load offers the best stability; low-density load is the least stable. The aerodynamic load exhibits both instability and stability (instability inherent until load streamlining occurs). An aviator must determine category, size, and weight of the load during preflight.

Cargo Nets and Slings

2-59. Cargo nets and slings are an essential part of the external-load operation and must be given the same attention during the preflight inspection as the cargo receives. Nets and slings with frayed or cut webbing will not be used for external loads. Due to critical strength requirements, field sewing of nylon should not be attempted, nor should nonstandard parts be substituted in assembling slings. The sling assembly must be commensurate with load requirements and must meet requirements in the operator's manual.

Aircraft Performance and Operator's Manual

2-60. It is imperative aviators consult the appropriate operator's manual to ensure a successful operation. Performance charts in this manual include gross-weight limitations, airspeed limitations, and endurance charts. The gross-weight chart provides a rapid means of determining load-carrying capabilities within safe operating limits. This performance planning data is crucial to successful sling-load operations.

2-61. The operator's manual also gives a complete operational explanation of sling-release systems. During preflight, aviators must inspect emergency-release systems and make operational checks of all normal release modes. Emergency procedures for any nonstandard occurrence experienced during external-load operations are outlined in the operator's manual.

Coordination with Flight and Ground Personnel

2-62. Preflight is not complete until the aviator briefs the flight and ground crews on their duties and mission to be performed. Essential criteria for a safe operation are predetermined prior to takeoff. Signaling procedures, unit standing operating procedures (SOPs), and emergency procedures are in the brief.

EXTERNAL-LOAD PICKUP PROCEDURES

Pickup Techniques

2-63. Pickup technique varies according to the helicopter in use, type and weight of the external load, terrain involved, and wind and weather conditions at time of pickup.

Approach Procedure

2-64. The approach to hookup (also release) should be conducted into the wind, yielding best aircraft stability and performance. Even if the load is light and there is excess power, the wind could be the critical factor during emergencies. A slow forward hover allows the aviator to receive directions from flight crew and ground personnel without jeopardizing the aircraft or hookup person's safety. When directions are received solely from ground personnel, a signalman must be in a plain view position of the aviator and give appropriate visual signals throughout the operation. The cargo-release switch is placed in the arm position as the aircraft approaches the load.

Hover Altitude

2-65. The appropriate hover altitude depends upon variables such as type of helicopter, terrain and ground effect, size of load, and safety of the ground crewmen. Once an altitude is decided, it should be kept constant to prevent false perception and possible load strike. To assist the pilot in maintaining a constant position and hover altitude, references should be selected in the front and to the sides of the helicopter.

Hookup Procedure

2-66. Hookup commences with final positioning of the helicopter over the load. In cargo helicopters, this normally is conducted through verbal coordination with a flight crewmember that is in a position to closely observe the helicopter's movements over the load. In helicopters where flight crews are unable to observe the helicopter's movements over the load, a signalman located on the ground and in plain view of the aviator must be used. In all cases, the signals (verbal or visual) must be standardized among the persons involved prior to the operation (see TM 4-48.09). The load is attached to the helicopter's cargo hook by the hookup crew when the helicopter is stabilized over the load.

Emergency Actions

2-67. In the event an emergency condition occurs while hovering over the load and the helicopter must be landed, the helicopter normally will land to the left of the load. Hook-up personnel must move in the opposite direction (to the right of the helicopter) to avoid injury. The unit SOP establishes this procedure and the aviator must brief all personnel before conducting external-load operations. The hookup man will approach from the helicopter's right and exit to the helicopter's right. When possible, ground personnel should not position themselves between the load and the helicopter during hookup. The load is to be attached according to the appropriate operator's manual, TM 4-48.09, and the unit SOP. Hookup personnel notify the pilot immediately when the load is attached to the cargo hook. Any emergency procedure following attachment must include cargo release.

Takeoff Procedure

2-68. There are two distinct phases when taking off with an external load—lifting the load to a hover and takeoff.

Lifting the Load to a Hover

2-69. Once the signalman indicates the load is hooked up and the hookup man is clear, the aviator initiates a slow vertical ascent until the sling becomes taut and centered. The aviator, flight crew, and/or ground crew closely coordinate ensuring the aircraft does not drift from over the load. The load is then slowly lifted to an appropriate hover altitude (normally about 10 feet above the ground). While picking up the load to a hover, the aviator must determine whether the helicopter has sufficient power to continue the operation. Security and proper rigging of the load are also reconfirmed.

Takeoff

2-70. After receiving the takeoff signal from the signalman and if all criteria have been met for flight, smooth acceleration and takeoff are initiated. Sufficient power (not to exceed maximum allowable) is applied on takeoff ensuring the load clears all obstacles by a safe altitude. Once established at a safe altitude, power is adjusted to maintain safe airspeed and altitude. The cargo-release switch is placed in the off or safe position after passing through above ground level (AGL) altitude as directed by the operator's manual and/or SOP. During flight below this altitude, the cargo-release switch is left in the on or arm position. Aviators should avoid flight over populated areas.

Note. A safe climb altitude is the altitude wherein the load is unquestionably clear of the highest barrier, usually 50 to 100 feet above the tallest immediate obstacle.

EN ROUTE PERFORMANCE

2-71. The weight and density of the load may determine airworthiness (steadiness in flight) and maximum airspeed at which the helicopter may be safely flown. Low-density, light loads generally tend to shift farther aft as airspeed is increased and may become unstable. When the load is of greater density, more compact, and balanced, the ride is steadier and airspeed may be safely increased. Any unstable load may jump, oscillate, or rotate resulting in loss of control and undue stress on the helicopter. This requires reducing forward airspeed immediately, regaining control, and steadying the cargo load. If an external load begins oscillating fore and aft, the helicopter should be flown into a shallow bank while decreasing airspeed. This normally shifts the oscillation laterally which can easily be controlled by further decreasing forward airspeed. At the first indication of a buildup in oscillation, it is mandatory to slow airspeed immediately. The oscillation may endanger the helicopter and personnel. This situation may require jettisoning the load. For a complete explanation of the cargo release system for the helicopter to be flown, see the appropriate operator's manual.

Chapter 2

TERMINATION-AND-RELEASE PROCEDURE

2-72. Termination and subsequent load release must include approach to the termination point, hovering to the load-release point, and releasing the load.

Termination Point Approach

2-73. The approach to the termination point should not be initiated until the appropriate termination point is identified. At the appropriate altitude, the cargo-release switch is placed in the arm position.

Load-Release Point Hovering

2-74. Procedure to the release point (RP) will be accomplished in the same manner as described earlier in external load pickup procedures. The procedure, however, reverses over the RP.

Load Release

2-75. Stabilize the aircraft over the load and descend to allow slack in the sling. If possible, slide the aircraft laterally to where the clevis will not fall on the load to prevent damage. When the aircraft is clear of the load, open the cargo hook to release the load. Usually the cargo hook is opened through the normal release modes of operation, in accordance with appropriate aircraft operator's manual. Manual and emergency release methods will be used in accordance with the appropriate operator's manual and the unit SOP when normal modes fail to function properly. Ground personnel, in accordance with SOP and other directives, may use any means necessary to free the load if the cargo cannot be released from the helicopter by the flight crew. These methods might include use of knives, bayonets, or blade-like instruments to cut nylon or rope components of the sling assembly. When metal components must be cut to free a load, devices such as diagonal cutters, bolt cutters, pliers, or cable cutters are appropriate.

HAZARDOUS MATERIALS

PLANNING REQUIREMENTS

2-76. Aviators and aviation planners must be aware movement of hazardous material by aircraft has different requirements. The following factors must be addressed when moving hazardous materials:
- Compliance with special procedures.
- Unique packaging and handling requirements exist for most items of hazardous cargo.
- Some items cannot be carried in aircraft unless specially trained escort personnel are aboard and particular security requirements have been met.
- Some items cannot be carried with other types of hazardous cargo and certain items of hazardous cargo may not be carried aboard the same aircraft with passengers.
- Regulations also prescribe items of information which must be provided to en route and destination airfields prior to an aircraft's departure.
- To carry some items of hazardous cargo, aircrews must be provided with protective clothing and special equipment.
- Additionally, there are some hazardous materials which may not be accepted for air shipment.
- Aviators must also be aware compliance with special in-flight emergency procedures may be required for aircraft carrying dangerous materials.
- Procedures, responsibilities, and guidance for handling, storage, and transportation of hazardous material are discussed in regulations and TMs.

2-77. While it is impracticable to discuss procedures for transporting all types of hazardous loads in this section, an overview of key publications is provided below. These publications should be reviewed to develop hazardous load SOPs appropriate to the unit's mission.

Dangerous Materials

2-78. Dangerous material is defined as any flammable, corrosive, oxidizing agent, explosive, toxic, radioactive, nuclear, unduly magnetic, or biologically infective material. Dangerous material also includes any other material that may endanger human life or property due to its quantity, properties, or packaging.

Publications

2-79. Following is a partial list of publications providing guidance for transportation of dangerous materials aboard aircraft. The applicability of procedures to tactical wartime operations normally is addressed in each publication.

Chemical Surety

2-80. Army Regulation (AR) 50-6 describes the Chemical Surety Program and provides guidance and directives for safe, secure, and reliable life-cycle management of chemical agents and their associated weapon systems. Included is guidance for transportation of chemical surety material by Army aircraft.

Flight Regulations

2-81. AR 95-1 prescribes procedures and rules governing command, control, and operation of Army aircraft. The following portions of this regulation pertain to transportation of dangerous materials:
- Procedures for packaging, handling, and air transportation of dangerous materials are described in AR 95-27 and TM 38-701. Aircrews assigned to move dangerous materials in Army aircraft will comply with the requirements listed in these publications.
- Aircraft must be grounded during refueling, arming, oxygen servicing, and loading or unloading of flammable or explosive cargo.
- At least one pilot seated at the controls must wear a protective mask when fused items filled with toxic chemicals are carried in aircraft. Other crewmembers will have protective masks readily available.
- When incapacitating or toxic chemicals with no arming or fusing systems are carried in an aircraft, pilots need not wear a mask; however, it must be readily available.
- All personnel aboard will wear a protective mask when incapacitating or toxic chemicals are dispensed and until the chemical safety officer or other crewmember reports the aircraft "clear" of the dispensed agent.
- Personnel who are not essential to the mission will not be carried in an aircraft with incapacitating or toxic chemicals on board.

Operational Procedures for Aircraft Carrying Hazardous Materials

2-82. AR 95-27 specifies special procedures applying to aircraft carrying nuclear, chemical, or biological research materials. Actions to be taken by PCs, aircrew members, and technical escorts during in-flight emergencies involving such materials are listed in this document. It applies to nuclear cargo, toxic chemical ammunition, highly toxic substances, hazard division 1.1 through 1.3 explosives, and infectious substances (including biological and etiological materials). In addition, it applies to Class VII (radioactive materials), which require a yellow III label, inert materials, and all other hazard classes or divisions, except Class IX and other regulated material-domestic, when shipped in quantities of 1,000 pounds or more aggregate gross weight. The following are a few of the many PC responsibilities:
- Brief crewmembers, couriers, and technical escorts on mission requirements, procedures governing hazardous cargo, notification requirements, and emergency procedures.
- Enter "hazardous cargo," "inert devices" (or both), and mission number and prior permission request number in the other information or remarks section of the flight plan unless prohibited by directives governing the area of operations (AO).
- Refuse to accept any clearance containing noise abatement procedures, in the PC's judgment, interfering with flight safety.
- Ensure compliance with in-flight notification procedures given in AR 95-27.

Storage and Handling of Liquefied and Gaseous Compressed Gasses and Their Full and Empty Cylinders

2-83. While AR 700-68 does not address transportation of gas cylinders by air, it provides excellent information on storage, handling, and inspection of gas cylinders. Information provided in this regulation should be reviewed by aircrews involved in air transport of gas cylinders.

Military Explosives

2-84. ATP 4-35.1 provides guidance for handling, storage, and transportation of ammunition and explosives. It includes operating regulations for aircrews, aircraft loading and unloading procedures, electrical-grounding requirements, quantity-distance standards, fire protection requirements, and considerations for establishment of ammunition and explosive sling-load pickup areas at ammunition resupply points.

Preparing Hazardous Materials for Military Shipments

2-85. TC 4-13.17 contains information useful to units preparing SOPs on hazardous loads. It provides instructions for personnel who prepare hazardous material for air shipment, labeling requirements, instructions for transporting passengers with hazardous materials, and instructions for notifying the PC of hazardous materials on the aircraft. It also contains the protective-equipment requirement quoted below.

Protective Equipment

2-86. Aircraft operators ensure appropriate equipment is available to protect aircrew and passengers when transporting materials whose vapors are toxic, irritating, or corrosive. Aircraft must have a closed oxygen system or protective mask for each person aboard. The shipper will provide any required special equipment to meet unique cargo safety requirements. While the exact equipment required depends on the materials being transported, the following are recommended minimums (or equivalent substitutions):

- Two pairs of rubber gloves.
- One pair of asbestos or leather (with wool inserts) gloves.
- One plastic or rubber apron.
- A 5-pound (2.3 kilogram) package of incombustible absorbent material.
- Three large plastic bags.
- One oxygen or protective mask.

STANDARDIZATION AGREEMENT REQUIREMENTS

2-87. Requirements of North Atlantic Treaty Organization (NATO) Standardization Agreement (STANAG) 3854(Edition 3) for carriage of ammunition and fuel as cargo by helicopter are outlined below. These requirements are applicable to operational conditions during both peacetime and wartime.

Ammunition

2-88. Ammunition is classified as explosives by national regulations, STANAG 3854, or International Air Transport Association (IATA) regulations, respectively.

2-89. Ammunition must be technically suitable and compatible for carriage by helicopter in accordance with national regulations. If not packed in its original packing material, extra care must be given to the labeling.

2-90. Ammunition shall not be considered cargo when needed by Soldiers on board immediately after landing for fulfilling their combat mission. Also, ammunition is not considered cargo when it is part of the equipment of the helicopter or crew.

2-91. When a helicopter is carrying ammunition, the landing place is classified as an in-transit storage place. Therefore, it becomes a risk to vulnerable locations such as residential areas, public roads, barracks, taxiways, parking lots, and aircraft parking areas. A helicopter loaded with ammunition is vulnerable to accident, interference, or hostile action and needs to be protected.

2-92. The required safe distances are to be determined according to corresponding regulations of the nation where the transfer of the load takes place.

Fuel

2-93. Fuel, petroleum, oils, and lubricants is classified as highly flammable liquid or flammable compressed gas and labeled in accordance with STANAG 3854 or IATA regulations, respectively. Fuel is to be carried only in approved containers or jerricans which meet regulations of the originating nation. The content of the containers or jerricans must not exceed 90 percent, unless specifically cleared for a safe higher content. The closure shall be leak proof. Carriage of fuel in gasoline containers of vehicles is determined by IATA regulations, but stationary internal-combustion engines may hold a limited amount for immediate operational requirements after off-loading. Types of carriage, such as internal or external load, are governed by regulations of the nation providing the helicopters.

Helicopter Safety

2-94. There is no smoking either within 25 meters of the helicopter or aboard the helicopter when it is carrying ammunition or fuel. The use of open flame light is prohibited within 25 meters of the helicopter or in the cargo hold.

2-95. Helicopters scheduled for carriage of ammunition or fuel should be refueled, if required, prior to loading. Defueling of helicopters loaded with this cargo is prohibited.

2-96. Static electricity of the helicopter shall be discharged prior to loading and unloading, as well as, pickup of sling-loads of ammunition and fuel. A non-conducting attaching device should be positioned between the load and hook.

2-97. Where possible, all loading and unloading procedures must be carried out with equipment authorized for this purpose and under supervision of qualified personnel. The cargo shall be loaded and lashed in such a manner as to be stationary during flight and checked at regular intervals. The cargo shall not be loaded near such potentially hazardous installations as heat conduits, heaters, or airborne electrical installations.

2-98. Prior to takeoff, helicopter crews in charge of transporting cargo are briefed by the supported unit on special handling measures. The aircraft must be well ventilated at all times. Unauthorized persons are to be kept away from helicopters carrying ammunition and fuel and nonessential personnel will not be transported on the same aircraft.

2-99. Service and maintenance work constituting a fire hazard are not performed on helicopters loaded with ammunition or fuel as cargo. This cargo must be off-loaded prior to such work being performed.

2-100. Whenever a helicopter loaded with ammunition and fuel as cargo takes off or lands at an airport, the air traffic control (ATC) service of that airport shall be notified, by the pilot, of the quantity, type, and classification of the cargo. If a fire breaks out in the cargo compartment during a flight, attempts to extinguish it using aircraft fire extinguishers are made. A landing will then be conducted in the closest area clear of obstructions. The cargo will be inspected before further flight is attempted. If during flight, or due to an emergency situation, a sling load has to be jettisoned and/or if it is believed a large quantity of fuel has leaked out, it is to be reported to the ATC service. In case of radio failure, the crew shall inform local authorities at the first opportunity.

2-101. In peacetime, flying helicopters loaded with ammunition and fuel as cargo over residential areas is prohibited. Whenever possible, avoid flying over houses, public means of transportation, or groups of people.

2-102. Several factors shall be considered when helicopters carrying ammunition and/or fuel as cargo are temporarily parked. Parking helicopters in aircraft hangars should be avoided. If they must be parked in a hangar, they are to be properly grounded and other aircraft should be removed from the hangar. Helicopters should be parked in the shade at a minimum safe distance of 275 meters from objects to be protected. Helicopters must be properly grounded, and where necessary, the area secured by guards. The minimum safe distance between parked helicopters is 25 meters between rotor disks.

Chapter 2

HAZARDOUS LOAD STANDING OPERATING PROCEDURE

2-103. An SOP on hazardous loads is extremely useful in utility and cargo aircraft units. The SOP normally is developed by examining unit mission requirements and referring to the publications discussed in this chapter and local regulations to determine appropriate procedures. The SOP is tailored to the unit and provides aviators with a one-source document to answer questions similar to the following:

- Should filled 5-gallon gas cans be carried internally or externally?
- Is it permissible to carry acid-filled automotive batteries inside an aircraft?
- Must a protective mask be worn while carrying gas grenades?
- Should mortar rounds and charges be carried on board the same aircraft?
- Can radio batteries be carried on board the same aircraft with dynamite or blasting caps?
- Must the aircraft be shut down while loading or unloading ammunition?

These are only a few questions aviators may have while performing routine resupply missions. The unit SOP should answer these questions as well as others.

LOADING AND STORAGE CHART

2-104. Air Force Manual (AFMAN) 24-204 shows which explosives and other hazardous articles must not be loaded, transported, or stored together. This publication specifies items not accepted for air shipment and provides classification, loading, and storage group codes and labeling and packaging requirements for most known hazardous materials.

Chapter 3
Rotary-Wing Environmental Flight

This chapter addresses the unique environments affecting aircraft performance and mission accomplishment. These environments may include desert, jungle, mountain, maritime, hot, and cold (see Field Manual [FM] 3-04 for vignettes). This overview helps prepare aircrews for mission execution. It does not replace available information; rather, it should supplement unit SOPs and enhance the knowledge of units assigned to perform missions in these locations. Units tasked to operate in these environments should, in addition to reviewing appropriate doctrine and technical manuals, seek guidance and necessary information to train and prepare their aircrews and personnel. Units with experience operating in these various environments have established training programs and 3000-series tasks not included in individual aircrew training modules (ATM), which are essential to mission accomplishment. Copies of these tasks and programs should be acquired to train aircrews for operations in unique environments.

SECTION I – COLD WEATHER OPERATIONS

ENVIRONMENTAL FACTORS

3-1. Aircrews may encounter cold weather flying conditions in many parts of the world. Extreme conditions vary according to latitude and season. Extreme cold and blowing snow pose special problems and difficulties in ground operations, preflight, and actual flight conditions.

CLIMATE AND WEATHER

3-2. Rapidly changing weather is one of the greatest hazards to cold weather operations and presents difficult flying for both inexperienced and experienced aircrews. Various factors—such as temperature range, snow conditions, and icing potential—are subject to rapid (within a fuel load) and dramatic changes and require crewmembers to be prepared at all times.

Temperature

3-3. In the arctic, sub-arctic, or any other region of the world subject to this type of weather, summer temperatures above 18 degrees Celsius are common. Winter temperatures can sometimes drop to -57 degrees Celsius, with typical temperatures as low as -40 degrees Celsius without wind chill. Within the continental United States (CONUS), these temperature ranges are common and should be expected and trained for. Aircrews should not only prepare for these flying conditions but should also ensure they carry necessary survival gear and aircraft maintenance equipment such as rotor head covers (maintenance concerns are covered later in this section).

Precipitation

3-4. Many areas of the far north receive less rain and snow precipitation than the southwestern United States. The average annual precipitation in the arctic, except near seacoasts, is equal to 10 inches of rainfall.

Snow

3-5. Like so many elements of winter weather, snow is most dangerous to the aircrews that rarely fly in it and do not adequately prepare for its effects. Snow varies widely in its characteristics. It may range from dry, fluffy flakes to a wet, heavy consistency clinging to every surface. The National Weather Service (NWS) categories for visibility restriction caused by snow are somewhat misleading. For example, light snow is defined as visibility greater than one-half statute mile or more. By this definition, light snow could be a serious restriction to visibility. The effects of snow and its inherent dangers to flight operations are covered in more detail later in this section.

Fog

3-6. Rapid temperature changes associated with winter are ideally suited to creation of fog. One form of fog unique to cold regions of the world is ice fog. It is most common in the arctic and sub-arctic. However, it can occur whenever the temperature drops to about -25 degrees Celsius or below. Ice fog consists of ice crystals suspended in the air. It is more common around cities and airfields. When there is little or no wind, it is possible for aircraft exhaust, combined with air disturbance caused by the rotor system, to create enough ice fog to halt operations. Ice fog can also be caused by aircraft flying low over an area—such as an LZ—for a low reconnaissance, leaving a trail of fog in the flight path. Ice fog is marked by near-zero visibility up to an altitude of only a few hundred feet AGL with clear skies above.

Icing

3-7. The most hazardous condition associated with flying in cold weather (excluding aircraft preflight) is aircraft structural icing. Icing accounts for loss of aircraft and personnel each year and must be a critical consideration. Aviators must review AR 95-1 for specific rules concerning flight in icing conditions. Icing is most common in temperatures ranging from 0 degrees Celsius to -20 degrees Celsius, accompanied by visible moisture such as clouds, drizzle, rain, or wet snow. Icing rarely occurs in areas maintaining temperatures of -20 degrees Celsius or below. Icing typically exists at altitudes well above the surface but can occur at any altitude all the way to the surface. Aviators must consider temperature inversions are common where surface temperature is too low for icing. An altitude difference of only a few thousand feet can place the aircraft into airspace where icing exists. Army helicopters equipped with anti-icing and deicing systems are not continuously operated in icing conditions. Systems are used to allow transitions, approaches, and departures.

3-8. The following weather conditions normally cause icing (see figure 3-1, page 3-3):
- Stratiform clouds indicate stable air in which minute water droplets/ice crystals are suspended. Water droplets may become super cooled at or below freezing and still be in a liquid state. Super-cooled droplets freeze on contact with aircraft and form layers of ice. The suspended ice crystals are not hazardous to flight as they do not adhere to the aircraft.
- Icing in cumuliform clouds with high moisture content can occur rapidly. Unstable air with currents may carry large super-cooled droplets that spread before freezing, causing rapid accumulation of ice.
- Icing in mountainous terrain occurs mainly when moist air is lifted over high peaks. Ice-producing areas usually occur on the windward side of peaks to about 4,000 feet above the peak and possibly higher when the air is unstable.
- Icing in frontal inversions also can be rapid. Although temperatures are normally colder at higher altitudes, when air from a warm front rises above colder air, freezing rain may occur. Rain falling from the upper (warmer) layer into a colder layer is cooled to a temperature below freezing but remains a liquid. This liquid rain freezes upon contact with the aircraft and can accumulate rapidly. This is the most hazardous type of icing.

Weight, Balance, and Loads

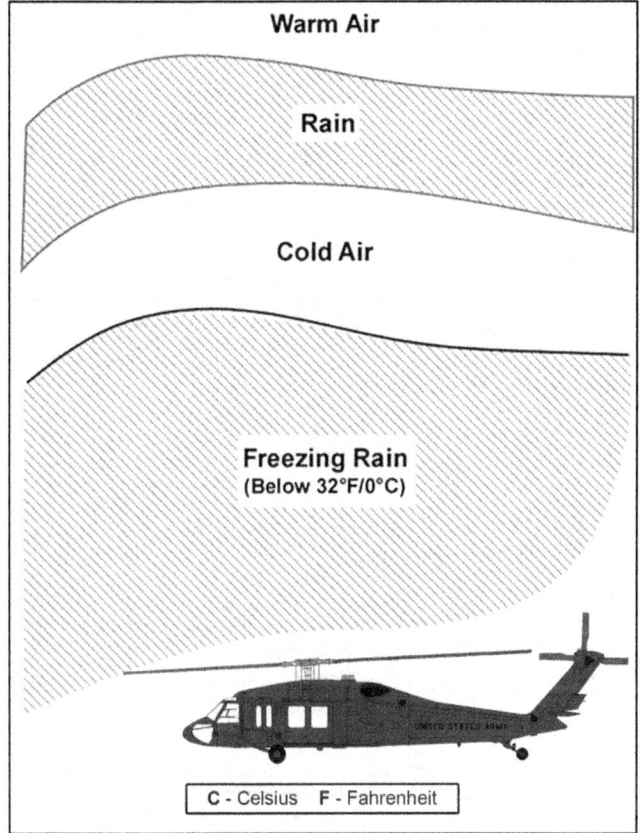

Figure 3-1. Weather conditions conducive to icing

Weather Rules for Cold Weather Operations

3-9. The following weather rules apply to cold weather operations:
- The aviator flies at altitudes below freezing level or clear of any visible moisture. Remain under visual flight rules (VFR) and stay clear of clouds.
- When flying near a warm front, the aviator determines; if temperatures in the cold air mass are in the ice-producing range (0 degrees Celsius to -20 degrees Celsius) and altitude of the inversion layer. These are critical elements for determining the potential conditions for icing.
- Ice on the aircraft windscreen usually occurs first on the wiper blades and arms accumulating at the edges of the windshield. Normally, the side windows do not ice over and usually provide some visibility.
- Rotor blade icing begins near the blade root. This ice buildup may cause loss of lift, which requires additional engine power.
- Asymmetrical ice shedding occurs when ice peels off the rotor blades in an uneven manner. This leaves the rotor out of balance and causes the helicopter to experience severe vibration. Shaking

the cyclic or other control input intended to shed ice will not help the problem and could worsen the condition. This shedding can also cause foreign object damage (FOD) from ice ingested into the engine. When landing the aircraft, be wary of asymmetrical ice shedding. Park the aircraft a safe distance away from other aircraft and ensure rotor blades are stopped before allowing passengers to depart the aircraft.

- When icing is encountered, aviators should descend or climb as appropriate to an altitude clear of clouds or out of the temperature range for icing. If icing conditions exceed aircraft limitations, the aviator must immediately exit these icing conditions and should land as soon as possible. Autorotational capability may be lost in minutes if ice is allowed to form on the blades.
- In freezing rain, it is important to know the altitude of the inversion layer and the freezing level. If the freezing level is at the surface, the best solution may be to climb through the inversion layer to the warmer air above.
- Aviators exercise great care when an obscuration (partial or full) is noted in a weather briefing. Ceiling and visibility are measured by taking vertical and horizontal measurements. Aviators are primarily concerned with slant range visibility, which cannot be accurately determined by a weather forecaster. Visibility conditions may be worse than indicated by weather personnel.
- As mentioned previously, snow visibility estimates and frequency/duration of snowstorms are difficult to forecast. Therefore, flights are carefully planned and include an alternate route or airfield whenever forecast accuracy is in question. Flight into snow conditions is very disorienting and can easily lead to inadvertent instrument meteorological conditions (IIMC).
- Ice and snow on runways or landing surfaces are dangerous for both fixed- and rotary-wing aircraft. Snow is particularly slippery when temperatures are near freezing. Snow accumulation is a hindrance to straight-line control when accelerating or decelerating.
- Aviators ensure all necessary actions regarding aircraft operation have been completed. According to the operator's manual, for example, aviators may need to remove air inlet screens and check the operational capability of the heater/defroster.
- Aviators remove all snow and ice from the airframe before any operations. Such accumulation adversely affects all aspects of flight performance in varying degrees. Before any flight operations, aviators move all control surfaces to confirm full freedom of movement.

3-10. More detailed information on icing can be found in chapter 8 and the appropriate operator's manual.

CAUTION
Do not remove ice from an aircraft by striking the aircraft with blunt objects (such as the hand or a hammer) or by using sharp-bladed objects. These methods can cause external and internal damage to certain aircraft components. The only effective way to remove ice is to apply heat using such techniques as ambient temperature change or a Herman Nelson heater. Care must be exercised in heat application as damage could result. Deice fluids can also be hazardous to aircraft components and the environment.

TERRAIN

3-11. Landscape varies widely in sub-arctic and arctic lands and includes nearly every possible type of terrain from mountain peaks to glaciers to plains. Nearly all surface types and conditions are found, including tundra, bogs, hard, soft, wet, and dry. In winter, freezing conditions open many areas—such as lakes and rivers—providing avenues of travel inaccessible during warmer months.

NAVIGATION

3-12. There is a marked difference in the way terrain appears in winter (with a snow covering) when compared to summer. Even familiar terrain looks very different and can easily lead to disorientation.

Weight, Balance, and Loads

Navigation in arctic regions may be hampered by rapidly changing and sometimes uncharted variation, mountainous terrain, snow-covered landmarks, and a lack of navigational aids (NAVAIDs). Under these circumstances, a combination of radio navigation, dead reckoning, and pilotage may have to be used to navigate to the destination. Time should be allocated for aircrew members to train in the changing conditions.

STATIC ELECTRICITY

3-13. During cold weather, especially when the air is very dry, static electricity creates serious problems. It is generated by activities such as moving an aircraft through the air, brushing snow and ice from the aircraft, and dragging steel ground cables over the snow. This is a particularly hazardous during refueling and rearming operations. It cannot be emphasized enough as to the importance of having aircraft properly grounded and bonded to prevent injury and reduce the potential for an explosive reaction. In addition, aircraft external load operations also present a serious potential for static electricity. Preparation should include measures such as ensuring static probes are available for use and verifying personnel are properly trained.

AMBIENT LIGHT CONDITIONS

3-14. Summer in the far north (above 55 degrees latitude) provides almost continuous daylight. At Fort Wainwright, AK (64 degrees latitude), for example, there are no night vision device (NVD) operations from the second week of May until the second week of August. During winter months, there are only 3 to 4 hours of daylight with extended sunrise/sunset periods (up to 1 hour each) as transition times between day and night. This presents some unique problems in such areas as mission planning and crew selection, and becomes a major consideration when planning for operations in this area.

3-15. There is great terrain contrast when conducting night operations over snow-covered terrain. While it can be much easier to see details of the terrain at night, certain aspects such as slope of terrain or landing area obstacles are not easily seen. When flying in mountainous terrain, it is also very difficult to accurately interpret details of terrain. Aircrews can easily lose visual reference or provide misleading information. Prevailing ambient light presents unique problems most aircrew members are unaccustomed to; this leads to frequent accidents or incidents (figure 3-2).

Figure 3-2. Ambient light conditions

Flat Light

3-16. Flat light is a variation of the height-depth illusion, also known as sector or partial whiteout. It is not as severe as whiteout, but the condition causes crewmembers to lose their depth-of-field and contrast in vision. Flat light conditions are usually accompanied by overcast skies inhibiting any visual clues. Such conditions primarily occur in snow covered areas but can occur anywhere in the world (dust, sand, mud flats, or glassy water). Flat light can completely obscure features of terrain, creating an inability to distinguish distances and closure rates. As a result of this reflected light, pilots may be given the illusion of ascending or descending when they may actually be flying level. However, with good judgment and proper training and planning, it is possible to safely operate an aircraft in flat light conditions.

Whiteout

3-17. As defined in meteorological terms, whiteout occurs when a person becomes engulfed in a uniformly white glow. The glow is a result of being surrounded by blowing snow, dust, sand, or water. There are no shadows, no horizon or clouds, and all depth-of-field and orientation are lost. A whiteout situation is severe in that there are no visual references. Flying is not recommended in any whiteout situation. Flat light conditions can lead to a whiteout environment quite rapidly. Both whiteout and flat light conditions are insidious, occurring quickly as visual references slowly begin to disappear. Whiteout has been and continues to be the cause of several aviation accidents.

FLYING TECHNIQUES

3-18. Conducting flight operations over snow-covered terrain is a difficult task, even for experienced aircrew members. Certain specialized techniques must be applied to fly safely during cold weather operations. Helicopter operations are emphasized because adverse effects of snow on rotary-wing aircraft are more critical than on FW aircraft.

OPERATIONAL PROCEDURES

3-19. Problems occurring when operating in extreme cold are related to preparation for flight, ice and snow, cold weather engine starts, taxiing, takeoff, en route, and landing. Problems presented by ice, snow, or freezing rain are such that provisions must be incorporated into flight planning to eliminate or reduce their effects.

3-20. A primary rule in any aircraft movement under winter conditions is to think before acting. This environment demands a thoughtful approach to every task. For example, an aviator does not bring the helicopter to a hover and then determine where to go. This will usually result in a whiteout, mandating an instrument takeoff (ITO) type maneuver to climb above the snow cloud and return to visual meteorological conditions (VMC). On an airfield, this results in traffic complication and a safety hazard. Each phase of flight requires a plan, which is announced to the other crewmembers according to sound aircrew coordination techniques. Crewmembers clearly establish and announce intentions before acting. The airfield itself can be a troubling place for winter operations.

TAXIING AND TAKEOFF

GROUND TAXI

3-21. Helicopters produce the greatest amount of rotor wash when hovering; thus, it is best to ground taxi whenever possible. This is more difficult with skid type aircraft. Ground taxiing is performed as a very deliberate movement accompanied by ground guides, if needed, to ensure all appropriate clearances are maintained. Loose snow conditions make this action much more difficult than first appearance as conditions are slippery and stopping will require more distance than normal. It is easier for wheel-type helicopters to ground taxi on a snow-covered airfield. However, this is often difficult or impossible due to snow accumulation, and the aircraft may be forced to hover. This places the helicopter in a mode of flight much more challenging in these weather conditions.

Pickup to a Hover or Takeoff

3-22. The execution of this task depends on snow conditions. If the snow is heavy (water saturated) or packed, there is little difference between hovering in this environment and in a no-snow condition. If the snow is dry and easily blown, this task can become extremely difficult requiring special techniques and specific avoidance procedures. If there is minimal accumulation, this dry snow condition is a small problem;

however it is worsened by a larger accumulation or crusted snow conditions, where the crust can shatter and reveal loose snow underneath. The rate of collective pitch application will vary for the same reasons. Sometimes a slower, more methodical collective application results in a more controlled climb or descent as opposed to the collective being adjusted rapidly in the belief it will alleviate the problem sooner. Hovering in snow can quickly result in a complete and persistent whiteout requiring the aviator to execute appropriate recovery procedures. The essential rule is to expect the worst when preparing to hover in snow conditions. Always assume a whiteout will result from your actions. This mind-set coupled with proper preparation will make for a safer flight. A takeoff should be performed into the wind as this will assist in keeping any snow cloud to the rear of the aircraft.

Note. Aircrew members experienced in desert environment flight, with its sand and common brownouts, have an advantage when confronted with snow conditions and whiteout. However, the two environments neither are not exactly alike nor require exactly the same technique. For example, snow crystals have a tendency to dissipate slower, which keeps the snow cloud suspended for a longer period. Only experience, accompanied by a conservative approach, ensures safe operations.

Hover Taxi

3-23. Aviators hover taxi the helicopter faster or higher than the existing snow cloud. This choice depends upon snow conditions. When forward visibility is essentially unrestricted and the snow cloud is visibly positioned aft of the cockpit, the hover taxi speed is correct. This requires continuous evaluation, depending on factors such as snow and wind conditions, and task to be performed. No one airspeed or rate of travel is correct as this procedure require evaluation of existing conditions each time. Air taxi is preferred over hover taxi. Air taxi allows the aircraft to fly at an airspeed/altitude that will not generate a snow cloud.

EN ROUTE

3-24. In a nontactical environment, aircraft are normally flown at an altitude and airspeed where rotor wash will have no effect upon loose snow.

TERRAIN FLIGHT

3-25. With its inherent low altitudes, terrain flight in a snow-covered environment can create a signature from the rotor wash. This signature can be seen on treetops in the disturbed snow path left by passing helicopters. It can also be seen as a snow cloud hanging in the air in the wake of passing helicopters.

3-26. Maintain at least 40 knots of airspeed to minimize signature effects caused by the rotor wash. Avoid flights at less than 40 knots over forested areas. Snow in the trees is more easily disturbed and may create a visual obstruction to following aircraft (in multihelicopter operations) and/or a signature.

3-27. Avoid flying in close formation over snow. Depending on the nature of terrain and condition of the snow, additional spacing will aid in reducing blowing snow as an obstruction to visibility. To prevent those aircraft behind the lead aircraft from landing into near whiteout conditions, aircraft separation must be increased before beginning final approach to landing. This increased distance between aircraft provides additional time for the snow cloud to settle.

3-28. Avoid flying through narrow valleys during multihelicopter operations, where aircraft must follow the same ground track, requiring aircraft to fly through any existing snow cloud.

LANDING

3-29. When landing a helicopter in snow-covered terrain (including an established runway), expect to be engulfed by a snow cloud unless a proper landing procedure is used. In loose snow conditions, there are essentially two types of approaches: to the ground or to a hover. Either type of approach should be performed into the wind. This assists in keeping any snow cloud to the rear of the aircraft.

Chapter 3

Approach to the Ground

With Forward Speed

3-30. This type of approach demands aviators maintain sufficient forward speed to keep their aircraft in front of the snow cloud ensuring contact with the ground before being engulfed by blowing snow. While this is often the preferred technique for landing, it is frequently avoided due to problems such as obstacles or too little space in the landing area. Although no two approaches are the same and any approach technique will vary by aircraft type, the basic technique remains the same. Because this approach involves touching down with some forward speed, the crew must be familiar with the landing surface and any potential obstacles that could damage the aircraft. The essential elements of this approach are the following:

- Sufficient forward airspeed is maintained to ensure the aircraft is traveling slightly ahead of the snow cloud being created by the aircraft.
- A shallow approach is generally used.
- The entire crew is prepared to call out the position of the snow cloud; for example, "at the tail" or "at the cabin."
- Both rate of closure and rate of descent are minimized ensuring the most controlled touchdown possible.
- This flight attitude is maintained until the aircraft contacts the ground and the collective is reduced to flat pitch.

Note. The optimum approach is marked by aircraft touchdown just before any snow cloud engulfs the cockpit. If the snow cloud engulfs the aircraft before contact is made with the ground, the rate of closure was too slow. Conversely, if the aircraft is completely in contact with the ground and sliding, and the snow cloud has not yet engulfed the cockpit, the aircraft was traveling too fast; an unnecessarily fast rate of closure was maintained.

With No Forward Speed

3-31. This landing is similar to the termination to the surface with forward airspeed, except this termination should be made to landing areas where slopes, obstacles, or unfamiliar terrain precludes a landing with forward speed. It is not recommended when new or powder snow or fine dust is present because whiteout conditions will occur. The termination is made directly to a reference point on the ground with no forward speed. Both the angle should be slightly steeper and the approach speed faster than a normal approach. After ground contact, slowly lower the collective to the full down position, neutralize the flight controls, and apply brakes as necessary to ensure no forward movement.

Approach to the Ground from a Hover

3-32. This technique generally requires termination over the designated landing point at an OGE altitude. This higher altitude is necessary due to the potential snow cloud and is an integral part of this approach technique. The increase in altitude minimizes effects of the snow cloud. It also allows the crew to maintain visual contact with the ground even while the snow cloud is dissipating. The crew can then begin descent to the ground. Termination at a lower altitude (for example, 10 feet) will not permit this visual contact; the crew will likely find itself in a whiteout. This technique works well and may be the only option in certain cases; for example, in a confined area or landing beside a slingload for hook-up. Caution must be used with this technique due to the snow cloud building under the aircraft.

LANDING SURFACE

Aviators should consider what is beneath the snow during all landings. While the snow appears level, the ground beneath could be sloped or covered with rocks, logs, holes, and other hazards. Treat all landings as possible slopes and be prepared if one side or both breaks through the surface. Snow-covered frozen bodies of water have the appearance of a good LZ.

Formation Landings

3-33. Formation landings pose special hazards when landing in this environment. An aircraft could easily be engulfed in the snow cloud of another aircraft during the land sequence. Careful consideration must be given to the appropriate landing formation. Landing distance separation must be increased. The essential elements of formation landings are—

- Increase rotor disk separation, especially for staggered formations. Trail formation is not recommended.
- If tactical and environmental conditions allow, echelon formations decrease the likelihood of being engulfed in another aircraft's snow cloud.
- During the landing sequence, the flight lead should land into the wind that allows the snow cloud to be blown away from the formation.
- All aircraft in the formation should land at the same time.
- Carefully plan go-arounds. If tactical and environmental conditions permit, one sound go-around technique is the forward go-around. This prevents the aircraft from landing back into the same snow cloud that was just departed.

DEPTH PERCEPTION

3-34. The ability to judge height and determine the contour of terrain is difficult when it is snow covered (figure 3-3). The normal tendency of an aviator is to estimate altitude as being higher than it actually is and view sloping terrain incorrectly.

Figure 3-3. Depth perception

3-35. Aircrews can use the following procedures to overcome depth perception difficulties:

- Use terrain features (trees, vegetation, and large rocks) as references. Knowing the approximate dimensions of these features produces a more accurate estimate of height and distance.
- Use a person, animal, or vehicle on the ground as a good reference.
- Improve depth perception by viewing terrain through the side window and comparing this perspective to the view through the windscreen. Maintaining a good scan pattern, similar to that used in night flight, is essential.
- Drop something on the landing surface to serve as a point for comparison when existing landmarks or features cannot be used to determine altitude and distance. An example is a length of pine bough or an item easily seen against the white background and not able to sink into the soft snow.
- Make frequent reference to flight instruments ensuring level flight, adequate altitude AGL, and appropriate airspeed. The information is correlated with current visual information. This continual process requires aviators to scan inside and outside the cockpit.

Chapter 3

- Use aircraft landing lights to assist in depth perception. Lights are adjusted to reduce reflection off the snow.

MAINTENANCE

3-36. Not all maintenance functions will be performed within heated hangars. Units operating in remote areas or in a tactical environment rarely have access to hangars. Although normal cold can be uncomfortable, arctic cold can be extremely dangerous and difficult to work in. Contact frostbite (frostbite caused by simply touching metal objects during very low temperatures) is a real possibility. Danger of such conditions is continually present during winter months. Knowledge gained by those units working in such environments is invaluable and should be sought by units deploying to a winter location.

Note. Refer to the appropriate aircraft operator's manual or local directives for the requirements to leave an aircraft parked outside during cold weather (below 0 degrees C).

AIRCRAFT COMPONENTS

Flight and Engine Instruments

3-37. Gyro-operated flight instruments, such as the directional gyro, turn-and-slip indicator, and attitude indicator, may be unreliable due to increased bearing friction caused by cold, congealed lubricants. Cabin heaters may be used to warm and keep these instruments at operating temperature. During engine start, engine and transmission oil pressures may indicate near-maximum requiring the engine to run at idle until the pressures are in the normal operating range. Transmissions are especially susceptible to cold and take longer to warm up thoroughly. Transmission temperatures and pressures should be carefully monitored.

Plastics and Protective Covers

3-38. Plastics may become brittle and crack from sitting outside or when the aircraft is moved from a warm hangar to the outside. Check for small cracks at the edge of mounting frames, bubbles, windscreens, windows, and doors, as cracks may lead to disintegration in flight. Protective covers provide adequate protection against rain, freezing rain, sleet, and snow when installed on a dry aircraft before precipitation. An unsheltered aircraft cannot be completely covered. Those portions of the aircraft left exposed should be carefully inspected/preflighted before operation of the aircraft occurs. In case of blowing snow, even the covered portion of an aircraft can be penetrated by the elements; these aircraft deserve the same level of attention. In addition, if an aircraft is pulled from the hangar with any accumulation of water, the water will quickly freeze and cause possible catastrophic damage to parts such as drive shaft joints and internal engine components. It is better to leave the aircraft outside unless it can be sheltered long enough to dry completely.

Synthetic Rubber

3-39. Synthetic rubber, when used for oil and fuel lines or to coat electrical wiring, may become stiff and easily broken. For example, the crew chief's microphone cord should be coiled and kept in the warmth of the helicopter until needed for the aircraft start or stop sequence. Lines and wiring should not be bent when cold.

Tires

3-40. Cold weather can cause tires to stiffen, leaving a flat spot until the tire is sufficiently heated through movement and friction from taxiing the aircraft. When beginning the taxi sequence, aviators should move slowly and minimize side load or excessive tire turning. Tires and tire pressure require frequent attention during these cold periods.

Hydraulic and Pneumatic Leaks

3-41. Leaks may appear more often due to contraction and expansion of fittings and lines during temperature extremes. Cold weather aircraft starts are completed according to the operator's manual and current maintenance directives. Static leaks may tend to disappear with increasing temperature. A close evaluation

Petroleum, Oil, and Lubricants

3-42. Aircraft are serviced with fuel upon landing to prevent condensation accumulating in the fuel tanks (when moved into a heated hangar). However, tanks should not be topped off as subsequent parking in a hangar will result in expansion and some fuel spillage. The fuel boost pump operation is checked before flight due to possible freezing or damage. Defueling an aircraft in winter conditions may be difficult and hazardous.

3-43. The viscosity of oil and grease used is very important in cold weather operation. Use only grades of oil and grease specified by the manual. Oil levels must be checked after operational temperatures have been reached (postflight is a good time), and any oil needed is preheated and added while the system is hot.

Control Cables

3-44. Adjust control cables to manufacturer's specifications to allow for contraction and expansion caused by temperature changes. Cables can freeze if moisture is allowed inside housing.

Batteries

3-45. Both dry and wet cell batteries require special consideration during cold weather.

Wet Cell

3-46. If the airplane must be parked outside, wet cell batteries should be kept fully charged or removed from the aircraft to prevent loss of power caused by cold temperatures and guard against battery freezing.

Dry Cell

3-47. Dry cells are usually associated with aircraft in only two applications—emergency lights and portable radios (including emergency locator transmitters). Manufacturer recommended batteries for this type equipment are resistant to power loss by freezing.

Operational Checks

3-48. Short engine ground run-ups must be avoided. Engine run-ups must be long enough to bring the engine to operating temperature. Any shorter period will cause water vapor to condense. This water could freeze and split oil coolers, block oil lines, and increase possibility of engine failure.

Aircraft Towing

3-49. Aircraft are towed at a slow rate of speed since control is difficult while turning or stopping. If the parking area is on an incline, aircraft will tend to push the towing vehicle. Caution must be used to avoid turning too short when marshaling aircraft out of or into a parking area. A disproportionate number of incidents occur due to aircraft damage resulting from sliding into a hangar doorway during towing operations. Enough personnel must be positioned around aircraft to monitor the operation. For example, a CH-47 is towed into the hangar with six personnel—one person in the cockpit manning the brakes, one person manning the tow vehicle, and one person at each corner of the aircraft equipped with whistles to alert the vehicle driver of potential collisions so the towing operation can be halted until clearance can be ensured. While this seems excessive, hangar entrances are often icy with aircraft subject to sliding.

TRAINING

3-50. Units qualifying aviators in cold weather operations are responsible for conducting a well-organized training program. Training programs are geared to instill confidence and develop skills in all areas of cold weather operations. Instructor pilots (IPs) and supervisory maintenance personnel must be highly qualified and skilled in all areas of cold weather operations.

3-51. Emphasis must be placed on safety. Snow conditions, wind velocity and direction, and aviator proficiency levels are factors instructors must evaluate to determine if safe training can be conducted. The professional judgment of the instructor to discontinue training due to unsafe conditions must be accepted and not criticized.

3-52. The flight training program allows each aviator to advance at an individual rate. Initial training is conducted under less challenging conditions. As aviator proficiency increases, conditions should be made more demanding until the most challenging mission can be performed.

RECOMMENDED PROGRAM OF INSTRUCTION

3-53. A recommended program of instruction for qualifying aviators in cold weather operations is provided in the following paragraphs. Additional academic subjects may be required, based on the specific mission and location of the unit.

Academics

3-54. Suggested topics include—
- Human factors associated with cold weather flying.
- Environmental factors that affect cold weather operations.
- Aircraft preparation for cold weather.
- Aircraft operational procedures in cold weather.
- Cold weather survival.
- Techniques to improve depth perception and determine snow condition.

Flight

3-55. Flight training may be limited by conditions at the unit's home station. Some areas may not be able to replicate snow conditions adequately for training in whiteout conditions. Instructors can demonstrate techniques and procedures to some extent. Crews are evaluated on these procedures during their annual proficiency and readiness test (APART) or no-notice evaluations. Flight simulators are also a great device in training for this environment.

3-56. Suggested maneuvers include—
- Snow landings (most crucial).
- Go-around procedures.
- Taxiing over snow covered areas.
- Snow takeoffs.
- En route flight techniques.

SECTION II – DESERT OPERATIONS

ENVIRONMENTAL FACTORS

3-57. Figure 3-4 illustrates desert regions around the world. The typical desert region is a dry, barren region, largely treeless, and sandy. A region of environmental extremes, it has violent and unpredictable changes in weather and contains terrain not conforming to any particular model. While frequent clear days offer unequaled visibility and flight conditions, a sandstorm can quickly halt all operations. Therefore, desert operations require special training, acclimatization, and a high degree of self-discipline. The lack of water makes this environment nearly inhospitable without a solid support structure. The desert environment is one of the most severe environments in which Army aviation must operate. Training and preparation are paramount.

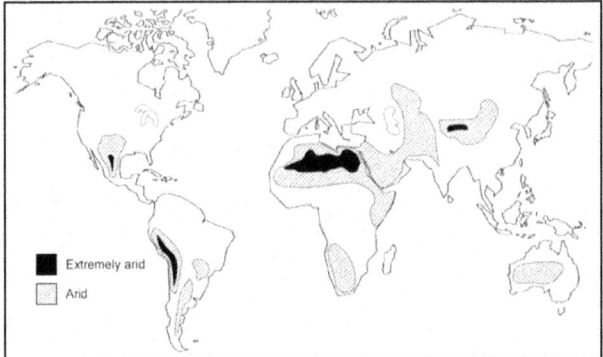

Figure 3-4. Desert areas of the world

CLIMATE AND WEATHER

3-58. At low altitudes, extremely high temperatures have been recorded in some desert areas. High daytime temperatures severely restrict lift capabilities of aircraft. This restriction may be overcome by conducting major operations during the cooler part of day or at night. High, violent winds are common to desert regions; aviators must be thoroughly briefed and prepared for these conditions.

Temperatures

3-59. Desert heat creates serious problems for humans and equipment and requires special consideration. Heat protective clothing is required when working on aircraft. Temperatures in Southwest Asia may exceed 55 degrees C while temperatures in a closed aircraft or vehicle can be substantially higher. Severe burns can result when bare skin touches any exposed metal parts. Extreme heat can cause electronic gear to malfunction or cease functioning. High temperatures can also cause lubricants to break down and seals and gaskets to distort, which may result in leaks or deadlined equipment. When these problems are compounded by sand accumulation, equipment experiences more difficulty. High temperatures can also cause softening of plastics, higher stress on pressurized containers, and shortened battery life.

Wind

3-60. Desert winds generally slow down around sundown and remain relatively calm until sunrise when they begin to increase again. Winds can achieve near-hurricane force and vary almost consistently with no distinct or predictable pattern. In all deserts, rapid temperature changes invariably follow strong winds. Strong winds may raise towering dense clouds of dust and sand. This condition is more common in sandy areas but can exist in any semiarid or arid region.

Precipitation

3-61. Annual rainfall varies between desert regions. Some regions in the world receive as much as 10 inches per year. The Sahara might receive 4 inches annually, with no rainfall for 8 to 10 months. When rain comes, it may be in the form of a deluge, with mostly surface runoff, and supply little help to the landscape.

Sunlight and Moonlight

3-62. Low cloud density results in bright conditions during the day and generally clear, moonlit nights. However, when moon illumination is low or during the new-moon cycle, the desert presents a formidable challenge to night flying. It is probably the most difficult environment in which to interpret terrain relief and elevation, especially while using NVDs. Unaided night flight and operations are far more difficult and not recommended.

TERRAIN

3-63. Large areas of open and relatively flat terrain create special flying problems. Distances and altitudes are difficult to accurately estimate in desert environments. The lack of definable terrain features makes navigation difficult, especially at night and over long distances. The likeness of the terrain can lead an aviator toward inattentiveness. Low flight requires constant observation, attention, and concentration. Desert terrain is also extremely rough with features such as rocks and gullies, which create problems and delays with equipment. There are three primary types of deserts—sandy, rocky plateau, or mountainous. Most deserts are rocky plateaus or mountainous, but some may be a combination of all three.

Sandy Desert Terrain

3-64. Sandy desert terrain (figure 3-5) consists of extensive basins completely filled with deep shifting sand, largely the product of wind erosion. The most familiar example is found in Saudi Arabia, where Desert Shield/Desert Storm operations occurred. This area and operation were marked by constantly shifting terrain and talc-like sand that permeated the tightest seals and made maneuvering difficult.

Figure 3-5. Sandy desert terrain

Rocky Plateau Desert Terrain

3-65. Rocky plateau desert terrain (figure 3-6, page 3-15) consists of relatively slight relief, interspersed with extensive sand-filled basins. This plateau, due to recurring floods, is cut by dry, steep-walled valleys. These valleys are filled with torrents of water during infrequent rains.

Weight, Balance, and Loads

Figure 3-6. Rocky plateau desert terrain

Mountain Desert Terrain

3-66. Mountain desert terrain (figure 3-7) consists of scattered ranges of barren hills or low mountains, separated by dry basins. Some rainfall occurs in the highlands during violent showers. The water runs rapidly over the surface and erodes deep ravines and gullies. Floodwaters rush from the mountains into the basins where sand and gravel are deposited. Evaporation results in dry salt or salt marshes.

Figure 3-7. Mountain desert terrain

NAVIGATION

3-67. Lack of terrain features and poor reference points make navigation difficult. Aviators generally rely on dead reckoning for navigation augmented by global positioning system (GPS)/Doppler equipment. During Operation Desert Shield/Storm and Operation Enduring Freedom/Operation Iraq Freedom, maps often proved to be of limited value due to ever-shifting sand dunes and constantly changing terrain. GPS was an invaluable tool.

SAND AND DUST

3-68. No discussion of desert environmental influences would be complete without reviewing adverse effects of sand and dust. The density, or consistency, of sand and dust vary throughout the world and even within the same desert region. All have a drastic effect on the operation of aircraft, especially helicopters. The large quantity of loose sand and dust creates serious erosion problems for rotor blades, turbine compressors, and windscreens–any moving part or parts in contact with other components. The corrosive effects of sand and dust in any desert region can cause severe damage to Army aviation equipment, as documented during Desert Shield/Storm. This information should be carefully reviewed and incorporated into training, planning, and preparation before any deployment. Hovering in loose sand and dust and nap-of-the-earth (NOE) flight continually expose the helicopter to the erosive effects of the desert. Even when parked, helicopters are exposed to blowing sand and dust from wind, vehicles, and especially rotor wash from other helicopters. Sand and dust particles collect on all exposed surfaces of the aircraft. These penetrate almost any crack or crevice accumulating inside the helicopter making daily cleaning a necessity.

FLYING TECHNIQUES

3-69. Conducting flight operations over desert terrain is a difficult task, even for experienced aircrew members. Certain specialized techniques must be applied to fly safely and effectively during desert operations. Primary emphasis is given to helicopter operations due to effects of the desert environment are more critical on rotary-wing than on FW aircraft. High temperatures, which cause inadequate engine cooling and reduced payloads, also hamper desert operations. An obvious effect of the high temperatures is the resulting performance degradation from high-density altitude conditions. This is compounded when the operational area is at higher altitudes. The temperature variation within a day also presents unusual problems in performance planning. This variation can often be 70 degrees F or more. FW aircraft experience longer takeoff and landing rolls and overheating of brakes caused by the reduced air density. Density altitude is a critical factor in desert flying techniques and mission capabilities.

OPERATIONAL PROCEDURES

3-70. Minimizing effects of sand and dust dictate, when possible, certain preventive measures. In this environment, the preferred aircraft position is a hardstand or other area with minimal sand and dust. The aircraft must be thoroughly inspected to remove as much sand as possible before flight. Aircraft performance, especially during hot summer months, may be adversely affected by temperature and altitude. This environment demands strict attention to PPCs and operator's manual.

3-71. Aircraft should not depart into a sandstorm or dust storm. A primary rule in any aircraft movement in desert conditions is to think before acting. For example, an aviator does not bring the helicopter to a hover and then decide where to move. This usually results in a brownout, mandating an ITO type of maneuver to exit the dust cloud by climbing above the dust to return to VMC. On a makeshift (soft-surface) airfield, this results in traffic complications and a potential safety hazard. Each phase of flight requires a plan be announced to the other crew members according to sound aircrew coordination techniques. As always, absolute control of the aircraft is paramount. Any techniques that differ from normal operations should be well-established procedures, backed up by a written task, condition, and standard. Such techniques will be trained with an IP.

TAXIING AND TAKEOFF

3-72. Avoid ground taxiing in areas with moderate to severe brownout conditions. Whenever possible, air taxi to minimize dust. In well-established areas or in areas with minimal dust, it may be possible to ground taxi. Takeoff should always be made into the wind to minimize brownout conditions. A normal takeoff utilizes an ITO technique. In prepared areas and operating near maximum performance, a running type takeoff may be used to minimize sand ingestion. Caution should be used when making abrupt takeoffs due to increased rotor thrust and the effects of sand and other debris. For formation flights in areas of heavy dust, the preferred technique is for aircraft to depart single ship and conduct in-flight join ups.

EN ROUTE

3-73. Flying through sandstorms, dust storms, and dust devils must be avoided whenever possible. The en route portion of a flight in a desert environment is especially dangerous over undulating terrain that shifts daily due to wind. This is especially true when flying with night vision goggles (NVG). Aerial perspective is difficult, especially at night, creating high-depth perception illusions. Poles and wires are hard to see even during the day; when covered with dust, these items blend into the background terrain.

TERRAIN FLIGHT

3-74. Terrain flight in a desert environment can leave a signature if the aircraft is flown too low and too slow. This potential signature is a major consideration. The exact flight altitude and airspeed must be carefully selected. With flight safety paramount, the crew must balance altitude, airspeed, and signature. With difficult terrain interpretation, flying too low and/or too fast can easily result in a catastrophic accident. Training to prevent this is crucial.

LANDING

3-75. When landing a helicopter in desert terrain, aviators should always expect severe brownout condition to occur. Soil conditions can change rapidly. All landings should be into the wind to assist in keeping any dust to the rear of the aircraft. The landing technique is mission, enemy, terrain and weather, troops and support available, time available, civil considerations (METT-TC) dependent for each site. The aviator should conduct high and low recon techniques to determine the best landing procedure. Hovering and low-altitude, low-speed flight modes are avoided if possible. Removal of doors and windows may increase visibility during the landing sequence. Aviators must expect dust and grit to enter the cockpit if the doors/windows are removed. If installed, all doors and windows should be closed and vent blowers turned off. Aircrew coordination is critical when landing. Crewmembers should call intensity and location of the dust cloud.

APPROACH TO THE GROUND

With Forward Speed

3-76. The normal approach technique is a VMC approach to the ground. This usually involves establishing aircraft in a landing attitude, with an appropriate and constant rate of closure and rate of descent fitting the condition and state of the terrain. Usually, the approach angle is greater than normal to minimize effects of the dust cloud. In each case, the aviator will have to determine several factors to achieve the proper rate of closure and rate of descent. These factors include the amount of dust, wind direction, slope, and roughness of the terrain (including obstacles). The approach proceeds with this predetermined information, intentionally touching down with some forward speed. The last portion of the approach, however, sees the aircraft engulfed in the brownout but under a controlled descent and with the aviator watching for the ground. When the aircraft contacts the ground, collective reduction and brakes, according to conditions, will stop any forward motion.

With No Forward Speed

3-77. This termination should be made to landing areas where slopes, obstacles, or unfamiliar terrain precludes a landing with forward speed. It is not recommended when fine dust is present because brownout conditions will occur. The termination is made directly to a reference point on the ground with no forward speed. The reference point is critical especially when utilizing NVG. The landing involves establishing aircraft in a landing attitude, with an appropriate rate of closure and rate of descent fitting the condition and state of terrain to the reference point(s). The angle should be slightly steeper and the approach speed faster than a normal approach. After ground contact, slowly lower the collective to the full down position, neutralize the flight controls, and apply brakes as necessary to ensure no forward movement.

APPROACH TO THE GROUND FROM A HOVER

3-78. This technique generally requires termination over the designated landing point at an OGE altitude. This higher altitude is necessary due to the potential dust cloud and is an integral part of this approach technique. It also allows the crew to maintain visual contact with the ground even while the dust cloud is dissipating. The crew can then begin the descent to the ground. Termination at a lower altitude, such as 10 feet, will not permit this visual contact, and the crew will likely find itself in a brownout. This technique works well if there is a limited amount of sand to be dispersed. It may be the only option in certain cases; for example, in a confined area or landing beside a slingload for hook-up.

Depth Perception

3-79. The nature of desert terrain with its open areas, relatively flat and without unique terrain features, makes it difficult to judge distances and altitudes. Radar altimeters provide the most effective reference for pilots to estimate altitude, day or night, over expanses of desert terrain–whether it is sandy, rocky plateau, or mountain. However, be aware that the radar altimeter is not terrain following and therefore does not guarantee terrain clearance when the aircraft is moving. Many aircraft have been lost due to inadvertent flight into such hazards as sand dunes and ridges. This phenomenon must be trained ensuring aircrew awareness. Systems, such as the heads-up display (HUD), greatly enhance SA and allow the aviator to concentrate more fully outside the aircraft.

MAINTENANCE

3-80. Most maintenance is performed outdoors and in the elements; that is, in heat and in the midst of the accumulating sand. This is a particularly hostile environment for equipment. It is made more difficult by temperature extremes. With the dramatic cooling period at night, accumulating moisture becomes condensation. This condensation, combined with accumulated sand, is a damaging mixture for equipment and causes a dramatic shortage of its service life as discussed earlier in this section.

Aircraft Components

Instruments and Avionics

3-81. The service life of avionics and electronic components is usually reduced due to heat, especially with windows and doors closed. If doors and windows are left open, however, the avionics' service life is reduced as a result of dust accumulation. This factor must be evaluated continuously to preserve the life of all the components. Use covers along with more frequent cleaning.

Protective Covers

3-82. Any windscreen or window surface, especially plastic, is covered to prevent the effects of blowing sand and minimize effects of the sun. Covers, such as canvas and condemned parachute canopies, are used. The crew ensures the window surface and cover are as grit-free as possible minimizing any abrasive elements between them. This accumulated grit can act like sandpaper as the cover moves around in the wind.

3-83. Aircraft are secured with the appropriate tie-down kit, supplemented by available tie-downs, ensuring tie-down points are secured deep enough to be effective. This tie-down procedure should be a daily function. Severe wind conditions occur in the desert with relatively little notice. Usually, the crew does not have enough time to tie down the aircraft once a storm begins.

Tires

3-84. Desert landings can cause considerable damage to aircraft wheels/tires. Crews should supplement stocking of these items prior to deployment.

Blades

3-85. Main and tail rotor blades and tip caps are subject to erosion. Anti-erosion kits should be installed on aircraft prior to extended deployment to desert environments. Also, look at paint or other materials that can be applied to blades between flights to extend their life.

Petroleum, Oil, and Lubricants

3-86. Oil should be changed more frequently to minimize internal component wear, including engines and transmissions. Oil should be checked for signs of sand accumulation which can clog filters. Oil and hydraulic fluid are added directly from their original unopened containers to help stop sand and dirt from entering the systems. Partially used containers are disposed of properly, a critical element of preventive maintenance, and helps preserve and extend the life of aircraft components in a desert environment.

3-87. Fuel contamination caused by sand/grit accumulation is a major consideration in proper maintenance procedures. When transferring fuel, exercise great care to ensure filters and screens are continuously used and changed frequently. Fuel tank caps and openings are closed whenever possible. Major damage from ingestion of the elements can occur when a cap is left open and the wind blows, or another aircraft hovers nearby.

3-88. During refueling and rearming operations, it is important to maintain a maximum aircraft separation to minimize blowing sand. Use pressure or closed circuit refueling equipment to avoid contamination.

3-89. Excess grease is wiped off each time lubricant is applied. This ensures sand and dirt are not attracted, forming a paste that can grind and wear lubricated parts.

Control Cables

3-90. Adjust control cables to manufacturer's specifications to allow for contraction and expansion caused by temperature changes.

Engines

3-91. During maintenance checks, engines are operated as little as possible. Sand ingestion increases erosion on compressor blades and decreases engine performance. Engine flush and wash intervals are reduced. As soon as the engine is shut down, appropriate covers are installed. Power checks may be deferred to areas permitting operation without damage.

Other Considerations

3-92. Other maintenance considerations include—
- Wipe clean exposed components during daily inspections and prior to flight. Sand collects on nearly every surface of the aircraft and in dead spaces. This causes additional wear on exposed actuators pistons, bearings, struts, and seals. Aircraft should also be washed frequently if water is available.
- To minimize pressure buildup in the system, do not set brakes when the temperature is expected to rise dramatically.
- Inspect filters more frequently than normal to ensure they are not clogged.
- Clean optical equipment, such as FLIR optics, and cover and protect these items when not in use. These items are subject to scratching or damage due to dust and grit in the environment. The care of such equipment is paramount.
- Weapons are particularly susceptible to accumulation of sand and dust. The light coat of lubricant, often covering weapon parts, invites an accumulation of retained grit more able to inflict damage and create jamming.
- Replace damaged seals around doors, windows, and access panels.
- Adjust aircraft prescribed load list to include additional supplies of filters, bearings, actuators, windshields, or other parts subject to wear in a desert environment.

TRAINING

3-93. Administering a desert weather training program to qualify crewmembers is a unit responsibility. The program outlined in this section is a suggested guide requiring modification by the commander to suit specific situations. Basic preliminary needs should include emphasis on physical fitness and careful maintenance to offset increased stresses and lowered efficiency of personnel and aircraft in the desert environment.

RECOMMENDED PROGRAM OF INSTRUCTION

3-94. The program starts with training completed routinely while at home station as part of the normal training cycle. This training includes academic and flight training, and defines the train-up of personnel upon notification of deployment. Experts from outside the unit may conduct this training.

Chapter 3

Academics

3-95. Suggested topics include—
- Human factors associated with hot weather operations.
- Environmental factors affecting desert operations.
- Aircraft preparation for hot weather.
- Aircraft operational procedures in hot weather.
- Principal difficulties during desert operations.
- Hot weather survival.
- Performance planning.
- NVD and night operations with zero illumination.

Flight

3-96. Flight training may be limited by conditions at the unit's home station. Some areas may not be able to replicate sand and dust conditions adequately for training in brownout conditions. Instructors can demonstrate techniques and procedures to some extent. Crews are evaluated on these procedures during their APART or no-notice evaluations. Flight simulators are also a great device in training for this environment.

3-97. Suggested maneuvers include—
- Sand/brownout landings (most crucial).
- Power management.
- Taxiing over sand covered areas.
- Sand takeoffs.
- En route flight techniques.

SECTION III – JUNGLE OPERATIONS

ENVIRONMENTAL FACTORS

3-98. Figure 3-8 illustrates jungle areas around the world. The jungle is an area located in the humid tropics. In this area, land is covered with dense growth of trees and other types of vegetation. This dense growth stifles military operations in many ways, such as restricting communication and travel. Jungle areas are characterized by heat, humidity, monsoon seasons, and other weather phenomena imposing particular restrictions on Army aviation. As with other environments, specific training and preparation are necessary.

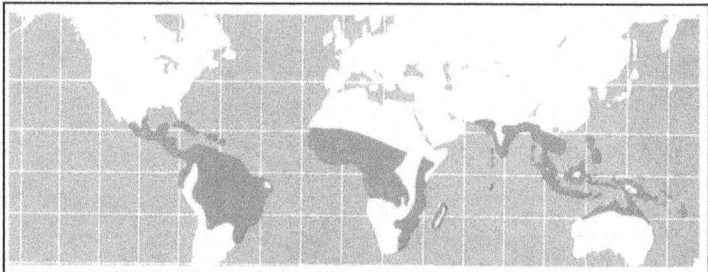

Figure 3-8. Jungle areas of the world

CLIMATE AND WEATHER

3-99. Tropical weather environments, marked by extreme heat and humidity, pose unique problems for aircrews.

Temperatures

3-100. Equatorial temperatures commonly reach the 35 degrees C to 40 degrees C temperature range. These temperatures, combined with extreme humidity, create a formidable work environment. This environment decreases human endurance due to the combination of heat and humidity. However, equatorial temperatures rarely reach the range of temperatures experienced in desert environments.

Wind

3-101. Southeast trade winds are prevailing winds south of the equator. Northeast trade winds prevail north of the equator. Doldrums, located along the equator, are marked by calms, squalls, and light and shifting winds.

Precipitation

3-102. Generally, the most abundant and regular rainfall is found at the equator. This rainfall diminishes progressively as distance (north or south) from the equator increases. Rainstorms can be brief and violent, and are often accompanied by intense thunder and lightning.

Sunlight and Moonlight

3-103. Since jungle environments are positioned at or very close to the equator, the length of day and night varies little throughout the year. This is markedly different from the arctic environment with its wildly varying cycle of day and night. At the equator, the split between day and night remains essentially 12 hours of each day throughout the year. This eases planning considerations for mission performance. Moonlight varies as it does throughout the world. Jungle terrain typically provides little reflective quality for any light source. Flying above the jungle canopy provides little reference to aid in depth perception. This makes terrain flight more dangerous with a greater chance for accidents such as tree strikes to occur.

TERRAIN

3-104. Jungle terrain is often rugged and swampy with deep valleys and steep ridges. Due to heavy rainfall, streams and rivers are plentiful, and the soil is usually very soft. Foot travel, especially on steep slopes, is frequently difficult. A steep slope may encourage rock or landslides. Trails tend to follow ridges, detouring to avoid low ground and deep valleys. The jungle landscape consists mostly of the following types of terrain.

Rain Forests

3-105. Rain forests consist of dense, high trees, often more than 100 feet high, with a rich undergrowth of smaller trees and bushes covered in greenery and vines. The tree trunks are usually straight, slender, and without branches for the first 50 feet but the undergrowth make travel difficult. Branches near the top of the trees spread out and interlock to form the upper layer of the rain forest, commonly known as the canopy. In some jungles, the canopy is composed of two or three successive levels of vegetation which are generally filled with foliage. At a height of 20 to 40 feet above the ground, this thick canopy nearly blots out the sun from the forest floor, which consequently, supports relatively little undergrowth. As with other environments, the jungle environment varies considerably.

Mangrove Swamps

3-106. Mangrove swamps consist of dense forests of trees and shrubs from 10 to 30 feet high, supported on tall, stilt like roots arching outward to anchor in murky water and mud.

Savannas

3-107. Savannas consist of vast areas of grass, shrubs, and isolated trees. Savannas range in size from small areas of a few miles to vast regions encompassing several thousand square miles.

Palm Swamps

3-108. Palm swamps are found in salt and freshwater areas. Movement is limited to foot traffic and, sometimes, small boats. Observation from air or ground is difficult.

NAVIGATION

3-109. Navigation in jungle terrain is often difficult due to lack of significant terrain features. The top of the jungle canopy is nearly devoid of distinguishing landmarks and makes determination of exact location very difficult. Dead reckoning is combined with GPS and Doppler navigational systems to assist aviators.

FLYING TECHNIQUES

3-110. High temperatures affect jungle operations. In determining jungle-flight techniques, density altitude becomes a major consideration. High-density altitude degrades aircraft performance.

OPERATIONAL PROCEDURES

3-111. Operations conducted in a jungle region will nearly always involve an exceptionally hot and humid atmosphere. This high humidity often results in condensation throughout the aircraft including fogging of instruments; rusting of steel parts; growing of fungus in tight, confined areas; and malfunctioning of electrical equipment. When operations are conducted in predominately high temperature conditions, engine operating temperatures must be closely monitored. As the ambient temperature increases, engine efficiency decreases and power availability, especially at high altitude, becomes limited. Performance planning is a critical factor for safe mission completion. Updates may be required as the day progresses, especially if conditions worsen. Jungle operations demand a planned and efficient use of the aircraft. In many situations–such as in high-altitude or high-density altitude conditions–the aircraft will be operating near its maximum operating capability. Any circumstances should be optimized. The aviator plans to terminate all approaches to the ground (conditions permitting), hovers as low as possible, and assumes tall grass and other obstacles will worsen the situation forcing hover OGE. The worst scenario is expected and performance planning calculated ensuring power is available for such situations. The aviator recognizes (before attempting the maneuver) when to avoid certain aircraft maneuvers.

TAXIING AND TAKEOFF

3-112. In jungle regions, aviators make use of available wind, the long axis of the LZ, and ETL during takeoff maneuvers. They select the most advantageous terrain, such as short grass or fewest obstacles, and avoid multiple aircraft departing simultaneously when planning taxiing or takeoff. A flight of aircraft may have to land and takeoff singly to avoid operating in the disturbed air of preceding aircraft and rejoin once airborne.

EN ROUTE

3-113. The primary consideration during en route flight is the potential for navigation problems.

TERRAIN FLIGHT

3-114. Terrain flight in a jungle environment is not unique. As always, the primary concern is safety. Due to the lack of a forced landing area, the need for safety of flight is emphasized. Jungle environments do not have deciduous trees with bare limbs which are difficult to see or may damage the aircraft during flight.

LANDING

3-115. Due to high trees and small LZs, a threat to landing in a jungle environment is the potential lack of power available for a steep-angle approach. Proper attention to PPC and available power helps prevent a steep approach landing with insufficient power while surrounded by tall trees that block a go-around.

MAINTENANCE

3-116. In the jungle environment, all equipment is subject to damage caused by corrosion and fungus.

AIRCRAFT COMPONENTS

3-117. High humidity in a jungle environment can cause problems throughout the aircraft. Avionics and electrical equipment malfunctions are common as humidity condensates in the equipment causing electrical anomalies or even failures and should be inspected often. Electrical connectors and cannon plugs are more susceptible to corrosion and require frequent cleaning. Fungus and mold are very common in the jungle environment and grow rapidly on fabrics in the aircraft. All fabrics in the aircraft, to include sound proofing panels, seat covers, and tie down straps, should be washed often and thoroughly dried to minimize fungus growth. Mold and fungus grow rapidly on rubber covered items and require frequent cleaning.

PETROLEUM, OIL, AND LUBRICANTS

3-118. Petroleum, lubricants, and fluids in the aircraft may also become contaminated by water or condensation. Condensation is minimized by keeping fuel tanks full, ensuring fuel samples are performed daily, and draining all water from the fuel tank In addition, aircraft fuelers should conduct frequent testing for moisture use and use water separators when servicing aircraft. Approved fungicides should be added to fuel in accordance with the appropriate operator's manual to inhibit fungi contamination. Hydraulic fluid and oil should be tested frequently for moisture contamination. Contaminated oil and hydraulic fluid will often have a milky appearance. Aircraft components contaminated with moisture should be drained, flushed, and serviced.

TRAINING

3-119. Administering a jungle training program to qualify aviators is a unit responsibility. Basic preliminary needs must include emphasis on physical fitness and careful maintenance to offset increased stresses and lowered personnel and aircraft efficiency in a jungle environment. Jungle-terrain training instills confidence, develops skills, and emphasizes safety.

RECOMMENDED PROGRAM OF INSTRUCTION

3-120. The program starts with routine training completed at home station as part of the normal training cycle. This training includes academic and flight training, and defines train-up of personnel upon notification of deployment. Outside experts may conduct unit training.

Academics

3-121. Suggested topics include—
- Human factors associated with jungle flight.
- Environmental factors affecting jungle operations.
- Aircraft preparation for jungle operations.
- Principal difficulties during jungle operations.
- Jungle survival.
- Performance planning.

Flight

3-122. Flight training may be limited by conditions at the unit's home station. Some areas may not be able to replicate conditions adequately for training in jungle environments. Instructors can demonstrate techniques and procedures to some extent. Crews are evaluated on these procedures during their APART or no-notice evaluations. Flight simulators are also a great device for training in this environment.

3-123. Suggested maneuvers include—
- Power management (Steep approaches and takeoffs).

Chapter 3

- En route flight techniques.
- IIMC procedures.

SECTION IV – MOUNTAIN OPERATIONS

> **CAUTION**
> This section should be trained, along with section I. Since mountain operations usually involve cold weather and snow, sometimes unexpectedly, this is the worst situation. Units should take additional time and concurrently train for both environments.

ENVIRONMENTAL FACTORS

3-124. Mountains are generally characterized by rugged, divided terrain with steep slopes and few natural or manmade lines of communication. Mountain weather is seasonal ranging from extreme cold, snow, and ice during winter months to extreme heat during summer months. Some mountain regions have snow and ice–including glaciers–throughout the year. Although these weather extremes are important planning considerations, varying weather within a compressed time also influences operations.

CLIMATE AND WEATHER

3-125. Rapidly changing weather is one of the greatest hazards to mountain operations. It presents difficult flight operations for experienced and inexperienced aircrews. Mountain flight affects aircraft performance, accelerates crew fatigue, and requires special flight techniques.

Temperature

3-126. The range of temperatures is wide; within some areas it can vary from -40 degrees C (during winter) to +30 degrees C (during summer). Additionally, as with cold weather environments, the temperature variant within a day or single flight can be significant. This must be expected and prepared for with appropriate performance planning completed and proper survival equipment carried on the aircraft.

Precipitation

3-127. Precipitation in mountain regions increases with altitude. Both rain and snow can be expected in these areas. Rain presents the same challenges as in lower altitudes, but snow drastically affects aviation operations. Refer to section I for information regarding cold weather operations.

Snow and Icing

3-128. Refer to section I for information regarding environmental factors.

Fog

3-129. The effects of fog in the mountains are the same as for lower regions. The topography, however, causes fog to occur more frequently in the mountains. Thus, fog becomes a more significant planning consideration.

Wind

3-130. Wind associated with mountains can be broken down into the following three main categories (figure 3-9):

- **Prevailing wind** is the upper-level wind flowing predominantly from west to east in the CONUS.

- **Local wind** is also called valley wind and is created by convection heating and cooling. This wind flows parallel to larger valleys. During the day, it tends to flow up the valley and flows down the valley at night.
- **Surface wind** is the layer of air lying close to the ground. It is less turbulent than prevailing and local wind.

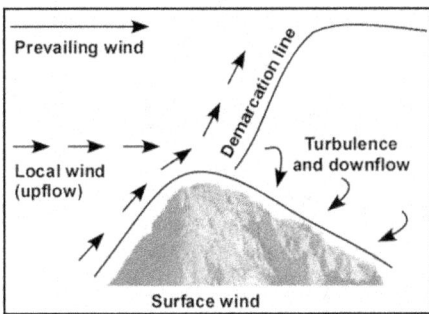

Figure 3-9. Types of wind

Demarcation Line

3-131. The demarcation line is the point separating upflow air from downflow air. It forms at the mountain's highest point and extends diagonally upward. The velocity of the wind and steepness of the uplift slope determines the position of the demarcation line. Generally, the higher the wind speed and steeper the terrain, the steeper the demarcation line. The effects of varying wind velocities on the demarcation line are described in the following paragraphs.

Light Wind

3-132. A light wind (figure 3-10, page 3-26) is 1 to 10 knots. It accelerates slightly on the upslope, giving rise to a gentle updraft. It follows the contour of the terrain feature over the crest. At some point over the crest of the hill, it becomes a gentle downdraft.

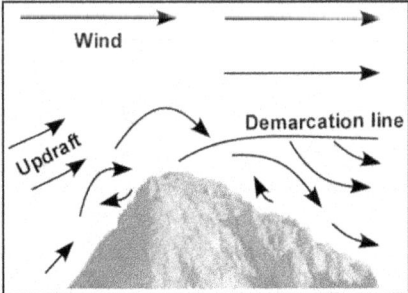

Figure 3-10. Light wind

Moderate Wind

3-133. A moderate wind (figure 3-11) is 11 to 20 knots. It increases the strength of updrafts and downdrafts and creates moderate turbulence. A downdraft will be experienced on the leeward side (sheltered from the wind) near the mountain's crest. The demarcation line forms closer to the crest and is steeper.

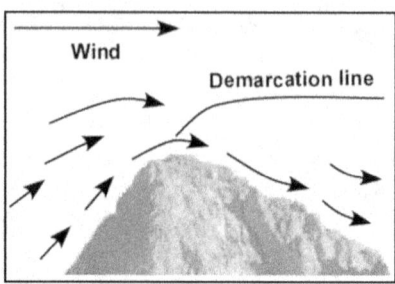

Figure 3-11. Moderate wind

Strong Wind

3-134. As wind increases above 20 knots, the demarcation line moves forward to the crest's leading edge (figure 3-12, page 3-27). It then matches the slope's steepness. The severity of updrafts, downdrafts, and turbulence also increase. Under these conditions, the best landing spot is close to the forward edge (windward side) of the terrain feature.

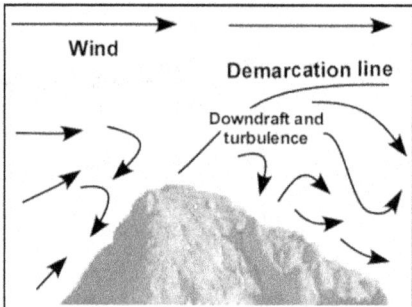

Figure 3-12. Strong wind

Mountain (Standing) Wave

3-135. A mountain, or standing, wave is a phenomenon occurring when airflow over mountainous terrain meets certain criteria and causes a complex weather pattern (figure 3-13). That pattern creates relatively smooth and strong, lifting winds on the windward side while progressing toward the leeward side very abruptly (at the crest) and dramatically. It thrusts the aircraft into an area dominated by downdrafts having sustained recorded values of at least 3,000 FPM.

Weight, Balance, and Loads

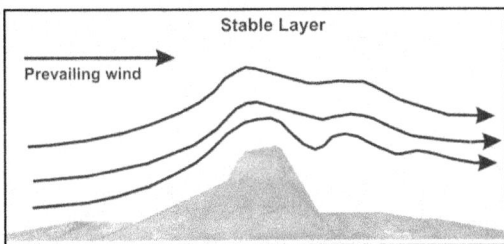

Figure 3-13. Mountain (standing) wave

3-136. The resulting turbulence of a mountain wave is determined by the following:
- Wind speed.
- Stability of the air mass.
- Degree of the slope.
- Height of the mountain.

3-137. Mountain waves are likely to occur when the following conditions are present:
- A low-level layer of unstable air.
- A stable layer of air above the lower levels.
- Wind direction fairly constant with altitude.
- Wind speed increasing with altitude–a larger increase in wind speed produces a stronger wave.
- A mountain or mountain range lying perpendicular to the airflow.

3-138. Near a mountain wave, the following conditions can exist:
- Vertical currents of 2,000 FPM are common, with more severe currents up to 5,000 FPM.
- Turbulence varies from moderate to severe.
- Wind gusts up to 22 knots per hour exist between waves. This condition is most severe near the mountain where the waves are closer together.
- Altimeter errors of as much as 1,000 feet may be experienced when penetrating a mountain wave.
- Icing can be expected in clouds when the temperature is below freezing.

3-139. When airflow meets the criteria for mountain waves, clouds form providing visible indications of the existence of a mountain wave (figure 3-14). The three types of clouds that may form due to a mountain wave are cap clouds, lenticular clouds, and rotor clouds.
- **Cap clouds** consist primarily of vertical updrafts, yet they develop updrafts and downdrafts passing over the mountain. The major part of the cloud extends upwind with finger-like extensions running down the slope on the ridge's downwind side.
- **Lenticular clouds** are lens-shaped clouds found at high altitudes, normally 25,000 to 40,000 feet. They may form in bands or as single clouds, located above and slightly downwind from the mountain's ridge. A mountain wave may exist without formation of lenticular clouds. Although airflow through the cloud is layered, aviators may encounter turbulence when flying under the cloud.
- **Rotor clouds** are located downwind from the ridge, sometimes in several rows lying parallel to the ridge. The bases may be at or below ridge level. The tops sometimes extend to the base of the lenticular cloud. They can produce updrafts and downdrafts of more than 5,000 FPM. Rotor clouds are of short duration and tend to disappear as rapidly as they build.

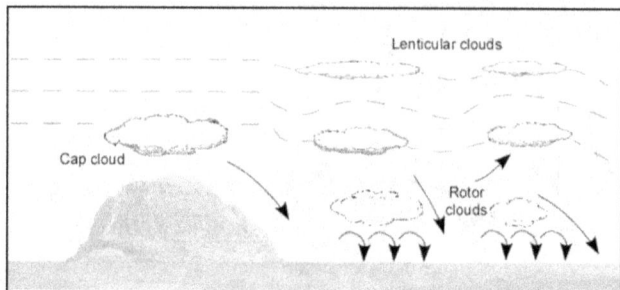

Figure 3-14. Cloud formations associated with mountain wave

Rotor Streaming

3-140. Rotor streaming (figure 3-15, page 3-29) is a comparatively rare occurrence. However, it does produce severe turbulence and has certain similarities to a standing wave. The conditions necessary for its formation are as follows:
- Unstable air in the lower level above ground.
- A stable layer (isothermal layer of inversion) at two to three times the mountain barrier's height.
- A strong surface and gradient wind decreasing noticeably with height in and above the stable layer.
- A hill mass providing upward deviation of airflow.

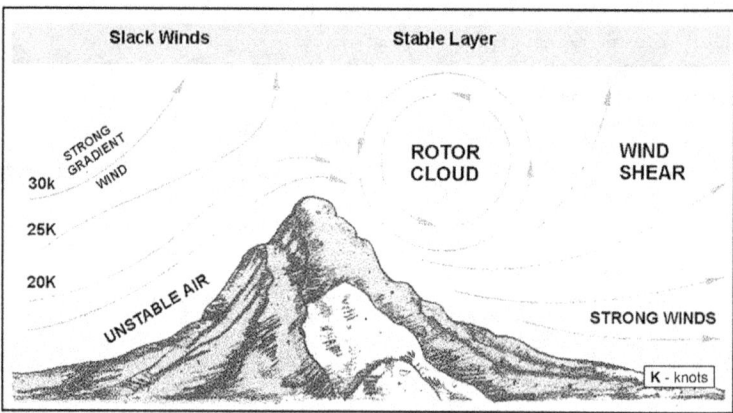

Figure 3-15. Rotor streaming turbulence

3-141. With these conditions, strong airflow begins to wave upward on the leeward side, meets the slack wind, and shears back on itself forming a rotary circulation. The stable air above acts as a lid holding down upward flow and assists circulation. This rotary circulation causes an increase in wind strength downwind with violent updrafts, downdrafts, and severe turbulence. This turbulence can cover a wide area to the leeward of the range and will lie in a roll downwind of the range. If enough moisture is available, a roll of clouds–known as rotor clouds–can form through the circulation's axis. The clouds will be rolling around their axis and are typified by broken, straggling tendrils around their outer edges. They also have pronounced vertical movement.

Terrain

3-142. When flying in mountainous regions, aviators will encounter many terrain variations. Each type affects the flow of air in its own way. Aviators must understand the different types and their effects to operate safely in a mountainous area. Mountain peaks are often warmer than the ground at the mountain's base due to the sun. This causes an uneven heating of air masses, triggering changing updrafts and downdrafts, and varying wind velocities.

Ridgeline

3-143. When the flow of air is perpendicular to the ridgeline and the ridge is characterized by gentle slopes, smooth air and updrafts will be experienced on the windward side of the ridge and downdrafts on the leeward side. Updrafts will be more severe when the updraft slope is steeper and the wind velocity is higher. If the air mass is unstable before the lifting action occurs or convection heating causes the air to become unstable, turbulence will be encountered. As the air flows over the crest, a Venturi effect is created; an area of low pressure develops on the leeward side of the mountain. When operating in this area, the altimeter will read high. Where the ridgeline is irregular, air funnels through the gaps causing a mixing of air on the leeward side. This condition tends to increase turbulence. Wind striking the ridge at less than 90 degrees produces fewer updrafts and downdrafts (figure 3-16, page 3-30).

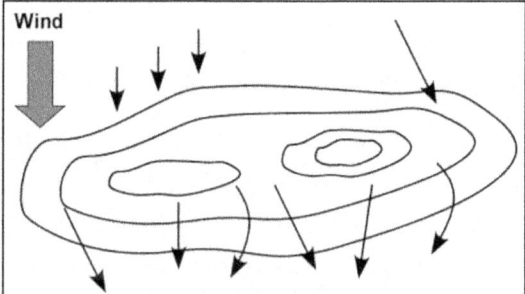

Figure 3-16. Wind across a ridge

Snake Ridges or Multiridges

3-144. The characteristics of wind flow over a ridge apply to snake ridges or multiridges (figure 3-17). However, downdrafts and turbulent air may be encountered on the windward slope of succeeding ridges. The severity of these conditions will be determined by the distance between the ridges, depth of the valley, and angle at which the wind strikes the slope. When the wind is perpendicular to the slope and the ridges are closer together, updrafts and turbulence are more severe. Greater turbulence will be experienced on the downdraft slope of succeeding ridges due to turbulent air flowing over the ridge.

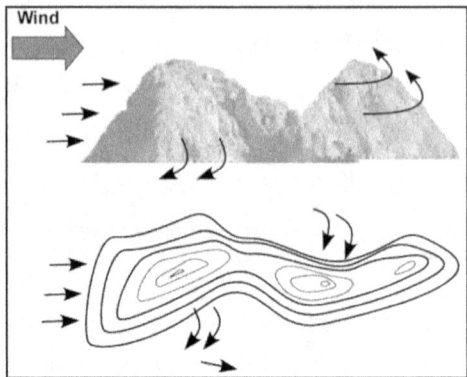

Figure 3-17. Snake ridge

Saddles

3-145. Saddles are formed by erosion of soft rocks. The turbulence severity in and around a saddle is determined by the saddle width and the slope's angle. Deep saddles, where terrain rises rapidly on each side, have the effect of a Venturi. The Venturi principle is when air is forced to flow through a constriction, velocity increases, and static pressure is reduced. This pressure reduction creates an altimeter error, causing the altimeter to read higher than the aircraft actually is. The wind flow over a shallow saddle with gentle slopes is much less severe.

Crown or Pinnacle

3-146. A crown or pinnacle is the highest point on a hill (figure 3-18). Because of its small size and separation from other terrain features, the effect on the wind normally is less severe. Usually, airflow near a crown is lateral around its outer edges and over the top. Turbulence will develop on the leeward side of the hill but does not extend too far out from the crown.

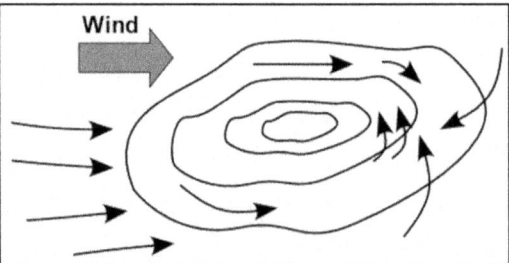

Figure 3-18. Wind across a crown

Shoulders

3-147. Shoulders are terrain features coming off higher ground (figure 3-19). The higher ground may be behind, in front of, or to the side of the shoulder. The airflow around a shoulder is extremely turbulent, regardless of wind direction. Extreme downdrafts may be experienced if the shoulder is located on the leeward side of the mountain. Rotary turbulence may be experienced on the uplift side of the shoulder.

Figure 3-19. Shoulder wind

Cliff (Bluff)

3-148. A cliff or bluff is a vertical or a near-vertical terrain feature. Extreme turbulence can be anticipated in front of, above, and below the cliff. This is caused by wind striking the face of the cliff and rebounding rearward. Eddies of airflow form above and below the top of the cliff. The air on the leeward side of the cliff is turbulent.

Canyons

3-149. Canyons are deep valleys with steep sides (figure 3-20, page 3-32) and enclosed on three sides. Usually, the lower winds flow parallel to the canyon floor. The degree of turbulence in the low areas of a canyon depends on the width and depth of the canyon and the wind speed. In a narrow canyon, the most severe turbulence is in the low area. However, the low area in a wide canyon may be relatively free of turbulence.

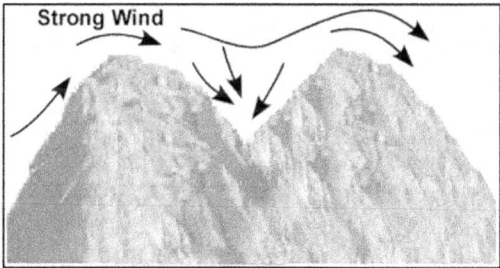

Figure 3-20. Wind across a canyon

NAVIGATION

3-150. Dead reckoning is the primary means of navigation. Maps of mountainous regions are usually easier to read because of greater relief, clearly defined features, and significant information from the contour lines. There is, however, a transition period necessary for aviators, especially if they are normally stationed in nonmountainous terrain. GPS and Doppler are extremely helpful and an invaluable part of the aircraft's avionics.

FLYING TECHNIQUES

3-151. Conducting flight operations over mountainous terrain is a difficult undertaking, even for an experienced aviator. The mountain environment is probably the least forgiving environment in which Army aviation operates. The following information, while extensive, is necessary and explains proven techniques.

Chapter 3

OPERATIONAL PROCEDURES

3-152. The mountain environment–combined with its effects on personnel and equipment–requires some modification of techniques and procedures. Important physical characteristics influencing mountain operations include peaks, steep ridges, deep ravines, valleys, limited communication capability, and continuously changing weather. When flying in the mountains, an aviator's senses are often unreliable. The natural tendency is to judge airspeed as too slow and altitude as too high. In addition, aviators tend to decelerate when flying upslope and accelerate when flying downslope. With constantly changing terrain and visual input, flying in the mountains demands an aviator's constant attention, which is divided between the outside environment and the flight instruments. The incorporation of an instrument scan into VFR flying assists aviators in maintaining appropriate airspeed, altitude (MSL and AGL), and rate of climb or descent. This demanding environment requires extensive training and practice. Mountain flight is a perishable skill. During high-altitude missions, one aviator should continually update the PPC to compensate for changes in the mission profile such as gross weight changes, CG change, temperature, and pressure-altitude change.

TAXIING AND TAKEOFF

3-153. Before takeoff, aviators conduct a hover power check. They obtain necessary information from the PPC, tabular data, or operator's manual including maximum torque available and go/no-go torque. These essential elements are calculated and verified by aircraft performance before departing and attempting the maneuver.

3-154. The primary difference between a mountain takeoff and a nonmountainous takeoff is the importance of gaining airspeed instead of altitude (figure 3-21, page 3-33). A nonmountainous takeoff emphasizes a combination of the two, accelerating as the aircraft climbs; mountainous takeoff emphasizes accelerating instead of climbing. When performing a mountain takeoff, an aviator applies torque, as necessary, to gain forward airspeed while maintaining sufficient altitude clearing any obstacles until climb airspeed is reached. Where a drop-off is located, the aircraft may be maneuvered downslope to gain airspeed. If an OGE takeoff is required to clear obstacles, angle of climb is minimized to conserve power. After clearing the obstacle, the flight attitude is adjusted to gain forward airspeed.

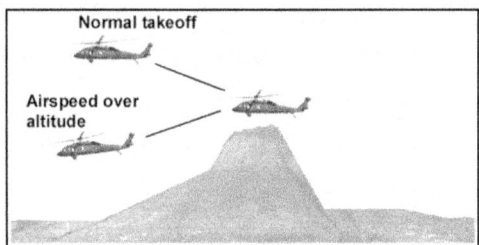

Figure 3-21. Mountain takeoff

EN ROUTE

Takeoff

3-155. Before takeoff, aviators identify the route of flight on a map. Although flight route and altitudes may change, the altitude for nontactical flight is generally considered to be at or above 500 feet AGL.

3-156. The following en route considerations apply to flight in mountainous terrain:
- When flying in a valley, aviators fly the aircraft in smoother up-flowing air on the lifting side of the valley. This technique requires less power and provides the aircraft a safer flight path. Wind velocity dictates how far away from the valley walls to fly. Aviators avoid flying too close to the valley walls during strong winds to avoid turbulence caused by irregular terrain. Under light

Weight, Balance, and Loads

winds, the aircraft is flown closer to the valley walls to facilitate a 180-degree turn if the valley is narrowing or there is rapidly rising terrain or a low cloud base.

- Terrain clearance is increased when strong winds exist. It may be necessary, however, to descend if flight is conducted below the rim of a large valley. Turbulence develops in the upper levels of the valley but diminishes closer to the valley floor.
- If a downdraft is encountered, full power is applied and best rate-of-climb airspeed (V_y) maintained. If unable to stop descent, a turn away from terrain is made. A maneuver downhill toward an area in the valley floor is attempted while maintaining airspeed. Near the valley bottom, downdraft begins to decrease in severity. Aviators continue to maintain full power and turn into the wind. If it appears the helicopter will be forced into the ground, a flat landing area is selected and an approach to the area planned. If the landing area is not level, land upslope.
- Techniques for crossing ridges vary depending on wind strength and direction of crossing, leeward to windward or vice versa. The basic rule is to cross the ridge diagonally. This procedure facilitates turning away from the ridge should the helicopter be carried below the crest by a downdraft. In strong winds, ample clearance is allowed above the top of the ridge when crossing from leeward to windward side of the ridge. Crossing from windward to leeward can lead to uncontrolled descents on the leeward side; this may require large applications of power to remain above terrain at a safe altitude. The clearance itself will be assisted by updrafts, but if a low cloud ceiling exists the aircraft may be carried up into the cloud even with minimum power applied. If the crossing is made at terrain flight altitudes, turbulence may be encountered on the ridge's leeward side.
- Maneuvering in narrow valley bottoms is easier at low airspeeds, thus reducing the requirement for large power demands. Airspeeds should not be reduced below ETL. Turns will be flat and, if possible, in the direction of torque reaction (right turn for U.S.-made helicopters).
- Flight on the sunny side of the valley has more turbulence; however, turbulence caused by solar heating is not as severe. An updraft is formed allowing the aircraft to fly with less power. On the shady side of the valley, downdrafts may be present and cause a need for additional power.
- As gross weight and altitude of the aircraft increase, the maximum allowable airspeed decreases. Failing to reduce airspeed according to the PPC may result in blade stall or simply running out of power.
- When turbulence is anticipated, airspeed is reduced to the recommended turbulence penetration airspeed for the type of aircraft being flown.
- Due to danger of encountering downdrafts, descents to follow terrain will be executed at less than 1,000 FPM. When following terrain in a descent, aviators often need to reduce airspeed keeping the aircraft within operating limits.
- When approaching a ridge, aviators will have trouble determining if the aircraft's altitude is sufficient to clear the ridge. When terrain beyond the ridge becomes progressively visible, the aircraft will clear the ridge. If this is not the case (if the aviator is not high enough to see beyond the ridge) then a climb must be established with a possible 360-degree turn made so the aircraft clears the ridge.
- When tactically possible, the flight is planned and flown using well-known routes facilitating a quicker rescue.
- When conducting multihelicopter operations during marginal weather operations, it is advisable to have an aircraft precede the flight determining actual weather conditions and ensuring satisfactory weather to accommodate the flight of aircraft. Mountainous terrain restricts avoidance maneuvers when aircraft are flying in multihelicopter formations.
- When conducting multihelicopter operations into a small LZ, aviators will use a greater than normal separation to provide sufficient reaction time and avoid forcing successive aircraft into the disturbed air of the preceding aircraft.

Determination of En Route Winds

3-157. While weather forecasters provide some information, it is impossible for them to know the exact wind at each point. Visual cues assist the aircrew in determining the wind. These cues are divided into two categories—ground indicators and aircraft indicators.

Chapter 3

Ground Indicators

3-158. Ground indicators work well for each isolated location and are subject to change a short distance away. Some examples are—
- **Bodies of water.** Upwind part of a small body of water indicated by a smooth surface. The downwind side indicates turbulence indicating an idea of the velocity. Whitecaps occur on an unprotected body of water at 20 miles per hour (MPH).
- **Smoke.** Indicator of wind and velocity. Rising smoke indicates a light wind while smoke moving laterally indicates a stronger wind.
- **Leaves.** Color of the leaves on a deciduous tree indicates wind direction. When leaves appear lighter in color, the aviator is flying downwind. Leaves appear darker when the aviator is flying upwind.
- **Tall grass.** Indicates wind direction and velocity. Direction is indicated by the movement of the grass and the frequency of that movement indicates wind velocity.
- **Manmade indicators.** Provides wind information accurately. Some manmade indicators used by aviators are wind socks or smoke grenades.

Aircraft Indicators

3-159. An experienced aviator can determine the wind by aircraft reaction and its apparent movement over the ground. When the aircraft drifts from the desired ground track, it indicates a crosswind. A difference between apparent ground speed and indicated airspeed suggests a headwind or tailwind. An increase/decrease of power from a previous setting for airspeed indicates downdrafts/updrafts.

Before Beginning an Approach

3-160. Before beginning an approach, an LZ reconnaissance is conducted to evaluate conditions around and on the proposed landing area. Aviators must assess subsequent takeoff conditions before landing in the area. This reconnaissance consists of two phases—high reconnaissance and low reconnaissance.

High Reconnaissance

3-161. During high reconnaissance, consideration must be given to determining the approach path. The three recommended flight patterns flown to conduct the high reconnaissance are figure eight, circular, and racetrack (figure 3-22). Regardless of the type flown, the following techniques are used:
- Flight altitude should be high enough to ensure safe operations in case a downdraft is encountered. Wind speed and the nature of the terrain are considered when selecting an altitude.
- Terrain should be observed while maintaining aircraft airspeed limitations.
- Flight pattern should be maintained relatively close to the landing area; however, aircraft maneuvers should be limited to bank angles of 30 degrees or less.

3-162. The high reconnaissance should assess the following:
- **Landing zone.** Determine slope, shadowed area, obstacles in and around the LZ, and any surface debris that could damage the aircraft.
- **Wind.** Determine direction, speed, and location of the demarcation line and any other variables of wind flow.
- **Takeoff route.** Locate takeoff direction (into the wind) and lowest obstacles, and identify escape routes and forced landing areas, if any.

Weight, Balance, and Loads

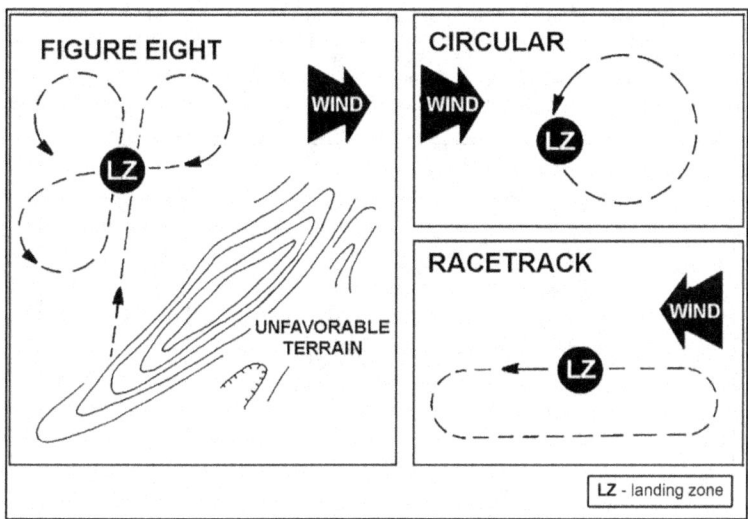

Figure 3-22. High reconnaissance flight patterns

Low Reconnaissance

3-163. Low reconnaissance may be performed to verify information gathered by high reconnaissance. If the information from high reconnaissance was sufficient, then low reconnaissance can be combined with the approach. Availability of power for approach and landing is determined from the PPC. When the aircraft is directly over the touchdown point, altitudes (MSL and AGL) are noted for use during the landing pattern. This knowledge aids aviators in establishing an appropriate traffic pattern altitude, especially when initiating the final leg. In mountainous terrain, the tendency–without this actual LZ altitude information–is to react to the surrounding terrain and fly the pattern too high or too low. If at any time during low reconnaissance it is determined conditions around the LZ are unsafe, reconnaissance is discontinued and the aircraft proceeds to an alternate LZ. The following specific conditions must be evaluated during low reconnaissance:

- Pinpoint wind direction and effects of the wind on surrounding terrain; wind indicators observed away from the LZ should be ignored as wind may be different over the touchdown point.
- Evaluate the touchdown point, size of the landing area, slope, type of surface, and any obstructions.
- Determine whether the approach should be terminated to the ground or to a hover.
- Evaluate the approach path.
- Identify escape routes.
- Evaluate the takeoff path.
- Determine air temperature and PA.

3-164. There are two recommended methods for conducting low reconnaissance and evaluating the wind condition. When performing either method, aviators fly the aircraft at an altitude slightly above the landing point. A portion of the flight path (for low reconnaissance) must be flown over the intended touchdown point to adequately assess the wind.

- **Time between two points** (figure 3-23). Aviator selects two recognizable terrain features (reference points) near the LZ. Separation between these points is approximately 300 meters and oriented in the same general direction as the wind. Constant airspeed is established and flown throughout reconnaissance. As the aircraft passes over the first point, position is noted and the

Chapter 3

clock started. When the aviator passes over the second point, time is noted in the seconds required to travel the distance between the two points. The course is reversed and the procedure repeated. Direction of flight requiring the shortest time indicates the approximate wind direction. If a crab is required to maintain ground track, the direction of crab indicates a more specific heading. Wind velocity is directly proportional to a larger time difference between the two points or a larger crab angle.

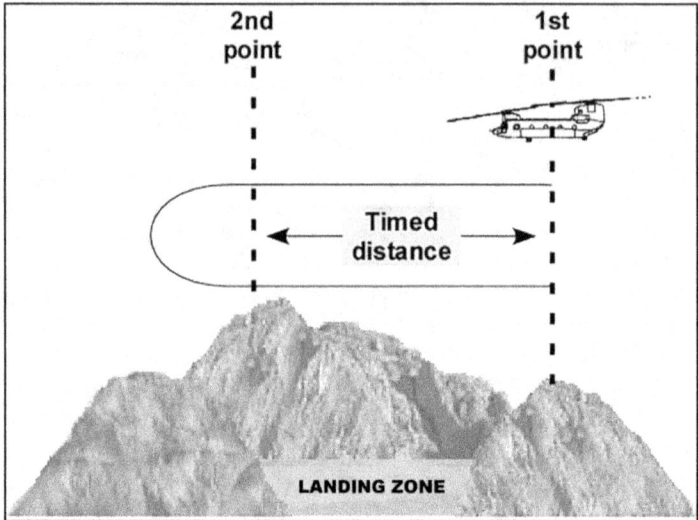

Figure 3-23. Computing wind direction between two points

- **Circle** (figure 3-24). The aviator selects a pivot point on the ground (LZ) around which a circle will be flown and identifies a starting point about 200 meters from that pivot point. As the aircraft passes over the starting point, the aviator notes the heading and stabilizes the airspeed. The aviator then starts a turn at the pivot point maintaining constant angle of bank and airspeed. As the aircraft passes through the original heading, the aviator should look at the pivot point, which indicates wind direction. Distance from the pivot point will indicate wind velocity.

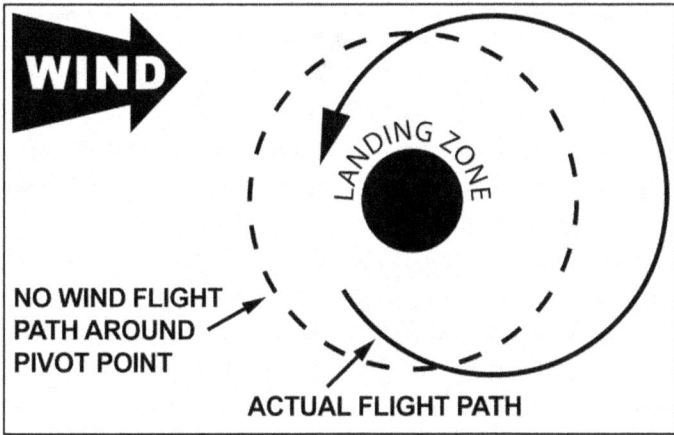

Figure 3-24. Computing wind direction using the circle maneuver

Approach Paths

3-165. Approach paths are common to both high and low reconnaissance. Figure 3-25, page 3-38, illustrates approach paths and areas to be avoided. The five basic factors to be considered for determining the approach path are—

- **Wind direction and velocity.** While it is desirable to land into the wind, the terrain and the effects of the wind may dictate a crosswind landing be made. Because of torque reaction and aircraft differences, the pilot should decide if a left/right crosswind landing should be made.
- **Vertical air currents.** The severity of the updrafts or downdrafts encountered may be more critical than landing into the wind and may require a downwind approach.
- **Escape routes.** Assess the escape routes by identifying where altitude can be exchanged for airspeed in case the aircraft experiences insufficient power or turbulence prevents a safe landing.
- **Terrain contour and obstacles.** Terrain and obstacles along the approach path should be low enough to permit a shallow approach angle into the LZ. When possible, select a landing point on or near the highest terrain feature.
- **Position of the sun.** Wind direction and nature of the terrain are the primary factors in selecting an approach path. The aviator must also consider, however, the position of the sun in relation to the approach path and the presence of shadows on the LZ. If the landing point is in the shadows, then the approach path should also be in the shadows to eliminate changing light conditions during the approach. An approach directly into the setting sun should be avoided because the distraction will not allow the aviator to see all of the LZ details.

Chapter 3

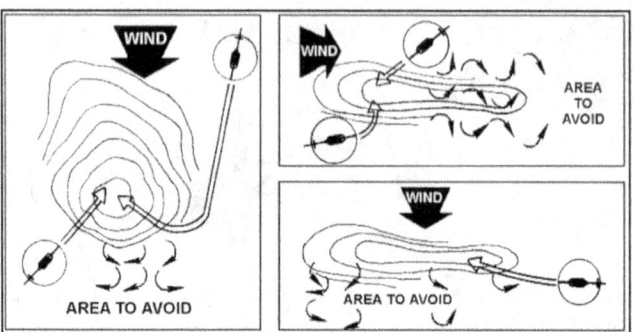

Figure 3-25. Approach paths and areas to avoid

APPROACH AND LANDING

3-166. There is no standard type of mountain approach. Ideally, it is made in a direction taking advantage of the wind to provide maximum tail-rotor control. The following are guidelines for successful approach and landing.

- In a light wind or when the demarcation line is shallow, an aviator uses a relatively low angle of descent or a flat approach. This type of approach requires less power; however, if downdrafts are encountered, the aircraft may lack altitude to continue the approach.
- As wind velocity increases and the demarcation line becomes steeper, the approach angle must also be steeper. This type of approach requires less power due to updraft and provides more terrain clearance if downdrafts are encountered.
- If torque is insufficient to make a normal or shallow approach and the landing area is suitable, a running landing may be performed. Before making this type of approach, alternate methods should be pursued such as reducing gross weight by flying longer to consume fuel or returning to the landing area after dropping off cargo or passengers. If insufficient power is available and an approach is executed, there may be insufficient power available to execute a takeoff. A running landing requires a smooth and long enough touchdown area. It is performed essentially the same as in a nonmountainous area except ETL is maintained until the aircraft contacts the ground.
- During a mountain approach, be aware uneven terrain surrounding the LZ can provide poor visual cues about the actual aircraft altitude (AGL) and rate of closure. When terrain slopes up to the LZ, the visual illusion leads an aviator to believe the aircraft is too high and the rate of closure too slow. When terrain slopes down to the LZ, the visual illusion leads an aviator to believe the aircraft is too low and the rate of closure too fast. When the LZ is a pinnacle and the surrounding terrain drops off sharply, it seems the aircraft is too high and the rate of closure initially too slow. As the aircraft closes the distance to the LZ, the rate of closure will appear excessive. These conflicting cues must be evaluated by cross-referencing visual cues with information on the flight instruments. The aviator must make appropriate adjustments.
- After low reconnaissance is completed, the aircraft is flown into a traffic pattern and the approach initiated. Where terrain conditions permit, the traffic pattern should be standard. When deviating from the standard pattern, an adequate distance on final approach is maintained to avoid descents greater than 300 FPM. Pattern altitude, depending on the terrain, will not normally be flown more than 500 feet above the LZ altitude. The pattern is flown over terrain where minimum downdrafts exist.
- The primary difference between the short final phase of a mountain approach and a flatland approach begins when the aircraft is approximately 50 feet above touchdown. To begin losing ETL, the aircraft slows but not to an OGE hover. Before reaching the near edge of the LZ, descent

is halted and airspeed reduced to a brisk walk. Torque is noted and a decision made to continue the approach or to abort.
- If it is determined the approach is unsafe in any way, the aviator initiates a go-around. Where escape routes exist, the aircraft is turned away from the mountain and flown appropriately. This normally consists of progressive acceleration, carefully-managed power (airspeed over altitude takeoff, if possible), and minimal bank angles established. The point is to execute appropriate control input predetermined on high and low reconnaissance, being careful to minimize control input, with minimum power application and precise handling of the aircraft. As always, an emergency may demand more aggressive action. Each situation is different and must be evaluated separately. The aviator should take nothing for granted and be prepared for the worst possible scenario, such as engine failure, loss of tail rotor effectiveness, and downdrafts. Potential problems must be discussed ahead of time and proposed actions considered and reviewed. In other words, the aviator should play the "what-if" scenario throughout the sequence, continuously considering and evaluating possible situations and reactions. During termination of the approach, if heading control is lost and the aircraft begins an uncommanded turn to the right (single-rotor helicopter), the collective must be lowered and the aircraft landed. This possible reaction should be anticipated, especially at high altitude/high gross weights. The most critical situation is when rotor RPM decreases and a go-around is not executed. In this situation, heading and power must be maintained. Emergency procedures for a hovering autorotation are followed if the aircraft is over open terrain. If the aircraft is over vegetation and descent angle will not allow the aircraft to clear the obstacles, the aircraft is decelerated to achieve minimum forward airspeed. Just before contact, full collective will be applied to minimize vertical descent.
- Mountain LZs are generally rough, small, and often sloped. To avoid possible aircraft damage, touchdown must be executed with zero forward airspeed, if possible. A slope landing may be required. After touching down in the LZ, the collective should be lowered gradually until it has been determined the aircraft is securely positioned on the ground. Slight control input assists in determining if the aircraft is secure. If the aircraft is rocking or wobbling in that position, it should be repositioned.

TERRAIN FLIGHT

3-167. Regardless of terrain, survival in a high-threat environment depends on flying missions at terrain flight altitudes. Terrain flight in a mountainous environment, with its inherently dangerous surroundings, should use the highest altitude flight mode possible—NOE, contour, or low level. Terrain flying in mountains imposes additional stresses on an aircrew with specific considerations and techniques required. Specific mountain regions may also present additional concerns. These are addressed in local SOPs. The following list presents those considerations common to terrain flight in any mountainous region including takeoff, en route, and approach flight techniques.
- Terrain flight in mountains is extremely fatiguing to aircrews. Demands of mountain flight and the level of attention it imposes cause most of the fatigue which is worsened by such conditions as high altitude and severe weather.
- Communications in a mountainous region are limited or restricted to varying degrees and must be planned for. This plan includes provisions for factors such as flight following, tactical communications, and search and rescue.
- Terrain flight in mountains restricts use of close formation flight. When conducting multihelicopter operations, free cruise, loose trail, or staggered trail formations are used for flexibility. Aircraft spacing may need to be increased accommodating conditions such as tight LZs, staggered landings, very steep terrain, and narrow valleys and crossings. It is common for a mountain LZ to accommodate only one or two aircraft at a time. Alternate flight routes and LZs should be planned. Each aircraft must prepare to assume flight lead during the mission.
- Downdrafts or turbulence may be a hazard and major consideration in the mountains. Aviators must be able to assess wind and evaluate its effect on flight at terrain flight altitudes. This evaluation is achieved using a combination of visual cues, flight instrument information, and experience in the environment. Emergency procedures must be well established and performed in

a minimum amount of time. Safe and successful mountain terrain flight is based on a thorough working knowledge of terrain flight and mountain flight procedures.

Terrain Flight Takeoff

3-168. Because terrain is steeper around mountain LZs, gaining airspeed is usually more important than gaining altitude. Airspeed over altitude takeoff is preferred whenever possible (figure 3-26).

- Aviators must consider wind condition and terrain to determine takeoff direction. Takeoff should be conducted into the wind and over the lowest obstacles and descending terrain. Mountainous terrain usually demands a compromise between these factors. Takeoff checks must include a hover power check and a thorough understanding of PPC information.
- A mountain takeoff is usually initiated from the ground. After lifting off, the aviator applies torque and change in pitch attitude to accelerate or climb as planned. Terrain flight often dictates descent to an appropriate altitude after clearing the LZ, using necessary control inputs. When terrain obstacles prevent airspeed over altitude takeoff, low level OGE takeoff should be performed with appropriate airspeed and altitude selected for continuing flight.

Figure 3-26. Nap-of-the-earth or contour takeoff (terrain flight)

Terrain Flight—En Route Flight Techniques

3-169. Any successful terrain flight includes detailed reconnaissance and a planned route with alternates. It should be an accurate reflection of the actual flight. During mission execution, the crew may discover the planned route does not offer the best concealment or terrain, or weather conditions prove the selected route is inadequate. Therefore, an en route change to the plan is common. The mode of terrain flight is dictated by the threat and terrain masking. The selected mode can be a hindrance in dealing with inherent hazards of mountain flight. It may dictate other compromises such as reduced airspeed or spacing of multihelicopter flights.

Nap-Of-The-Earth and Contour Flight

3-170. For this discussion, NOE and contour flight will be reviewed together. These flight modes are used when enemy detection is likely. Where noted, the following flight considerations apply:
- Aviators conduct flight at the bottom of the valley providing cover and concealment. However, this area may have a great deal of turbulence and more stable air might be found along the base of the lifting side of the valley. If adequate cover is available, plan the route in this area.
- If a downdraft is encountered, an aviator will apply maximum power to arrest any descent or maintain altitude. A deceleration may assist in making more power available to prevent ground contact while providing enough altitude to clear the tail rotor. If descent cannot be stopped and it appears ground contact will be made, the aviator continues to decelerate the aircraft to minimize forward airspeed at touchdown. Just prior to impact the aviator should establish a level attitude.
- If possible, aviators avoid flying close to abrupt changes in terrain where severe downdrafts may exist. If this is unavoidable, airspeed is reduced minimizing the effects of any downdraft.
- Aviators should avoid flying across the mouth of an adjacent valley where turbulence is often encountered. When crossing a ridge (figure 3-27, page 3-41), aviators fly a route over the lowest

point minimizing exposure. Aviators will adjust their heading to cross the ridge at a 45-degree angle to ridge direction. This technique increases the opportunity to turn away from the ridgeline in case of an emergency.

Figure 3-27. Ridge crossing at a 45-degree angle (terrain flight)

- Aviators fly parallel to the lifting side of and as close to terrain features as possible when climbing to a higher altitude for a ridge crossing or to approach an LZ. This allows the aircraft to receive the benefit of any updraft while reducing the chance for enemy detection.
- Aviators must avoid situations requiring rapid ascents from the valley. Climb should be initiated with adequate time to clear terrain and obstacles.
- Aviators must avoid making turns of more than a 30-degree angle of bank at low altitudes. A reduced airspeed may be necessary to allow a turn to be executed within the space available (figure 3-28). Extreme caution must be exercised if a turn is performed downwind and airspeed is reduced below ETL. If the valley is very narrow, it may be necessary to stop the aircraft and perform a pedal turn. If insufficient power prohibits this maneuver, a combination of cyclic and pedal turn should be performed while maintaining minimum forward airspeed.

Figure 3-28. Steep turns or climbs at terrain flight altitudes

- When approaching a terrain feature where terrain drops off rapidly on the reverse slope, aviators reduce airspeed before crossing. This deceleration allows the aircraft to follow the terrain more closely without silhouetting the aircraft against the skyline.
- Aviators must watch for wires in narrow canyons. Wires are difficult to see and may be stretched across the valley with no support in the middle.
- Aviators reduce airspeed during periods of low visibility to increase required reaction time.

- During day multihelicopter operations, aircrews normally fly a free cruise or staggered formation. Narrow corridors limit maneuver airspace and require all aircraft to follow essentially the same ground track. One option is staggering aircraft at varying distances to reduce enemy detection capabilities.

Low-Level Flight

3-171. Low-level flight is conducted where terrain features do not dictate lower altitudes in NOE/contour flight because sufficient masking exists at a higher AGL. Many of the same en route flight techniques that apply to NOE/contour flight also apply to low-level flight. Aviators should review these techniques in addition to the following considerations.
- When flying in a valley, an aviator's flight path should be as close to the lifting side of the valley as possible (figure 3-29). This technique allows more room for turning and exposes the aircraft to less turbulence while taking advantage of any updrafts.
- Aviators avoid making turns over terrain requiring an increase in altitude.
- A staggered-trail formation is normally flown for multihelicopter operations. Less separation is required than for NOE/contour flight, while aircrews have greater freedom of maneuver and can avoid the same ground track as the preceding aircraft.

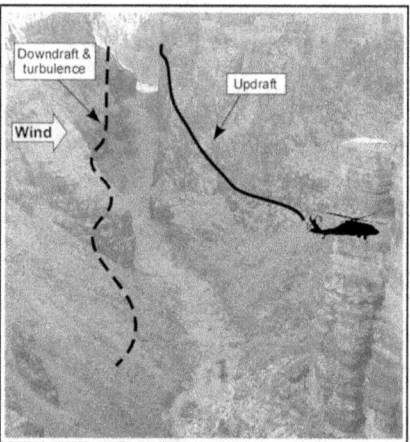

Figure 3-29. Flight along a valley (terrain flight)

Terrain Flight–Approach Techniques

3-172. During basic mountain-flight training, aviators are taught to use a shallow- or normal-approach angle, requiring the approach to be initiated from an altitude higher than the LZ. In terrain flight, this must be altered to avoid enemy detection. A terrain flight approach may be initiated from a point below the LZ. Additionally, traditional high and low reconnaissance cannot be performed due to threat of enemy detection. A straight-in approach from the inbound course is the preferred method but presents limitations. If turbulence/strong downdrafts are encountered, it may be necessary to conduct a low pass over the LZ confirming wind direction and velocity, turbulence, and any additional information. It also may be necessary to maneuver some distance from the LZ and approach from another direction.

Nap-Of-The-Earth/Contour Approach

3-173. This approach is the most difficult and subjects the aircraft to mountain hazards for the longest amount of time. Aviators must recognize if there is sufficient power, as any avenue of escape or opportunity for a go-around will be severely limited. Figure 3-30 depicts an approach beginning from the valley floor and terminating at an LZ on a ridgeline. As the approach is initiated, any downdrafts or turbulence will determine if a straight-in approach is feasible. If a climb is required, it should be started soon enough that a maximum 1,000-FPM climb and 40 knots indicated airspeed (KIAS) minimum allow the aircraft to ascend on the approach path without exceeding either of those parameters. If this is not possible, a go-around should be executed. As the aircraft approaches the LZ, the aviator selects a point along the approach path about 100 meters short of the LZ. This is the initial decision point for continuing the approach. The aircraft should be located about 50 feet above the highest terrain feature and at the desired airspeed. The decision to continue should include the wind direction and velocity, availability of sufficient power, and presence of any downdrafts forcing the aircraft into trees or the ground. A decision to go-around should be made and the approach discontinued as soon as possible.

Figure 3-30. Nap-of-the-earth or contour approach (terrain flight)

Low-Level Approach

3-174. This approach combines techniques used for NOE/contour and nontactical approaches. A low-level approach will normally be initiated from an altitude below the LZ, with the final leg of the approach path begun at an altitude 100 to 200 AGL above the touchdown point. In other words, the aircraft will initiate a climb from the en route altitude to an altitude appropriate for beginning the final leg of the approach path. At a predetermined point, this climb is started and should not exceed a 1,000-FPM rate of climb or be slower than 40 KIAS. As the aircraft approaches the LZ, the same considerations used for NOE/contour approaches should be used. The decision to continue or execute a go-around should be made as soon as possible.

MAINTENANCE

3-175. With power demands inherent to mountain operations, the focus will be engines and maintaining maximum performance. In addition, the aircraft is subject to the demands of any environment the mountainous region is located (for example, a cold weather region).

TRAINING

3-176. Units qualifying aviators in mountain operations are responsible for conducting a well-organized training program. This training instills confidence and maintains the aviator's interest. The IP should be experienced in mountain flight and preferably, a High Altitude Army Aviation Training Site (HAATS) graduate. The IP must be capable of initiating corrective action for any emergency that may occur.

Chapter 3

3-177. Mountain flight is very hazardous; therefore, greater emphasis should be placed on preflight planning. Wind velocity and the level of turbulence restricting flight training must be identified. The judgment of the instructor to discontinue training due to unsafe conditions is accepted and not criticized.

3-178. The flight training program allows each aviator to advance at an individual rate. Initial training should be conducted over less challenging terrain during non-turbulent conditions. As proficiency increases, conditions should become more demanding until the most challenging mission can be performed.

3-179. The program starts with training occurring routinely while at home station as part of the normal training cycle. This training includes academic and flight training, and defines the train-up of personnel upon notification of deployment. Outside experts may be brought in to conduct training for the unit. However, the best option is sending personnel to HAATS.

Academics

3-180. Suggested topics include—
- Human factors associated with mountain flight.
- Environmental factors affecting mountain operations.
- Mountain weather patterns.
- Aircraft operational procedures in mountainous areas.
- Principal difficulties during mountain operations.
- Mountain survival.
- Performance planning.

Flight

3-181. Flight training may be limited by conditions at the unit's home station. Some areas may not be able to replicate conditions adequately for training. Instructors can demonstrate techniques and procedures to some extent. Crews are evaluated on these procedures during their APART or no-notice evaluations. Flight simulators are a great device in training for this environment.

3-182. Suggested maneuvers include—
- Power management.
- En route flight techniques.
- Mountain approaches and landings.
- Go-around.
- Power checks.
- Mountain takeoffs.
- IIMC.

SECTION V – OVERWATER OPERATIONS

3-183. In nearly every major conflict and operation since World War II, Army aviation has been assigned missions in maritime environments, either basing off naval vessels for land attack or operating from ships for sustained overwater missions. Recently, the nature and complexity of those missions have changed dramatically, dictating aviation units complete specialized preparatory and sustainment training. Recent worldwide deployments have shown Army aviation has a versatile combination of equipment sophistication, deployability, and personnel to accomplish specific strategic missions requiring operations in the maritime environment.

ENVIRONMENTAL FACTORS

3-184. Army aviation units presently participate in many joint operations requiring proficiency in shipboard and overwater operations. Individual training, aircraft modifications, development of SOPs, and application of established policies are complex and necessary to ensure Army aviation can perform safely and effectively in an overwater environment. FM 1-564 is the primary reference for overwater operations and for working with the United States Navy (USN). It is imperative to reference this source and contact units routinely involved in overwater operations before an aviator performs missions in this environment.

CLIMATE AND WEATHER

3-185. Army aviators are accustomed to working in areas where a visible horizon exists with normal ceiling and visibility during all flight operations. In an overwater environment, the horizon over the water is the reference line for VFR attitude. The water and skyline often blend because of fog, rain, or other obscurations, eliminating any visible horizon. Overwater winds are not affected as much by the surface of the earth and generally remain steady and constant from a given direction. This can pose a challenge when conducting shipboard landings and takeoffs based on limitations and capabilities of the aircraft. At times, aircraft must depart and land with left or right crosswinds. Performance planning should be completed in-depth ensuring adequate power margins. The surface of the water can also create the illusion that an aircraft is higher than its actual altitude. In extreme calm seas with clear water, aviators can see through the water and often believe they are at a higher altitude than they actually are. During night operations at altitude, generally above 200 feet, waves blend and the actual surface becomes difficult to detect.

TERRAIN

3-186. Overwater flight is marked by a near-featureless surface.

FLYING TECHNIQUES

3-187. Conducting flight in overwater operations usually includes lack of visible horizon due to overcast skies, restricted visibility, difficulty in detecting altitude above water, water spray coating the windscreen, and the potential for spatial disorientation.

3-188. Individual training for aircrew members includes, but is not limited to—
- Swim testing and proficiency in drown-proofing.
- Dunker training.
- Use of helicopter emergency egress device or other approved emergency breathing system.
- Use of specific floatation devices.
- Egress procedures specific to the aircraft.
- Downed aircrew member extraction procedures.
- Academic training including FM 1-564 and appropriate SOPs including theater operations or unit SOP.

MAINTENANCE

3-189. The major concern in this environment is corrosion. TMs dictate special maintenance required for salt-water operations.

TRAINING

3-190. Administering an overwater training program to qualify aviators is a unit responsibility. Training in overwater operations instills confidence, develop skills, and emphasize safety. If operations are conducted aboard USN vessels, then FM 1-564 describes training requirements necessary for shipboard operations.

CREW/TEAM/SCENARIO TRAINING

3-191. Crew coordination training for overwater operations to ships needs to address several new areas and be incorporated into academic training. The USN uses different terminology and procedures for aviation operations. Cockpit communications require a thorough understanding of terminology to comply with instructions from USN vessels. Teams operating overwater for landing or takeoff need to be familiar with holding patterns and shipboard departure/arrival procedures. Units routinely flying overwater must

Chapter 3

incorporate established procedures into unit SOPs. An example overwater SOP is located in FM 1-564. It can be tailored to add specific unit capabilities and requirements. Overwater training scenarios can be included in simulator training programs and might be used as part of the initial qualification and sustainment program.

RECOMMENDED PROGRAM OF INSTRUCTION

3-192. The program starts with training done routinely while at home station as part of the normal training cycle. This training includes academic and flight training and defines the train-up of personnel upon notification of deployment. Outside experts may conduct training for the unit.

Academics

3-193. Suggested topics include—
- Human factors associated with overwater flying.
- Environmental factors that affect overwater operations.
- Principal difficulties during overwater operations.
- Overwater aviation life support equipment (ALSE) requirements.
- Water survival.

Flight

3-194. Flight training may be limited by conditions at the unit's home station. Some areas may not be able to replicate conditions adequately for training. Instructors can demonstrate techniques and procedures to some extent. Crews are evaluated on these procedures during their APART or no-notice evaluations. Flight simulators are also a great device in training for this environment.

Chapter 4

Rotary-Wing Night Flight

Vision is the most important sense used in flight. Day or night, instrument meteorological conditions (IMC), or VMC, vision is the primary sense providing crewmembers with awareness of aircraft position. Eyes can rapidly identify and interpret visual cues in daylight. During darkness, visual acuity decreases proportionally as the level of illumination decreases. NVDs improve the capability of the human eye to see at night and increases the survivability of the aircrew. This chapter covers NVD and provides a general discussion of night vision and techniques for completing missions safely.

SECTION I – NIGHT VISION

NIGHT VISION CAPABILITY

4-1. In today's aviation environment, night missions are seldom conducted completely unaided. There are times however, that aviators may determine unaided flight is preferable to aided flight. The basic principle is to choose the sensor that is most appropriate for the mission and flight environment. Generally, aircrews should only go unaided when the light level is very high (usually 75 percent or above), where the advantages of a full field of view and color perception outweigh the limitations of reduced or uneven light levels. This occurs most often when there is a significant amount of cultural lighting, such as in a city or at an airfield. This is a decision the PC of the aircraft must make in accordance with the unit SOP. One of the biggest problems with unaided flight is that dark areas can be difficult or impossible to see into. The level of available detail is inconsistent due to the highly variable light level. Obstacles can be very difficult to see. Additionally, the flight environment (visibility or weather) may change while the aircrew is effectively "heads down" in the lighted area.

TYPES OF VISION

4-2. The types of vision (photopic, mesopic, scotopic) are related to the light level. They relate to the level of detail available to the aviator under different conditions. These concepts are more fully explained from a medical perspective in TC 3-04.93.

COMBAT VISUAL IMPAIRMENTS

4-3. An aviator's eyes can be damaged during Army aviation missions. The following instances should be considered and preparation made to prevent such occurrences.

NIGHT LASER HAZARD

4-4. The eye is more vulnerable to laser damage at night as the iris of the eye opens more to accommodate lower levels of illumination. Laser damage to the eyes includes flash blindness, minor and major retinal burns, and impaired night vision. The effect of flash blinding is similar to the temporary effect of a flashbulb and can last seconds to minutes, possibly leaving colored spots in the field of vision that are distracting and potentially dangerous. Minor retinal burns can cause discomfort and interfere with vision. The injuries may involve internal bleeding in the eye, immediate pain, and possible impaired or permanent loss of vision. Night vision acuity may be lost due to undetected damage. Fovea damage may affect vision sharpness and color interpretation. Normal cockpit tasks, obstacle avoidance, and use of acquisition or targeting devices may become difficult or impossible. Aviation unit training must emphasize aircrew use of the aviator's helmet laser visor when performing missions in an anticipated or known laser environment. To reduce chances of laser injury, aviation support personnel must be trained to wear laser protective spectacles when performing aviation ground support functions.

NERVE AGENTS

4-5. Exposure of the eyes to minute amounts of nerve agents adversely affects night vision. When direct contact occurs, pupils constrict (miosis) and do not dilate in low ambient light. The available automatic chemical alarms are not sensitive enough to detect low concentrations of nerve agent vapor. The exposure time required to cause miosis depends on the concentration of the agent. Miosis may occur gradually as eyes are exposed to low concentrations over a long time. Conversely, exposure to a high concentration may cause miosis during the few seconds it takes to mask. Repeated exposure over a period of days is cumulative. Symptoms range from minimal to severe, depending on amount of exposure. Severe miosis, with reduced ability to see in low ambient light, persists for approximately 48 hours. The pupil gradually returns to normal over the next several days with full recovery taking up to 20 days. The onset is insidious as it is not always immediately painful, and personnel suffering from miosis may not realize they have suffered the condition. Units exposed to nerve agents must assume damage has occurred and crewmembers will suffer the described effects. Awareness and preparedness could prevent mishaps. Currently, no effective drug is available to counteract the effects without causing side effects. Aviators showing the effects of miosis may not be able to safely fly an aircraft. Aviators exposed to a nerve agent and exhibiting symptoms must be cleared by the flight surgeon.

FLASH BLINDNESS

4-6. While dark adaptation of the rods develops rather slowly over a period of 30 to 45 min, it can be lost in a few seconds of exposure to bright light. If the eyes are exposed to bright light after dark adaptation, their sensitivity is temporarily impaired. The degree of impairment depends on the intensity and duration of the exposure. Brief flashes from high-intensity, white xenon strobe lights commonly used as aircraft anti-collision lights have little effect on night vision because the energy pulses are of such short duration, lasting only milliseconds. Exposure of 1 second or longer to a flare or searchlight, however, can seriously impair night vision. Depending on brightness (intensity) and exposure duration or after repeated exposures, complete dark adaptation recovery time can range from 5 to 45 minutes or longer. Accordingly, during night operations aircrew members should avoid bright lights or protect one eye. Dark adaptation is an independent process in each eye. Even though a bright light may shine into one eye, the other eye will retain its dark adaptation if it is protected from the light. Aircrew members should avoid looking at flares, flames, or gun flashes to avoid temporary flash blindness. If a supplemental light must be used, it should be as dim as possible while still being readily usable (the lowest easily readable level) and should only be used for the shortest possible period.

AIRCRAFT DESIGN

DESIGN EYE POINT

4-7. The design eye point (DEP) is the point where the aviator's eyes should be. Proper seat adjustment is required for DEP. The aviator should be able to easily reach all switches and circuit breakers appropriate to the crew station and have unobstructed visibility of all appropriate instruments and gauges. The aviator must also be able to clear and control the aircraft during all modes of flight. A field expedient method of enhancing DEP is to position a reference individual directly to the front of the seat position for which the DEP is being determined. The reference individual should position themselves at a distance from the nose of the aircraft as designated below (table 4-1, page 4-3). The reference individual crouches down with fingers barely touching the ground as the aviator adjusts the seat until the individual's fingers touching the ground are visible. This seat position optimizes the aviator's position in the cockpit. DEP reference information is not available for the UH-72.

Table 4-1. Design eye point reference distance

Aircraft	Distance from Nose
UH-60	12 feet
CH-47	20 feet

SECTION II – HEMISPHERIC ILLUMINATION AND METEOROLOGICAL CONDITIONS

4-8. Ambient light is any atmospheric light, whether natural or artificial, providing useful illumination for the aircrew. Sources of ambient light include the moon, background illumination, artificial light, and solar light. Regardless of the ambient light source, meteorological conditions affect levels of light. Aviators can conduct night aviation operations more easily when ambient light sources provide the greatest amount of hemispherical illumination. The aviation unit operations officer, with assistance of supporting Air Force weather personnel, can develop a light-level calendar to predict when optimum levels of ambient light will exist. Computer programs can also be used for illumination and ambient light planning purposes.

LIGHT SOURCES

NATURAL

4-9. Moonlight is the greatest source of natural illumination at night. The sun also has a significant impact on illumination for approximately the first hour after sunset and the first hour before sunrise. Starlight provides the illumination used by NVG when moonlight is insufficient. Much of the illumination provided by the starlight falls into the near infrared range, and is not visible to the unaided eye.

Lunar Light

4-10. The moon angle changes about 15 degrees per hour (1 degree every 4 minutes). Ambient light level changes as the moon angle changes. Light from the moon is brightest when at its highest point or zenith. The time of moon's rising and setting changes continually. Detailed planning is required to determine ambient light levels during a particular night flight. The different phases of lunar light and illumination include—

- **New moon.** The new moon phase is completed in approximately eight days. Moonlight increases toward the end of the phase when about 50 percent of the moon is illuminated.
- **First quarter.** Nearly seven days are required to complete the first quarter phase. The percentage of moon illumination at the beginning of the phase is close to 50 percent and increases until slightly less than 100 percent of the apparent disk is illuminated.
- **Full moon.** The full moon phase begins when 100 percent of the disk is illuminated. It ends seven days later when almost 50 percent of the moon is visible.
- **Third quarter.** The duration of the last phase is about seven days. It begins when close to 50 percent of the moon is visible and ends when 2 percent or less is visible.

Solar Light

4-11. Ambient solar light is usable for a period following sunset and before sunrise. After sunset, the amount of available solar light steadily decreases until the level of light is not usable to the unaided eye. Solar ambient light becomes unusable when the sun is 12 degrees below the horizon or about 48 minutes after sunset. This is end evening nautical twilight (EENT). Before sunrise, solar light becomes usable when the rising sun is 12 degrees below the horizon or about 48 minutes before sunrise, which is begin morning nautical twilight (BMNT). In addition, end evening civil twilight (EECT) occurs when the sun is 6 degrees below the horizon, while begin morning civil twilight (BMCT) occurs when the sun is 6 degrees below the horizon. Civilian and law-enforcement agencies commonly use civil twilight for BMCT and EECT.

Starlight

4-12. Starlight provides the illumination used by NVG when moonlight is insufficient (about 1/10 the illumination of a half-moon). Much of the illumination provided by the starlight falls into the near infrared range, and is not visible to the unaided eye. This night sky near infrared (IR) energy matches the peak sensitivity of the NVG. It is possible to fly effectively with NVG under these conditions with a good training program and proper pre-flight mission planning. On a moonless night, about forty percent of the illumination is provided by emissions from atoms and molecules in the upper atmosphere known as air glow.

ARTIFICIAL LIGHT

4-13. Lights from cities, automobiles, fires, and flares are normally sources of small amounts of illumination. The lights of a large metropolitan area will increase the light level around the city. Artificial light is most pronounced during overcast conditions. Tactical employment of flares or illumination rounds may be used with either unaided vision or NVG operations. This increases apparent illumination contrasts in the target area and denies the adversary use of NVG in the immediate vicinity.

OTHER CONSIDERATIONS

METEOROLOGICAL EFFECTS

4-14. Because meteorological conditions vary, light levels cannot always be accurately predicted and weather elements can change slowly or rapidly. A flight may begin with clear skies and unrestricted visibility; however, these conditions could deteriorate rapidly within the span of one fuel load. In addition, adverse weather is difficult to detect at night. Often the decrease in visual acuity and a gradual loss of horizon are very subtle. As visual meteorological conditions deteriorate, aviators must decrease airspeed to reduce risk of flying into IMC. Heightened awareness of changing weather conditions better prepares the aircrew to evaluate available ambient light.

CLOUDS

4-15. Clouds reduce hemispherical illumination to some extent. The exact amount of reflection or absorption of light by different cloud types is highly variable; therefore, a common factor cannot be applied to each condition of cloud coverage. Aviators consider the amount of cloud coverage and the density, to subjectively evaluate light reduction. For planning purposes some illumination computer programs also give reduction by cloud coverage. Obviously, a thick, overcast layer of clouds will reduce ambient light to a greater degree than a thin, broken layer of clouds. Aviators can detect any reduction in the ambient light level with some basic cues. The following cues and a continuing awareness of present weather conditions are critical in avoiding IMC or unsafe situations:

- Aircrews continuously monitor the apparent light level, paying attention to any reduction, with an accompanying reduction in visual acuity and terrain contrast.
- Increasing cloud coverage obscures moon illumination/visible stars.
- Shadows caused by clouds obscure the effects of moon illumination and should be observed by the aircrew.

FOG

4-16. The effects of fog are similar to those of clouds. Generally, fog is distinguishable from clouds only in regard to distance from the ground. Since fog tends to stay close to the ground it is more a navigation hazard to rotary wing aircraft than to fixed wing aircraft. Fog can mask or partially mask ridgelines and other navigational features making it more difficult to navigate. One way to note an increase in the moisture content of the air while utilizing NVG is to observe a decrease in the intensity of ground lights. This is especially obvious when flying at an altitude high enough to compare ground lights in the immediate area to ground lights beyond the area of increased moisture content. Also, the halo effect noted around lights when viewed directly with NVG may become slightly larger and more diffuse in an area of increased moisture. The enhanced contrast in an area illuminated by ground lights will also be lessened or absent.

4-17. Fog droplets can produce almost 100 percent scattering for the FLIR. Larger particles such as raindrops or snow will absorb as well as scatter far IR energy. In both cases, the end result is an attenuation of the FLIR image. Even with these adverse effects, the FLIR can "see" through this atmospheric obscurant better than the unaided eye, and under the right conditions, can still identify "hot spots" such as fires, operating factories, etc. This information may help in detecting targets, but due to the lack of detail surrounding these hot spots, the information may be of little help for navigating.

Rain

4-18. As with clouds, the performance of NVG and thermal systems in rain is difficult to predict as droplet size and densities are variable. All previous discussions on water vapor, clouds, fog, absorption and scattering are applicable. Due to small droplet size and low density, light rains or mists cannot be readily seen with NVG. However, contrast, distance estimation and depth perception will be affected due to light scattering and the resulting reduction in light level. Heavier rains will be more discernible due to luminance blocking and more obvious signs such as rain on the windscreen. The effect of rain on the FLIR is similar to that of fog. Since the droplet size is larger than fog there will be relatively more absorption. But, like fog, there can be some information gained through a light rain or mist. However, with a heavier rain there will be significant attenuation. Also, constant rain over a period of time will cool surfaces to a more uniform temperature and thus decrease the thermal contrast and ultimate scene discrimination.

Snow

4-19. The density of the snowflakes will determine how much illumination and luminance is blocked and how much degradation occurs to the NVG image. Snow can reflect available light and thus enhance luminance when on the ground. Also, snow can add a slightly different texture that may aid in contrast discrimination. Due to the excellent reflectivity of snow, less illumination is required to give the same luminance for the subject without snow. Thus the NVG can see the terrain under lower light level conditions.

4-20. As with the other forms of moisture, the effect on the FLIR depends on the flake size and density. Due to the general size of snowflakes, scattering of the thermal energy causes most of the attenuation. Therefore, density of the flakes is of primary concern. For snow on the ground, the degree of attenuation will depend on the duration of the snow cover. Snow can cool surfaces to a reasonably uniform temperature and thus will attenuate the FLIR image. However, a fresh blanket of snow on the ground may be "invisible" to the FLIR, making it the sensor of choice if there is little texture / contrast for the NVG to work.

Sand, Dust and Other Obscurants

4-21. The impact of battlefield obscurants on NVG performance depends on particle size and density. NVG visibility "inside" or through these obscurants is usually poor. Hovering in a dusty environment can be very dangerous. Visual references are easily lost and disorientation follows rapidly due to the swirling dust. Use of aircraft systems such as the HUD or helmet display unit (HDU) is strongly recommended. Dust operations are normally trained at the unit. Aviators should use high contrast references closer to the aircraft. Aircraft position and anti-collision lights can interfere with the ability to see outside the aircraft to the point of jeopardizing the safety of the aircraft. The pilot in command should consider turning off aircraft lights IAW regulations and local SOP. Even small amounts of dust with light winds can obscure the horizon.

SECTION III – TERRAIN INTERPRETATION

4-22. The ability to interpret terrain during night flight is determined by a combination of the flight method employed, whether aided or unaided, ambient light level, and aircrew ability to effectively employ night vision techniques. Different conditions affect visual presentation of natural and manmade features during any mode of night flight. This section covers factors affecting night terrain interpretation and various techniques used to compensate for limitations imposed by the terrain.

Chapter 4

VISUAL RECOGNITION CUES

4-23. During night unaided and NVG flight, color vision is degraded or entirely absent. Aviators do not have the relatively high level of visual acuity to identify objects in their environment and must rely on cues such as the size, shape, contrast, and reflectivity of objects. The size and shape of objects should be unique in relation to the operating environment.

SIZE

4-24. Large structures and terrain features, such as towers, are easier to recognize at night than small objects. Small objects tend to get lost in the clutter of other objects (figure 4-1). To see and recognize small features often requires crewmembers to view an area several times. A shorter viewing distance also aids in visual recognition.

Figure 4-1. Identification by object size

SHAPE

4-25. Aircrews can identify objects at night by their shapes/silhouettes (figure 4-2, page 4-7). Some buildings are recognizable at night by their design. For example, a church may be marked by a steeple or cross on top of the structure. Religious buildings of other faiths may look markedly different. Aviators should consider these details in mission planning. Often, maps depict manmade features through symbology which aid in navigation. Aviators may have to reposition aircraft to see objects from different perspectives to recognize shape. A water storage tank/tower may be similar in shape to an oil storage tank requiring the aircrew to seek other viewing angles or supporting information. For example, storage tanks positioned in a group are likely oil tanks not water tanks. The shape of terrain features is also a means of identification at night. Landmarks, such as a bend in the river or a prominent hilltop, provide distinct shapes and assist in night terrain interpretation.

Figure 4-2. Identification by object shape

CONTRAST

4-26. The contrast between an object and its background can aid in object identification (figure 4-3). The degree of contrast depends on the type and amount of ambient light, texture of the object, background, and whether the object is illuminated. These items also serve as cues in identifying objects or features.

Figure 4-3. Identification by object contrast

Color, Texture, and Background

4-27. Color, texture, and background of a manmade or natural feature determine its reflective quality. Various reflective qualities of objects in a FOV help determine degree of contrast. For example, an unplowed

Chapter 4

field with no vegetation provides a good reflective surface. Conversely, an area covered with water provides less overall reflection and appears darker than adjacent foliage with aviator's night vision imaging system (ANVIS). Dense vegetation, however, actually provides very high reflectance of IR radiation with ANVIS. A crewmember familiar with the reflective quality of a feature may be able to identify it by contrast. This is one advantage of knowing the local area and its features. Manmade and natural features most identifiable by contrast include roads, water, open fields, forested areas, desert, and snow-covered terrain.

FACTORS

4-28. Factors affecting an aviator's ability to use cues for terrain interpretation include ambient light, viewing distance, flight altitude, moon angle, visibility restriction, terrain, seasons, and type of night vision sensor used.

AMBIENT LIGHT

4-29. NVG performance and the ability to interpret terrain are directly related to ambient light levels in the flying environment. Reduced light levels at night decrease visual acuity. This restricts the distance at which an object can be identified and terrain interpretation becomes more difficult as the light level decreases. Adjustments in airspeed and/or altitude may be required to improve visual interpretation and increase viewing and reaction time. Reduced ambient light levels can be detected by recognizing the following indications:

- Scintillation-Increase in video noise within the NVG image as a result of low ambient light levels.
- Increase in Halo Intensity-As ambient light levels decrease, halo intensity will increase.
- Loss of Celestial Lights-The moon and stars fade or disappear due to cloud cover or other obscurants.
- Loss of Ground Lights-Cultural lighting begins to fade or disappear due to obscurants.
- Loss of Scene Detail-As ambient light levels decrease, loss of scene detail will occur.

VIEWING DISTANCE

4-30. The viewing angle becomes smaller as the distance from the object increases (figure 4-4); therefore, large and distinctly shaped objects viewed from a great distance at night may become unrecognizable. Range is also difficult to estimate at night and can result in a miscalculation of object size. The distance at which interpretation of an object becomes unreliable also depends on ambient light level. Aviators may be able to identify an object by its shape and size at up to 1,500 meters during a high light condition; however, they may not be able to recognize the object at 500 meters during a low light condition.

Figure 4-4. Identification by object viewing distance

Flight Altitude

4-31. The altitude AGL at which an aircraft is flown affects the aircrew's ability to interpret terrain. The effects of high- and low-altitude flights are discussed in the following paragraphs.

High Altitude

4-32. Changes in viewing angle and distance at which an aviator is viewing an object will change the apparent shape of that object. The ability to identify manmade or natural features progressively decreases as flight altitude increases. This condition is affected at all levels of ambient light. When flight altitude increases, contrast between features becomes less distinguishable and features tend to blend. As terrain definition becomes less distinct, detection from altitude becomes difficult.

Low Altitude

4-33. Terrain becomes more clearly defined and contrast is greater when an aviator flies closer to the ground. This allows manmade and natural features to be more easily recognized and permits increased navigational capability. However, the viewable area of a crewmember at low altitudes is smaller than at higher altitudes. With NOE/contour altitudes, that area is even smaller, sometimes requiring an aviator to reduce airspeed to permit more accurate terrain interpretation. Objects can also be identified at low altitudes by silhouetting them against the skyline.

Moon Angle

High Angle

4-34. Higher moon angles produce greater levels of illumination and reduce shadows that cause distortion and loss of ambient light. This creates the best conditions for visual interpretation because increased ambient light levels improve visual acuity and contrast.

Low Angle

4-35. Terrain interpretation is more difficult when the moon is low on the horizon. This is due to the lower light level and the shadows caused by the low angle. If low-level flight is conducted toward the moon, with the moon at a low angle, glare may bother the aircrew causing distorted vision and a loss of dark adaptation. During aided flight, glare may also degrade NVD capability. However when the moon is low on the horizon, terrain features or objects on the skyline are more recognizable.

Azimuth Angle

4-36. With ANVIS and high moon illumination, trees increase in apparent brightness. When the moon is positioned behind an aviator, the contrast between the terrain and sky at the horizon may be reduced to a zero value. However, when the moon is positioned within the frontal 180 degrees of the flight path, the trees at the horizon will be shadowed, appearing darker, thereby increasing the contrast at the horizon.

Visibility Restriction

4-37. Weather conditions (dust, rain, fog, or snow) restrict visibility, reduce ambient light, and cause a loss of visual acuity. These conditions normally cause visibility to decrease gradually, beginning with reduced visual range, followed by loss of terrain definition. Eventually, the loss of visibility may impair night vision to the extent that terrain flight is not safe and should be discontinued. These weather conditions also complicate procedures such as hovering in a battle position (BP) and external load hook-up. Dust or snow particles reflect light from a searchlight and can become a major distraction. In addition, the swirling dust or snow can cause the illusion of relative motion when the aircraft is at a stable hover. A scan pattern should reference vertical fixed points such as bushes, rocks, and trees.

Terrain

4-38. The nature of terrain determines the amount of light reflected from the surface of the earth. Deserts, heavily vegetated rolling terrain, mountains, jungles, and arctic are some of the terrain types that reflect light differently.

Deserts

4-39. The amount of vegetation varies greatly from substantial numbers of shrubs and trees to sparse, sandy wastelands. In sparsely vegetated, sandy deserts, the texture and color of the soil on the desert floor is normally very uniform. This can make it difficult to identify changes in terrain elevation or locate individual features, such as ridges, valleys, wadis or ravines. The sandy soil provides optimum reflectivity of available ambient light and a useful background for identification of objects by contrast. Man-made objects in particular usually stand out well against their background. Vegetation that does exist will aid terrain interpretation by providing good contrast with the soil. Desert terrain can vary from relatively flat expanses of sand to mountains. The amount of detail available may change dramatically during a single mission. Gradually rising terrain may be quite difficult to detect, particularly in areas of lower contrast. Aircrews should exercise caution when transitioning to areas of lower contrast. Comparing MSL and AGL altitudes can assist crewmembers in identifying rising or descending terrain. Aviators can encounter blowing dust or brownout in this environment requiring a practiced technique to overcome. Low level winds can raise just enough dust to obscure the horizon without otherwise interfering with visibility. Lack of terrain features and reference points makes terrain flight navigation and concealment more difficult.

Vegetated Rolling Terrain

4-40. Dirt roads and farm structures provide the most distinguishable man-made features and contrast is good between forested areas and open fields. Rivers and terrain features which give distinct changes in elevation from surrounding terrain provide the most recognizable natural landmarks for navigation. Visibility of these terrain features depends on the amount of vegetation present; dense vegetation can mask terrain features and changes in elevation. Dense vegetation makes reconnaissance difficult and may mask thermal signatures. Airspeed may need to be reduced during these missions.

Mountains

4-41. Terrain interpretation can be enhanced due to large distinct terrain features and is aided by terrain silhouetting. When the moon is near the horizon, large shadows can severely restrict what can be seen in the shadowed areas. Varying lighting conditions may make it difficult to see folds in the terrain, such as ridges or valleys. Mission planners should allow additional time for reconnaissance or attack missions in mountainous terrain. The complex textures normally found in mountains can mask flight hazards such as towers, wires or other man-made structures. It is not safe to fly into areas too dark for obstructions to be seen. If you can't see into it, treat it as an obstacle.

Jungle

4-42. Jungles are similar to heavily vegetated rolling terrain areas. The canopy obscures the view of most features lacking significant vertical development. Precise terrain interpretation is more difficult as the dense vegetation may also mask changes in elevation.

Arctic/Snow

4-43. Visible vegetation and dark features provide good contrast. Similar to the desert environment, blowing snow can cause a "white-out" condition and high reflectivity may cause the NVG to gain down, reducing image clarity. Terrain interpretation may be difficult due to drifting or deep snow that can hide key features, fill in valleys, or create hills. Unlike desert terrain, there are very few differences between various areas of snow, making it very difficult to distinguish terrain features. Landing to snow must be considered with great caution, as obstructions, holes or partially frozen bodies of water may be impossible to detect. The ability to judge height and determine the contour of terrain is difficult when it is covered with snow. The normal tendency is to estimate altitude as being higher than it actually is and misjudge slope angles. Check instruments frequently.

Overwater

4-44. Poor contrast, minimal reference points, and a reduced sense of motion parallax make aviators operating over water susceptible to a variety of visual misperceptions and spatial disorientation. Long flights over water without a visible horizon should be avoided. A greater reliance on flight instruments and heads-up display systems will be required to maintain spatial orientation and situational awareness. Before flying over water, check the barometric and radar altimeters for proper operation. Aviators should set the radar altimeter low altitude indicator to the minimum acceptable altitude. As an aviator crosses from land over water, the aircraft may appear to stop in mid-air due to the loss of visual references. As a result, there is a tendency to lower the nose of the aircraft and enter an unintended descent. Aircrew not on the controls should maintain a cross-check of the flight instruments and other indications of altitude to prevent inadvertently flying into the water. Trail aircraft should monitor and advise the flight if any aircraft appears to be descending below the briefed altitude. Water is the most difficult surface over which to hover as there are almost no visual references. Loss of the visible horizon will have a significant impact on the ability to maintain aircraft orientation. Aviators display a tendency to drift in the direction of waves. If possible, the aircraft should be maneuvered near some object, such as a tree stump, or buoy to provide a reference point. Remember that objects floating in the water may move unexpectedly. At night under low effective illumination due to the poor contrast, maximum emphasis must be placed on crew coordination and being deliberate with flight control inputs. Changes in airspeed and altitude should be clearly announced and level-off acknowledged by the crew. Minimizing and briefing power setting for en route flight will aid in individual aircraft control while operating overwater and assist multi-ship flight.

SEASONS

4-45. Seasons of the year affect the amount of ambient light reflected from the surface of the earth; however, aviation focuses on two seasons—winter and summer. While significant differences are present between the two seasons, which season is easier to interpret terrain and detect visual cues is determined by that AO. Aviators must evaluate each location separately to avoid generalizations or assumptions.

Winter

4-46. Contrast improves during winter as many areas lack vegetation. Ground snow also improves contrast by increasing total illumination as it reflects ambient or artificial light. The light color of snow, compared with the dark color of structures and heavily forested areas, enhances visual interpretation.

4-47. The loss of foliage on deciduous trees makes ground features, such as small streams, easier to identify. Plants and grass in open fields change color and improve contrast between open fields and coniferous trees. Barren trees, however, reflect less light and become more difficult to see often causing an aviator to fly higher.

4-48. In winter, the orbital path of the moon is closer to the earth causing the ambient light level to be higher than at other times of the year. This improves visual acuity and enhances terrain interpretation.

4-49. Cloud cover and restricted visibility occur more often during winter than summer. Both conditions significantly reduce ambient light level, thereby decreasing visual acuity and making terrain interpretation more difficult unless sources of artificial light are nearby.

4-50. Heavy buildup of snow may conceal manmade and natural terrain features. Snowdrifts may obscure a road intersection normally used as a navigational checkpoint (CP). An aviator can still identify this obscured CP by associating it with other objects or terrain features such as a power line, fence line, or cut through a wooded area. In addition, heavy snow buildup combined with severe cold cause small rivers or lakes to freeze over and become unrecognizable. Aviators must identify these landmarks by associating them with a depression or tree line.

Summer

4-51. Identifying objects and terrain features by contrast in summer is less effective than during winter months. The increased amount of vegetation and abundance of growth on deciduous trees makes it difficult to recognize small rivers or streams and decreases the ability to recognize military targets when located in or near forested areas. Concealment and camouflage are much easier during summer months.

OTHER CONSIDERATIONS

TERRAIN FEATURES

4-52. Analysis of terrain is the most reliable means of orientation. Features unique in shape or providing a distinct change in elevation are excellent CPs. Man-made objects can be equal to or better than natural features as navigation aids, particularly because navigation is primarily conducted through aircraft systems and they may be more accurately plotted.

Silhouetting

4-53. This cue is best described as sighting the darkened shape of an object when positioned against a lighter background. Silhouetting is visually achieved during low-altitude flights. Aviators also use silhouetting to locate terrain definition as well as manmade objects. High terrain can create shadows hiding hazards or other important features.

Vegetated Areas

4-54. Deciduous trees appear different when compared with coniferous trees. With ANVIS, heavily forested areas reflect light well at the tops of the trees but may appear darker than open fields due to shadows, viewing angle, and altitude. An open field stands out in forested areas due to good contrast. Contrast, shape, and texture are cues as to which type of vegetation an aviator is viewing.

Fields

4-55. The amount of light reflected by a field depends on the season and amount and type of vegetation. The type of vegetation or a harvested or plowed field may also provide highlights due to good contrast. In addition, isolated fields make good CPs; however the surrounding trees may mask the field.

Hydrographic Features

4-56. Water generally provides little contrast unless wind disturbs the surface. Identification depends on the amount of reflectivity (moonlight) and ambient light levels. Small ponds and lakes generally make poor CPs; vegetation or terrain can easily hide them. Vegetation may also hide rivers and streams; however, deciduous trees generally grow in wetter areas while coniferous trees grow on ridges which can assist an aviator in locating rivers and streams.

Cultural Features

4-57. Manmade features are excellent NVG navigational cues. Flight altitude is important for recognition of these features. They include—

- **Roads.** A dirt road may provide excellent contrast between the surrounding terrain, vegetation, and its surface. Composition of the soil must be considered as it changes the degree of contrast the road will provide in comparison to the surrounding terrain. In addition, roads cutting through heavily forested areas are easily identifiable if visible through foliage. A concrete road is generally more reflective than an asphalt road but may or may not be more visible through NVG due to surrounding background reflectance. Freshly paved asphalt roads appear dark through NVG; however, roads reflect more IR energy as they age and wear. An asphalt road is usually difficult to identify as its dark surface reduces the contrast between it and the surrounding terrain. The exceptions are if the asphalt road is located in a desert or snow-covered environment or an area with open fields, which provides good contrast, making it easier to recognize. Although roads are not good CPs, certain features can serve as orientation cues or CPs. Roads normally make excellent barriers when associated with other CPs.

- **Intersections.** Intersections accurately plotted on maps can serve as orientation cues or CPs. Check the type of intersecting roads, road heading, and surrounding cues to ensure the correct intersection has been located.
- **Bridges.** Bridges can be good CPs if they have vertical development. A bridge is also a good CP if the bridge surface contrasts with the road surface or surrounding vegetation.
- **Railroads.** Aviators can easily identify railroads; however, surrounding vegetation or terrain often hides them. The viewing angle is important for locating railroads. They make poor CPs and barriers unless they are in open fields.
- **Buildings.** Isolated, large, or light-colored buildings provide excellent contrast. Aviators should not use small, dark-colored buildings as orientation cues.
- **Cemeteries.** Most cemeteries have light-colored, polished headstones contrasting well against a natural background and often reflecting a considerable amount of light.

SECTION IV – NIGHT VISION SENSORS

4-58. The purpose of using night vision sensors is twofold. First, night vision sensors enable friendly forces to sustain around-the-clock operations. Second, night vision sensors allow the command to conduct offensive and defensive operations against an enemy force with the element of surprise while increasing survivability of an aircrew and aircraft.

4-59. Night vision sensors are described as either night vision system (NVS), night vision goggle (NVG), or NVD. NVS refers to the night vision system that is attached to the aircraft systems (for example, the modernized target acquisition designation sight (MTADS)/modernized pilot night vision system (MPNVS). NVG refers to any night vision goggle image intensifier system, for example, the AN/AVS-6, an aviator night vision imaging system (ANVIS). NVD refers to both NVG and NVS.

ELECTROMAGNETIC SPECTRUM

4-60. The electromagnetic energy spectrum includes the range of wavelengths, such as gamma rays, X-rays, ultraviolet, visible light, IR, microwaves, and radio waves, or frequencies of electromagnetic radiation. NVDs make use of visible light energy bands and IR energy bands. These bands make up a small portion of the electromagnetic spectrum. Figure 4-5 highlights the portions of the electromagnetic spectrum used by NVDs.

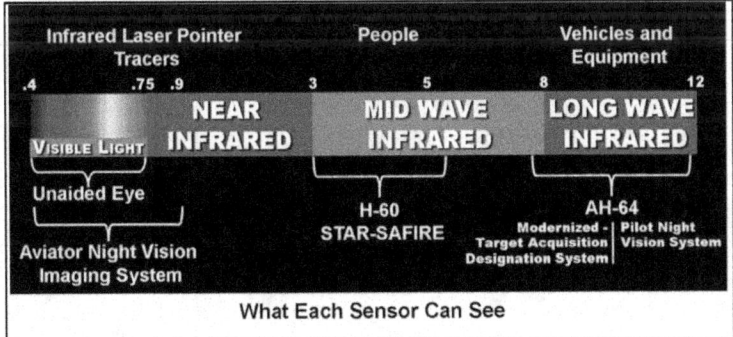

Figure 4-5. Electromagnetic Spectrum

Visible Light

4-61. The amount of reflected visible light determines what the human eye sees. The eye sees color due to the reflective or nonreflective properties of the object being viewed. In other words, a leaf appears green because it reflects mainly the green wavelength within the visible spectrum (0.52 to 0.57 micron) and absorbs most of the remainder. For the leaf to reflect visible light energy, it must have energy in the wavelengths between 0.4 to 0.7 micron incident upon it.

4-62. During daylight, the greatest source of visible light energy is the sun. The sun continuously emits energy and permits the eye to discern form and color. When the sun sets, most naturally occurring visible light energy is reduced and normal eye function makes the transition to scotopic vision decreasing visual acuity. Scotopic vision requires either naturally occurring night light sources or artificial lights. Image intensifier (I2) systems amplify natural and artificial visible and near IR energy.

Infrared Radiation

4-63. The sun emits energy across the entire electromagnetic spectrum, not just the visible light spectrum. As IR energy enters the atmosphere and penetrates to the surface, it is reflected or absorbed to produce stimuli for NVDs. I2 devices can amplify reflected IR light. When IR light is absorbed, temperature changes occur in those natural and manmade substances in the environment. As the sun sets, effects of this solar heating remain. FLIR is effective because this system can detect heat as the environment radiates it. IR radiation exists due to molecular activity within elements of substances. As molecules are stimulated, they vibrate, which radiates energy–including IR energy. The stimulus for molecular activity is heat. The intensity of molecular activity is directly proportional to temperature. The temperature of an object is caused by natural or artificial thermal sources or a combination. The amount of IR energy radiated by an object depends on the exposure amount and how much thermal energy is absorbed, reflected, or transmitted.

Infrared Energy

4-64. Reflectance, transmittance, absorptance, and emissivity determine the amount of IR energy an object will radiate when exposed to "x" level of thermal energy for "x" amount of time. The total amount of IR energy an object radiates is the sum of reflected, transmitted, and emitted energy (figure 4-6, page 4-15) which are defined as—

- **Reflectance.** The tendency of energy to bounce (reflect) off of an object is called reflectance. Objects with a high reflectance (such as a container with a shiny surface) tend to heat more slowly in sunlight.
- **Absorptance.** The tendency of an object to retain energy.
- **Transmittance.** The ability of an object to allow energy to pass through it.
- **Emissivity.** The relative power of a surface to emit heat by radiation. It is the ratio of radiant energy emitted by a body (due to its temperature only) to that emitted by a reference body (blackbody) at the same temperature.
    - This characteristic has considerable significance regarding object IR radiation. A blackbody is an ideal body or surface that completely absorbs all radiant energy falling upon it with no reflection making it the theoretical standard for laboratory comparison. A blackbody absorbs 100 percent of IR energy acting upon it and emits 100 percent of its IR energy.

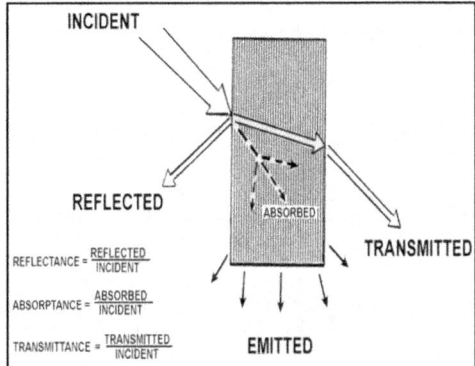

Figure 4-6. IR energy

NIGHT VISION DEVICES

OPERATION

4-65. An image intensifier (figure 4-7) is an electronic device that amplifies light energy. Light energy, consisting of photons, enters the objective lens, is inverted and focused onto a photocathode that is receptive to both visible and near IR radiation. Photons striking the photocathode are then converted to a proportionate number of electrons.

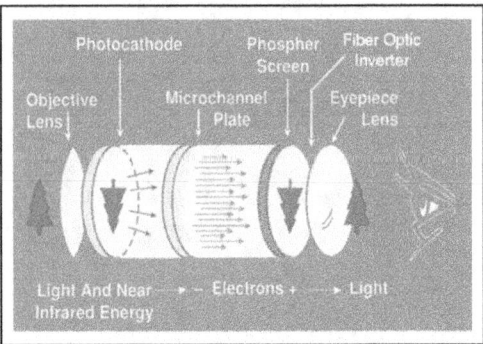

Figure 4-7. Image intensifier

4-66. Electrons are accelerated away from the photocathode to the microchannel plate (MCP) via an electrical field produced by the power supply. The MCP is a thin wafer of tiny glass tubes that are tilted about 8 degrees. Electrons enter these tubes and strike the walls, creating a reaction which exponentially increases the amount of electrons. These increased numbers of electrons are then accelerated to the phosphor screen. The phosphor screen emits an amount of photons proportional to the number and velocity of the electrons striking it creating a lighted image. The image is then passed through a fiber-optic inverter to rotate the image 180 degrees to correct the inverted image caused by the objective lens. The image is then focused onto the viewer's eye through an eyepiece lens. The power supply provides automatic brightness control (ABC) that automatically adjusts MCP voltage to maintain image brightness at preset levels by controlling the number of electrons that exit the MCP. Another feature is bright source protection (BSP) which reduces the voltage to the photocathode when exposed to bright light sources. This feature protects the I2 from damage and enhances its life; however, it lowers resolution. Exposure to bright light sources could result in damage to the photocathode, MCP, or the operator's eye.

AN/AVS-6

4-67. The AN/AVS-6 (figure 4-8) is a helmet-mounted, light-intensification device. This NVG and its variants allow aircrews to conduct operations at terrain flight altitudes during low ambient light levels, to include overcast conditions. It has a 40-degree FOV, which can enhance visual acuity from normal unaided night acuity of about 20/200 to approximately 20/25 under optimum conditions. The AN/AVS-6 amplifies light 2,000 to 3,000 times and provides sufficient imagery for pilotage from overcast starlight to moonlight conditions. In practical application, when illumination is below quarter moonlight conditions, artificial illumination (usually IR) may be required to light the flight path of the helicopter. The AN/AVS-6 is powered by batteries or aircraft interface. The dual-battery pack has a low voltage warning indicator on the visor mount consisting of a red light that flashes when the battery is at 2.4 volts or less. The AN/AVS-6 also incorporates a 10- to 15-G breakaway feature allowing the goggles to separate from the attachment point on the helmet preventing injury to the aircrew member during an accident.

CAUTION
An aircrew member's eyeglasses or protective masks can block the warning light. The low-battery light is often detected first by the other crewmember.

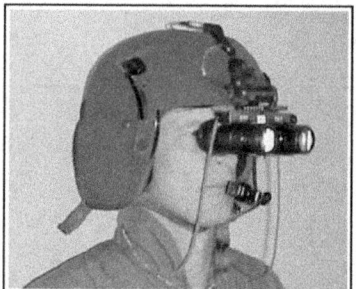

Figure 4-8. AN/AVS-6 in operational position

SYSTEM COUNTERWEIGHTS

4-68. The counterweight system consists of the weight bag and counterweights. The weight bag is locally constructed and is the responsibility of the maintainer. Attachment of the weight bag should be low on the back of the helmet with the battery pack mounted vertically above it. The recommended initial weight is 12 ounces; however, maximum allowable is 22 ounces. The aviator adds or removes weight, with the goggles attached and flipped down, to achieve the best balance and comfort. A Velcro™ patch on the back of the helmet is required to attach the counterweight system as well as the battery pack. The helmet can only be modified with the Velcro™ patch by a qualified ALSE technician. Using tire weights and like materials with sharp edges is discouraged as they can become missile hazards during a crash sequence. It is recommended buckshot in zip-lock pouches or rolls of pennies be used, allowing the amount of weight to be adjusted easily and the weight bag to conform to the contour of the helmet.

Rotary-Wing Night Flight

HEADS-UP DISPLAY

4-69. HUD systems are designed to display flight, navigation, and aircraft system information onto the NVG display. It enables an aviator to concentrate his or her vision outside the cockpit while maintaining the ability to view critical information. Depending on the system, the aviator has the ability to determine and display critical information and symbology into his or her FOV and is able to keep eyes outside the cockpit.

OPERATIONAL CONSIDERATIONS

Operational considerations for NVG systems include the following. **Magnification Versus Enhancement**

4-70. NVG systems do not magnify an image; they enhance the illumination of an object. An object viewed through an NVG system is the same size as if seen with the naked eye.

Lights and Lighting

4-71. Using NVG, an aircrew member can detect light sources not visible to the unaided viewer. Examples include certain lights on other aircraft, flashlights, chemical light sticks, cockpit supplementary lighting, and even cigarettes. As ambient light level decreases, aircrews can more easily detect these light sources; they will have greater difficulty correctly estimating distance. Performance of NVG is directly related to ambient light. During high light levels, resolution is improved and objects can be identified at greater distances. Conversely, lights too bright, such as searchlights, street lights, or moonlight, can adversely affect NVG.

4-72. Fixed pattern noise (honeycomb) is usually evident at high light levels or when viewing bright lights. Internal circuitry automatically adjusts output brightness to a preset level restricting peak display luminance. When an area with bright lights is viewed, display luminance of the background decreases. In addition to the halo effect around a bright light source, overall display luminance of the remaining scene also dims. The brighter light sources dim the viewed scene. This same problem is usually evident when an aviator is viewing in the direction of a full moon (usually at low angles above the horizon). The ability to see objects within a lighted area depends on the intensity of the light and distance of the object from the viewer. To prevent degrading NVG performance, a crewmember should minimize the time spent looking at bright light sources within the 40-degree FOV. In addition, when flying with the landing light, searchlight, or IR band-pass filter installed, an aircrew should avoid concentrating on the area illuminated by the light. They should also scan the area not illuminated for hazards or obstacles.

4-73. The sky above the horizon tends to activate the ANVIS ABC level to dim objects below the horizon when an aviator is flying in the direction of the setting sun before EENT or in the direction of the rising sun after BMNT. The more the sky above the horizon fills the NVG's FOV, the greater the dimming of the image and details below the horizon. NVG training flights during these periods are not recommended.

Depth Perception and Distance Estimation

4-74. NVG distort depth perception and distance estimation. The quality of depth perception in a given situation depends on factors including available light, type and quality of NVG, degree of contrast in the FOV, and user experience. The aircrew must often rely on monocular cues.

Color Discrimination

4-75. Color discrimination is absent when a crewmember views scenes through NVG. The picture viewed is monochromatic (single color) and has a green hue due to the type of phosphor used on the screen. The green hue may cause crewmembers to experience a pink, brown, or purple afterimage when they remove NVG. This is called monochromatic adaptation and is a normal physiological phenomenon. The length of time the afterimage remains varies with each individual.

Chapter 4

Scanning Techniques (Aided Flight)

4-76. The basic principles of scanning, flight techniques, and visual cues are the same for aided and unaided flight; however, a few specific items are considered when conducting operations with NVG. NVG use improves ground reference but significantly reduces FOV.

4-77. An NVG's FOV significantly reduces peripheral vision as compared with unaided flight. Crewmembers must use a continual scanning pattern to compensate for the loss. Moving the eyes will not change the viewing perspective; the head must be turned. Rapid head movement, however, can induce spatial disorientation. To view an area while using NVG, a crewmember's head and eyes must rotate slowly and continuously. The length of time and frequency of the scanning pattern is based on the type of terrain and obstacles, airspeed, and what is actually seen through the NVG. When scanning to the right, crewmembers should move their eyes slowly from the left limit of vision inside the device to the right limit while moving their head to the right. This enables a crewmember to cover a 70- to 80-degree FOV with only 30 to 40 degrees of head movement, minimizing head rotation. The crewmember should scan back to the left in reverse order avoiding rapid head movements. The crewmember must blend aided and unaided vision techniques to view the scene. After a few NVG flights, head and eye movements for proper scanning become intuitive and natural.

4-78. NVG are the primary source for detailed visual information. The intensity, distance, or color of illumination sources–such as aircraft position lights and ground lights–may not be accurately interpreted when using NVG. Unaided vision can provide this additional information. Inside the cockpit, an aviator can look under or around the framework of NVG. This technique is also used to view outside the cockpit to detect the true color of position lights or possible obscuration, or any distorted observation by NVG. Initially, unaided peripheral vision may be distracting until the crewmember develops adequate experience combining aided and unaided vision.

Obstruction Detection

4-79. Obstructions having poor reflective surfaces, such as wires and small tree limbs, are difficult to detect. The best way to locate wires is by looking for the support structures. However, aviators should review the most current hazard maps with known wire locations before NVG flights.

Spatial Disorientation

4-80. Maneuvers requiring large bank angles or rapid attitude changes tend to induce spatial disorientation. An aviator should avoid making drastic changes in attitude/bank angles and use proper scanning and viewing techniques.

Airspeed and Ground Speed Limitations

4-81. Aviators using NVG tend to overfly their capability to see. To avoid obstacles, they must understand the relationship between the NVG's visual range, forward lighting capability, and airspeed. This is especially true when flying in a terrain flight mode.

4-82. Different light levels affect the distance at which objects are identified and limit the ground speed flown at terrain flight altitudes. Ground-speed guidance is not specified due to continuously changing variables such as type of aircraft, supplemental lighting, visual obscuration, and ambient light conditions. Aviators should reduce ground speed to allow enough reaction time for detection and obstacle avoidance, especially during low ambient light levels or when visibility is poor.

4-83. Object acquisition and identification are related to ambient light levels, visibility, and contrast between the object and its background. Light levels appropriate for training may need to be different from operational conditions to ensure safe operation and reduce risk. Variables affecting the ability to see with NVG include—

- Type, age, and condition of NVG.
- Cleanliness of aircraft windscreen or sensor window.
- Moisture content in the air (humidity).
- Individual and collective proficiency and capability.

- Weather conditions (fog, rain, low clouds, or dust) and amount of ambient light.

Aircraft Lighting

4-84. Exposure to various sources of lighting not compatible with NVG, especially red, may degrade an aircrew member's ability to see. Adverse effects of lighting are greatest during low ambient light conditions.

4-85. The AN/AVS-6 is designed to be operated with blue-green cockpit lighting. Red cockpit lighting is not compatible and not authorized for use with NVG. While blue-green cockpit lights will not degrade system performance, the lights should be dimmed to a low readable level.

> **CAUTION**
> During tactical operations at night, cockpit lighting should be adjusted to the absolute lowest usable levels and crewmembers should be discrete in the use of supplemental lights to avoid detection by enemy forces.

4-86. NVG operations are degraded by aircraft external lights unless properly modified. The lights should be adjusted to the lowest level allowing detection by other aircraft or a control facility. Red navigation lights (left side of aircraft) produce more usable light with NVG than green lights. Aviators switching seats should anticipate this, especially before hovering or performing external load work.

4-87. Other aircraft external lights such as position lights, formation lights, anti-collision lights, or electroluminescent light panels should be turned off or subdued as appropriate for the operation. Compliance with all local requirements and any appropriate Federal Aviation Administration (FAA) exemptions must occur prior to conducting lights-out operations or modifying helicopters. Exterior lights of other aircraft do not degrade the vision of an aircrew using NVG if the lights are properly operated. Consult other publications to determine if an IR search/landing light must be installed before conducting NVG operations.

Weather

4-88. When using NVG, aviators may fail to detect entry into or presence of IMC. NVG enable crewmembers to see through obscurations, such as fog, rain, haze, dust, and smoke, depending on density. As density increases, aircrews can detect a gradual reduction in visual acuity as less light is available. Certain visual cues are evident when restriction to visibility occurs. The apparent increase in size and density of halos during bad weather is an illusion. The halos are due to the electron spread for bright light sources; size remains the same. Any reduction in visibility decreases light intensity and reduces density of the halo. While contrast decreases, video noise may increase. There may be a loss of celestial lights, while the moon and stars may fade or disappear due to overcast conditions. When these conditions are present, severity of the condition is evaluated and appropriate action taken. Actions include reducing airspeed, increasing altitude, reversing course, aborting the mission, or landing. If visual flight cannot be maintained, the crew must execute appropriate IMC recovery procedures.

4-89. Rain causes unusual effects when using NVG. Specifically, rain will not be detected on the windshield of an aircraft primarily because the NVG's depth of focus makes the windshield out of focus.

Weapons

4-90. During Hellfire missile engagements, NVG may momentarily shut down if the aviator looks directly at the motor during ignition. When firing the 2.75 inch folding-fin aerial rocket, 20- or 30-millimeter cannon, 7.62 millimeter, or .50-caliber machine guns, aircrews may briefly lose sight of the target. Although the bright flash resulting from the rocket launch lasts only milliseconds, the muzzle flash from the weapons may cause the aircrew to lose sight of the target throughout the entire firing burst. The recovery from bright flash illumination is more rapid with NVG than unaided. A greater concern is observing impact due to flash signature momentarily degrading the NVG.

Chapter 4

FORWARD-LOOKING INFRARED SYSTEMS

4-91. The operation of a FLIR differs from NVG systems. Thermal systems operate passively without regard to levels of visible light. These systems do not transmit energy; rather, they sense and display energy radiated from objects. FLIR provide aviators with an image of an IR scene. Thus, aviators can operate in environments that could restrict or prohibit unaided operations. Increased effectiveness of FLIR occurs when there is a large difference in detected IR radiation between an object and its background. Effectiveness is also improved when atmospheric considerations, such as obscuration, are minimized between the system and the object.

TYPES

4-92. Currently, Army attack and reconnaissance helicopters utilize the FLIR for target acquisitions during day and night operations. The AH-64D/E MPNVS, and MTADS, figure 4-9 are passive systems which sense and display various levels of IR energy radiating from objects. This allows operators to view objects regardless of visible light levels required for unaided and aided operations. The effectiveness of the FLIR depends on the difference in detected IR radiation between an object and its background. Effectiveness also depends on atmospheric considerations, specifically, the degree of obscuration present between the system and the object. FLIR is most effective when a large difference in IR radiation exists between an object and its background and when obscuration is minimal. The AH-64D/E MPNVS is currently the only FLIR designed for pilotage but crews may use the MTADS as a backup should MPNVS fail. Aviators should consult the appropriate aircraft operator's manual for specific operating instructions.

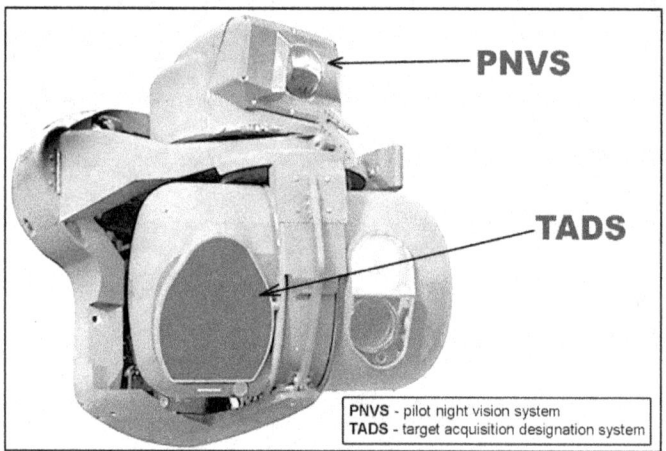

Figure 4-9. MPNVS/MTADS

OPERATIONAL CONSIDERATIONS

4-93. Operational considerations for FLIR operations include the following.

Minimum Resolvable Temperature

4-94. FLIR operations depend on the ability of the system to detect temperature difference. These differences are displayed in the cockpit by shading variations of the display screen. The lowest thermal difference that can be resolved is called the minimum resolvable temperature (MRT). An operating tank shows up well against a cool background, but a cold-soaked tank sitting out in a field with only a minor temperature difference may not show up at all unless the difference in temperature is greater than the MRT of that system.

A low MRT also provides more contrast and detail in the FLIR picture and allows operations in a broad range of environmental conditions.

Forward-Looking Infrared Radar Sensor Optimization

4-95. Detailed procedures for optimizing are found in appropriate operators manuals. Generally FLIR optimization is a combination of level and gain settings, which produces the most detail in the displayed image. It is equivalent to the brightness control on a cathode ray tube (CRT). From these settings, an adjustment of either control in either direction produces less detail and degrades the image quality of the FLIR. The level and gain controls on the display control panel are used to adjust the FLIR sensor. Proper adjustment of FLIR provides the highest possible resolution picture for the operating environment at time of adjustment. If the FLIR is operating properly, scene content—such as terrain and metal buildings, temperature, humidity, atmospheric conditions, and range to the viewed objects—will determine the image quality.

4-96. The level control regulates overall intensity or brightness of the total light-emitting diode array. An increase in level control uniformly increases intensity or brightness of the total light emitting diode (LED) array. Conversely, a decrease in level control uniformly reduces intensity of the total LED array. This is presented as a brightening or darkening of the total display. An aviator should increase or decrease level control, as necessary, to bring significant object signals (whether hot or cold) within the dynamic range of the LED array.

4-97. The gain control also affects intensity of the LED array but on an individual LED basis. It is equivalent to contrast control on a CRT. Increasing the gain decreases the range of visible temperatures in the scene, which allows detection of smaller temperature differences in the midrange. With excessive gain adjustment, cooler and hotter temperatures will appear either black or saturated white with few visible gray shades. With too little gain adjustment, overall contrast is reduced and small temperature differences are less apparent. Gain control regulates the response of each LED to the electrical signal produced by the IR detectors. Each IR detector in the detector array is electronically connected through preamplifiers and postamplifiers to one LED within the LED array. An increase in the gain control increases amplitude of the electrical signal, leaving the postamplifier to power an LED. Conversely, a decrease in gain control decreases amplitude of the electrical signal. If gain control is increased, an LED will be brightened or dimmed to a greater extent or degree than when gain is decreased. For example, an LED response to an IR detector signal is increased with an increase in gain and decreased with a decrease in gain. This is presented as a variation in intensity between the shades of gray within the total display. Aviators will perceive a reduction in gain on the display as a softening or clouding of the image. Increases in gain reduce the apparent cloudiness in the image until only black and white are visible with no shades of gray between them.

4-98. Considerations for FLIR optimization are the following:
- The FLIR should be allowed to cool to proper operating temperatures before optimizing.
- The aviator selects a scene that is potentially rich in detail or best represents the planned flight environment.
- The aviator selects the desired polarity.
- Only one control should be adjusted at a time; never simultaneously.

4-99. To accomplish FLIR optimization, the aviator fully decreases level and gain controls, which will completely darken the display. Level control is increased until the display just begins to brighten, then gain control is advanced until obvious variations in shading appear in the display and stop the advance. The aviator then makes minute adjustments in level and gain controls to complete the optimization process.

4-100. FLIR optimization described above is appropriate only for the scene viewed and existing atmospheric conditions at the time of the optimization. Generally, changes in atmospheric environment and scene content will require only minor adjustments of level and gain controls after FLIR is initially optimized. To ensure effectiveness of FLIR, aviators should continually optimize the FLIR image. Aviators must clearly understand and effectively practice the principles of FLIR optimization.

Atmospheric Effects

4-101. Atmospheric transmission pertains to signal reduction (attenuation) caused by the distance a signal travels through a given air composition or density. IR signal attenuation is directly proportional to changes in air composition or density. As moisture increases in the air, IR signal strength is attenuated. Condensing moisture forms clouds that may form heavy overcast conditions. These overcast conditions, especially over a period of days, prevent most solar thermal radiation from reaching the surface. The loss of thermal energy reduces molecular activity in substances beneath the overcast conditions and subsequently reduces IR radiation from those substances. Heavy concentrations of moisture between a FLIR sensor and the objects viewed tend to attenuate IR radiation from those objects. These particles of moisture generate their own molecular activity. In comparison, the radiation from these particles is very small. It may add, however, to the overall interference in the IR signal transmission. Elements other than moisture, such as dust, haze, or smoke, also affect the composition or density of the atmosphere and IR signal transmission. FLIR penetration of these substances depends on the size and amount of particulates between the sensor and the objects viewed.

4-102. FLIR performance for specific environments cannot be absolutely defined as they depend so much on the intensity and relative unpredictability of atmospheric effects and conditions. The differences in total radiation of objects relative to the existing backgrounds normally permit safe terrain flight operations. FLIR exceeds the capability of the human eye to operate in visual obscurations or adverse weather conditions and usually allows detection of any obscurations before penetration. This advance notice allows the aircrew the option of circumnavigation or penetration. The effect of atmospheric obscurations on thermal system performance varies in direct proportion to the quantity and density of the obscuration. The distance between the sensor, viewed scene, and IR signature strength of the objects also affects the performance level. A scene viewed with FLIR is rarely totally obscured.

4-103. Since 1979, MPNVS has been subjected to flight operations in weather phenomena including heavy rain, snow, sleet, fog, and haze. These conditions were encountered in various types of terrain including deserts, mountains, and densely foliated swamps. Data gathered during these environmental tests proved MPNVS FLIR permits safe NOE flight operations most of the time. As visibility was degraded, airspeed was reduced to avoid obstacles. Figure 4-10 illustrates atmospheric effects on IR radiation.

Figure 4-10. Atmospheric effects on infrared radiation

Infrared Energy Crossover

4-104. IR energy crossover is the final factor affecting IR radiation. Figure 4-11, page 4-23, depicts a specific location and shows temperature distributions of various substances during a 24-hour period. The effects of solar thermal radiation can be observed by tracing any of the substance curves from 0600 hours (assuming that to be sunrise) to 1400 hours. Point A depicts the time of day when soil, water, and concrete cross over—when thermal radiation of each is nearly equal. The ability of FLIR to discriminate soil from concrete or water would be based on MRT. The FLIR with the lowest MRT would experience the least effect from IR energy crossover and would increase the amount of time when FLIR is unaffected by IR energy crossover. Point B depicts the time of day when temperature differences among soil, water, and concrete are greatest. If this were always the case, the MRT of FLIR could be much higher and still permit scene definition. The temperature differences depicted in point B are generally the exception. The common condition lies somewhere between points A and B.

4-105. Soil, concrete, and water cross over twice daily. However, soil and concrete do not cross over with vegetation, while vegetation and water cross over twice daily. Figure 4-11 depicts conditions for a moment in time at one location. Given the effects of weather, it is unlikely these same conditions will recur, even in the same location. The variance occurs because changing weather patterns make the same day different from one year to the next. Geographically, terrain exhibits vastly different temperatures over a period of time. Crossover in the desert may occur several times in one day and not recur for several consecutive days.

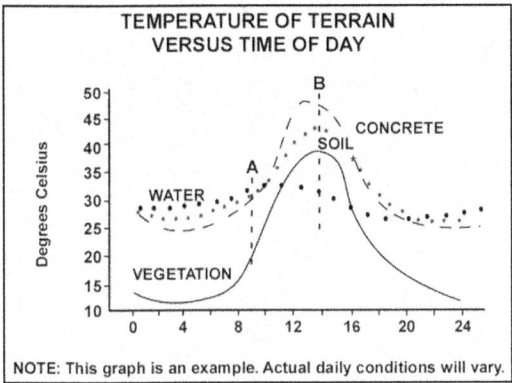

Figure 4-11. Infrared energy crossover

4-106. Predicting crossover is not an exact science. IR energy crossover has the greatest effect on FLIR operations when it occurs with haze, fog, or some other weather phenomenon and results in a poor image quality that cannot be predicted. Modern sensors have very low MRT, enabling them to detect very small temperature differences in a thermal scene. This significantly reduces the impact of IR crossover.

Lights

4-107. Lights visible to the unaided eye at night will not normally be visible through FLIR. Aviators can compensate for this by looking for lights with the unaided eye.

Depth Perception and Distance Estimation

4-108. The FLIR system greatly affects depth perception and distance estimation. To help overcome the loss of peripheral vision cues and the two-dimensional image, flight information is symbolically superimposed on the FLIR image. An aviator must rely on flight symbology and monocular cues for accurate depth perception and distance estimation.

Color Discrimination

4-109. Color discrimination of objects is absent due to the operational properties of FLIR. Color is based on energy which falls into the visible light spectrum. FLIR images are produced by detecting IR energy radiating from objects and do not require visible light. FLIR displays are monochromatic and shading is used to display different levels of detected energy. The unaided eye will be able to distinguish the color of lights that are bright enough for photopic vision.

Chapter 4

Parallax Effect

4-110. This occurs in a MPNVS due to the relative distance between the FLIR sensor and the HDU. The FLIR sensor is contained within the MPNVS turret located on the nose of the aircraft, while the HDU is positioned in front of the aviator's eye. The MPNVS turret is located approximately 10 feet forward and 3 feet below the aviator's design-eye position. In both crewmember positions, the thermal scene viewed on the HDU is obtained from the physical perspective point of the FLIR sensor. The aviator flying with the MPNVS views with the FLIR sensor, not with his or her unaided eye. Attempts to correlate the thermal scene viewed through the HDU with the actual scene viewed using the unaided eye can result in an apparent difference in the location of objects within the scene.

4-111. In figure 4-12, the aviator has turned his or her head 90 degrees to the right, and the MPNVS turret is pointed 90 degrees to the right. An object (tree) is located at A in the illustration. The FLIR sensor views the tree in the center of the FOV along the line-of-sight (LOS); however, the aviator's unaided eye would not see the tree in the center of his or her FOV, rather slightly left of his or her FOV. The parallax effect increases with the turret offset angle and the relative closeness of obstacles to the aircraft. The aviator must relate his or her view of the MPNVS scene between the origin of the image (MPNVS turret) and his or her seating position.

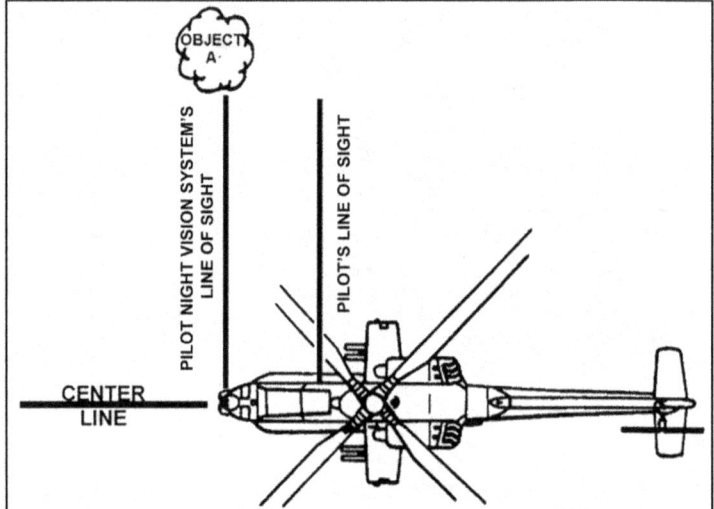

Figure 4-12. Parallax effect

Binocular Rivalry

4-112. Binocular rivalry describes the competition between the MPNVS aided eye and an unaided eye, while an aviator is flying with monocular-equipped MPNVS. This rivalry can be described as an undirected attention shift of an aviator's desired visual reference point (HDU display) to an undesired point or scene, or vice versa. The frequency and length of these occurrences depend on several variables including HDU luminance, ambient scene luminance, HDU scene complexity, ambient scene complexity, and to some degree, eye dominance at early stages in training. Aviators are accustomed to using both eyes while performing flight duties. The MPNVS monocular display is positioned in front of only one eye (aided eye), leaving the other eye unaided. Difficulty arises when an aviator is forced to manage the direction of both eyes, while maintaining a high degree of concentration with the MPNVS. This is coupled with the need to absorb any information observed by the unaided eye such as caution lights or a flare outside the cockpit. The goal is to prevent an uncommanded shift by either eye. To control or prevent binocular rivalry, an aviator should select, through experimentation, one of three cockpit lighting configurations—floodlights bright, dim, or off. During night flights, external light interferences are commonplace, so the aircrew should plan its flights to eliminate disturbances from known light sources. A high degree of concentration is required when managing visual perception with MPNVS. Even experienced aviators are susceptible to binocular rivalry. Adequate crew rest aids in overcoming this problem. Using a tinted visor periodically also assists in reducing visual distractions. If all else fails, an aviator may have to close one eye until the rivalry subsides.

Scanning Techniques

4-113. Proper scanning techniques are essential during FLIR night flights. To overcome limited FOV (30 degrees by 40 degrees) and loss of peripheral cues, an aviator must use a continual scanning pattern during terrain flight. With a moving aircraft, the scan is performed by looking left and right of the aircraft centerline while maintaining reference with the symbolic head tracker. An aviator bases the length of time and frequency of the scanning pattern on terrain type, obstacles, airspeed, and scene content quality of the MPNVS. Aviators use close cues to determine obstacle clearance altitude, airspeed, and closure rates, as well as midrange and far cues to evaluate route trends and patterns such as direction, turns, and obstacles. Avoid overflying close-in cues, maintain obstacle clearance, and use aided and unaided eyes to detect and avoid obstacles.

4-114. An aviator can overcome the inability of FLIR to see lights by incorporating the unaided eye into the scanning pattern. While scanning, periodically changing FLIR sensor polarity assists in distinguishing obstacles such as aircraft, tree branches, and power poles. These may be difficult to detect because of direct current (DC) restoration. DC restoration (horizon blooming) washes out the upper portion of the video image during rolling maneuvers when the bank angle is increased.

4-115. Finally, an aviator needs practice and experience to obtain maximum visual information from both the aided and unaided eyes. An aviator must learn to correctly interpret and use flight symbology for aircraft control. Reliance on imagery alone or dependence on unaided vision is not enough and will result in erratic aircraft control. Unaided peripheral vision in the aided eye may be distracting until an aviator learns to use primarily FLIR cues and symbology, and disregard unwanted peripheral distractions.

Spatial Disorientation

4-116. Aviators avoid maneuvers requiring large bank angles or rapid attitude changes. These maneuvers tend to induce spatial disorientation. An aviator flying with MPNVS may become disoriented and experience an unusual attitude when he or she has visual reference with the surface of the earth with the FLIR sensor. This also occurs when reference is lost due to FLIR image degradation or sensor failure. Proper scanning techniques–using a slow, purposeful head movement and positive aircraft control with the proper symbology mode–aids in preventing spatial disorientation. Adequate crew coordination should be preplanned and prebriefed to assist aviators in recovering from spatial disorientation. Regardless of the symbology mode being employed at the time of disorientation–hover, bob-up, transition, or cruise–the initial recovery steps are the same. The crew orients the MPNVS turret toward the aircraft nose and minimizes head movement during the recovery. The head tracker should be located and cross-checked with the LOS reticle. The crew can use the remaining flight symbology to complete reorientation and recovery.

Airspeed and Ground-Speed Limitations

4-117. Aviators using FLIR tend to overfly their capability to see. To avoid obstacles, aviators must understand the relationship among the system's visual range, atmospheric conditions, and airspeed. With the limited visual range of FLIR, aviators must exercise extreme care when using the systems during terrain flight modes. With poor atmospheric conditions and subsequent poor thermal resolution, a reduction in ground speed may be appropriate. Object acquisition and identification are related to atmospheric conditions and thermal contrast. The variables affecting the ability to see with FLIR include—

- IR crossover.
- FLIR sensor optimization.
- MRT.
- Aviator's proficiency and capabilities.

Chapter 4

- Humidity.
- Obscurations such as, dust, smoke, or haze.

Weather

4-118. FLIR systems have the ability to see through most obscurations; however aviators flying with MPNVS should be aware they may fail to detect IMC. If the aircraft has entered IMC, an aviator must be careful not to overfly FLIR capabilities and to continue slowing as conditions degrade. Increasing graininess and reduction in scene quality indicate deteriorating weather such as a denser obscuration or equalizing temperature in the viewed scene. When an aviator recognizes this restricted visibility, reoptimizing the FLIR is attempted using level and gain controls. If this does not work, an aviator must turn away from the weather conditions, land, or execute IIMC recovery procedures.

Target Detection

4-119. Detection of targets at night using the FLIR system is fairly easy; however, identification of those targets is often difficult. To help alleviate this situation, the pilot can assist the copilot gunner in detection by using MPNVS. The primary duty, however, is to fly the aircraft. Because the MPNVS has no magnification capability, the maximum range an aviator can detect during optimum conditions is 1,500 to 2,000 meters. The MTADS FLIR is the primary night acquisition source for the AH-64D/E. It is a passive night viewing device with four different FOVs. The copilot gunner's ability to optimize and operate the MTADS FLIR directly influences the capability to detect targets. The black/white hot feature allows pilots to change the way the temperature differences are displayed. This feature can enhance target detection and target recognition. The white-hot setting displays a thermal image of the scene in which hot objects are displayed with positive contrast, so appear bright. In white hot, hot targets such as tanks are generally more apparent than in black hot because the bright end of the gray scale is usually more noticeable than the dark end. Since targets are typically warmer than their background, white-hot produces bright targets that can generally be detected at greater ranges. Black hot tends to produce a more natural scene for pilotage and terrain recognition purposes (figure 4-13).

4-120. The use of "white hot" polarity can normally best be optimized for target detection (figure 4-13).

White Hot	Black Hot
Target Acquisition	Pilotage and Terrain Recognition

Figure 4-13. Polarity

Weapons

4-121. During rocket firing, the motor burn from the rocket illuminates the cockpit area letting the aviator see some sparkling effect to the front of the aircraft. Other than this momentary distraction to the unaided eye, the crew should not experience any adverse effects. When the aircrew member fires the 30-millimeter cannon, the muzzle blast may distract the unaided eye if the gun is fired off axis. Crew coordination and communication can minimize this temporary distraction. While firing Hellfire missiles, the crew will experience a temporary illumination of the cockpit area similar to rocket firings. This temporary distraction from the flight motor of the missile will not affect either crewmember's aided eye, which is already adapted to photopic vision.

SECTION V – NIGHT OPERATIONS

4-122. Flight operations at night use many of the same techniques as day flight; however night flight is inherently more dangerous due to visual limitations which affect mission planning and execution for aided and unaided flight.

PREMISSION PLANNING

MISSION BRIEFING AND DEBRIEFING

4-123. Aircrew mission briefings are conducted according to AR 95-1, appropriate regulations and directives, and the unit SOP. All missions are briefed. At the end of a mission, a thorough debriefing should be conducted and any post mission debriefing forms completed. The debriefing should include any problems, issues, recommendations, and lessons learned with a plan to notify necessary personnel.

COMMON TERMINOLOGY

4-124. Common terminology must be established among aircrew members and any other participants. Each aircraft ATM should identify standard terms used by aircrews during flight. Common terminology needs to be specific in its meaning to prevent confusion.

PREFLIGHT INSPECTION

4-125. Aircraft preflight inspection is a critical aspect of mission safety. It must comply with the appropriate aircraft operator's manual. Preflight should be scheduled as early as possible in the mission planning sequence, preferably during daylight hours, allowing time for maintenance assistance and correction. If a night preflight is necessary, a flashlight with an unfiltered lens should be used to supplement lighting. Oil and hydraulic fluid levels and leaks are difficult to detect with blue-green or red lens. Windscreens are checked ensuring they are clean and relatively free of scratches. Slight scratches are acceptable for day but may not be for night flight. The searchlight or landing light should be positioned for the best possible illumination during an emergency descent.

AIRCRAFT LIGHTING

4-126. The use of aircraft lights should be standardized to reduce adverse effects on night vision. AR 95-1, ATM, aircraft operator's manual, and the unit SOP will help define the standardization.

Cockpit Lights

4-127. During before-starting checks, cockpit lights are adjusted to the lowest usable intensity level. For aided night flight, aircraft interiors must be tailored according to modification work orders. Interior lighting, supplemental lighting, or flashlights assist in illuminating the cockpit and cabin area. If a particular light is too bright or causes reflection, it is modified or turned off. As ambient level decreases from twilight to darkness, intensity of the cockpit lights is reduced to a low, usable intensity level reducing any glare or reflection off the windscreen. A flashlight, with appropriate lens filter, or map light can supplement the available light in the cockpit. If an existing map/utility light is used, it should be hand-held or remounted to a convenient location. A separate flashlight, not the aircraft map/utility light, is required according to AR 95-1.

Chapter 4

> **CAUTION**
> During tactical operations at night, cockpit lighting should be adjusted to the absolute lowest usable levels and crewmembers should be discrete in the use of supplemental lights to avoid detection by enemy forces.

Anticollision Lights

4-128. In formation flight, anti-collision lights are turned off with the exception of trail aircraft. Operation of anti-collision lights can be a major distraction to succeeding aircraft within the flight and may hamper safe operation. Anti-collision lights are used according to AR 95-1, FAA directives, host country/theater directives, and appropriate SOP guidance. In addition, with installation of two anti-collision lights on some aircraft, the bottom light may be turned off eliminating vision restriction when conducting night or NVG operations.

Landing Light or Searchlight

4-129. Use of landing lights or searchlights is determined by factors such as crewmember experience and ambient light conditions as directed by ATC. Aviators who constantly rely on it might not develop techniques to fly without it; however, a crew striving to never use it may put the aircraft at risk. The landing light must be used with discretion and due consideration for other aircraft and safety. The use of the landing light may reduce the ability to see under certain atmospheric conditions such as fog or blowing snow. Each situation must be evaluated separately. During tactical operations, the landing light is only used to prevent a hazardous situation from developing, with due consideration of enemy threat. The unfiltered landing light can be used with NVG under emergency/administrative conditions, but aircrews must direct their scan and the light to prevent dimming the NVG and reducing their effectiveness. There are different types of bulbs available for use in the IR searchlight, chose the best bulb for the conditions and requirements of the mission.

Position and Navigation Lights

4-130. Inappropriate use of position and navigation lights can degrade night vision and increase the possibility of detection by an existing threat. Aircraft in formation flight can be distracted by position and navigation lights, thereby hampering safe operation. During formation flight, with the exception of trail aircraft, position or navigation lights should be dimmed or turned off according to AR 95-1, FAA directives, host nation/theater directives, and appropriate SOP guidance.

Supplemental Cockpit Lighting

4-131. Supplemental cockpit lighting is any light device not part of the aircraft lighting system. Examples include finger lights, lip lights, flashlights, and chemical light sticks. Light sources must be compatible with NVG, and checked according to current directives for compatibility with NVG.

4-132. The general procedure to conduct a light degradation check is as follows:
- At night, in an aircraft located in an area of low ambient light, with interior lighting set for NVG operations, and with ANVIS prepared for use, position a reflective material (map sheet, note card, vinyl checklist) approximately 12 to 18 inches from the eyes.
- Shine the supplemental light onto the material.
- With the unaided eye, look at the resultant reflection cast on the windscreen.
- Observe this same reflection through the ANVIS. An acceptable supplemental light source will allow NVD aided vision through the reflection. The reflection may even disappear. If the reflection, glare, or stray light interferes with the ANVIS aided vision of any crewmember the light source is unacceptable.

AIRCREW PREPARATION

4-133. Preparation of aircraft and ground facilities before a night flight contributes to mission success; however, crewmembers must be physically and mentally prepared to participate in the flight.

NIGHT FLIGHT TECHNIQUES

HOVER

4-134. Aviators may have difficulty hovering at night as visual ground references are not easily seen or identified. The surface type surrounding a hovering helicopter affects an aviator's ability to judge movement. The technique used varies with surface type and any available lighting.

4-135. It is impossible to overstate the importance of scanning during night flight, whether unaided or using NVG. With peripheral vision reduced or eliminated, it can be very difficult to detect aircraft drift. During night flight, aviators use motion parallax to detect drift. Pick an object to the front of the aircraft that is relatively close. This can be a bush, a tree branch of an unusual shape, or some other easily identifiable item. Also pick an object at a greater distance, such as a tower, a light on the horizon or a far tree. If the helicopter is stationary, these objects will not move in relation to each other. Remember that this only shows helicopter movement in two axes. It is important to follow the same procedure 30-60 degrees offset from the initial references, such as through a side window, to identify helicopter movement. It is helpful to select one set of references somewhat higher than the other to assist in detecting climbs or descents. The aviator must change his field of view every 3-6 seconds, looking forward, then off to one side, then forward again (figure 4-14 and figure 4-15). Use of HUD symbology will greatly help with detecting drift.

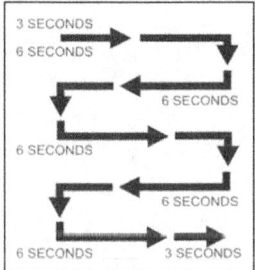

Figure 4-14. Stop-turn scanning

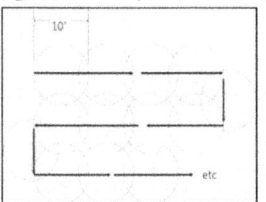

Figure 4-15. Scanning with ten degree circular overlap

Asphalt or Concrete

4-136. Estimating hover height over asphalt or concrete is difficult due to lack of visual cues. An aviator can use markings, such as taxiway lines or centerlines, to provide reference points. These surfaces lack contrast; however, a distinct contrast exists where a hard surface adjoins a soft surface. An aviator must continually scan to maximize detection of movement and avoid fixation.

Grass

4-137. Finding reference for precise hovering over grassy surfaces is difficult due to the lack of contrast and absence of visual reference points. Tall grass worsens the effect, making it more difficult to hover precisely creating an illusion of movement that may exaggerate or contradict actual aircraft movement. An aviator also tends to hover higher than normal or is necessary.

Snow or Dust

4-138. These elements present a very difficult surface over which to hover. Chapter 3 contains further information about these conditions.

Water

4-139. Water is the most difficult surface over which to hover as it is nearly absent of visual reference points. If possible, the aircraft should be maneuvered near some object, such as a tree stump, Chemlight, or buoy, to provide a reference point. If waves are present, the aviator tends to move laterally with the waves. Accurate height estimation requires use of a radar altimeter when hovering over water. Some operator's manuals contain directives mandating use of such equipment.

Lighting Types and Effects

Position Lights (Aided and Unaided Flight)

4-140. When hovering unaided with the help of position lights, aviators tend to stare at a single reference point on the ground. Reference points should be selected to the front and side of the helicopter to assist in scanning and detecting aircraft drift or movement. When hovering with position lights on "dim," there is a tendency to hover too low, especially with fewer visual reference points. To assist aircraft control, continue to scan and use all available information such as taxiway lights or shadows. Position lights also assist the aircrew when using NVG, keeping in mind the effects of red and green lights on NVG.

Landing Light or Searchlight (Aided or Unaided Flight)

4-141. When hovering unaided with the help of either of these lights, aircraft movement is easily detected, but the tactical situation is compromised. The position of the landing light can be critical to the aircrew's night adaptation. If the light is viewed directly, a dark adaptation period will again be required for the aircrew. If the light is positioned to provide adequate lighting without viewing the beam directly, dark adaptation can be partially preserved. When hovering at night, references are generally limited to the area illuminated by light.

TAKEOFF

4-142. If enough illumination is available to view obstacles, the aviator can accomplish takeoff as a day VMC takeoff. Figure 4-16, page 4-31, illustrates a night VMC takeoff. If illumination is insufficient, the aviator should make an altitude-over-airspeed takeoff until the aircraft reaches an altitude that clears obstacles. Takeoff may be performed from a hover or from the ground. The aircrew should treat visual obstacles, such as shadows, the same as physical obstacles. If the aviator applies more than hover power for the takeoff, that power setting should be maintained until about 10 knots before reaching the desired climb airspeed. At that point, the aviator adjusts power establishing the desired rate of climb and airspeed. The aviator not on the controls crosschecks the instruments. The lack of visual references during takeoff and throughout the climb may make maintaining the desired ground track difficult. Using the known surface wind direction and velocity assists in maintaining the ground track. Whenever possible, the takeoff heading should be in the direction of the first leg on the flight route as this helps in initial orientation, especially during low illumination. If the landing light is used during takeoff to detect obstacles, the illuminated area increases in size as altitude increases. As soon as possible, the landing light is extinguished to aid vision. When the landing light is turned off, the aircrew can expect some reduction in night vision. Takeoffs in severe dust or snow conditions are extremely hazardous as ground references will likely be obscured. The aviator performs an ITO until clear of the obscuration.

Figure 4-16. Night visual meteorological conditions takeoff

EN ROUTE

Unaided

4-143. After reaching the desired flight altitude, aviators allow time to adjust to flight conditions. This includes readjustment of instrument lights and orientation to outside references. During the adjustment period, the aircrew's night vision continues to improve until optimum night adaptation is achieved.

Aided

4-144. The viewing distance increases with altitude. However, depth perception and visual acuity decrease significantly at higher flight altitudes.

Overwater

4-145. Long flights over water without a visible horizon should be avoided without a radar altimeter. Before flying over water, check the barometric and radar altimeters for proper operation. Aviators should set the radar altimeter low altitude indicator to the minimum acceptable altitudes. The aviator not on the controls should maintain a cross-check of the flight instruments to prevent inadvertently flying into the water. Trail aircraft monitors and advises the flight if any aircraft appears to be going below the briefed altitude. The lower the visibility or ambient light, the higher the en route altitude should be over water.

LANDING

4-146. With reduced visual capability at night, night LZs should be larger than day LZs. Night LZs should be relatively clear of obstacles on approach and takeoff paths. During an approach without aircraft lights, aviators should observe the contrast between the dark trees and the lighter open area as this aids in identification of obstacles along the LZ boundary. Forward and lateral limits of the open area appear darker when contrasted with the open area.

4-147. Altitude, apparent ground speed, and rate of closure are difficult to estimate at night. Throughout an approach, other crewmembers provide information to the aviator on obstacle avoidance, altitude, airspeed, and approach angle. Maintaining a thorough scan, including the side windows, aids in estimating such information as the rate of closure. If approach is made to tactical lights, lateral movement can be detected by the relative position of the aircraft and lights. Except in blowing snow or dust, night approaches to an unlighted area should be terminated at a hover and followed by a slow vertical descent to the ground.

4-148. The approach can be made to the ground or terminated at a hover. Approaches to the ground require the most skill and proficiency. Field LZ approaches are normally planned to terminate at a hover because it is difficult to determine the surface condition. However, if the LZ surface can be adequately assessed during approach, an aviator may continue to the ground. Each approach must be separately evaluated. As the aircraft nears the ground, it is difficult to predict when ground contact will be made. To avoid slowing the vertical descent excessively and over-controlling the aircraft while waiting for touchdown, an aviator should reduce collective gradually and continuously.

4-149. An aviator executing a night landing to a field LZ must consider all aspects of the approach. The touchdown point should be selected before reaching the entry point on approach. When landing on a runway or taxiway, an aviator should select a specific group of lights or point on the runway while on the downwind leg or base leg or, if on a straight-in landing, as soon as the area is in sight. This timely selection aids the aircrew in determining entry point, approach angle, rate of descent, and rate of closure. The apparent ground speed and rate of closure are difficult to judge during night operations. The last portion of the night approach should be slower than during the day to avoid abrupt attitude changes at low altitudes and slow airspeeds. To avoid reducing airspeed too soon or too high, the aviator should maintain an instrument cross-check ensuring all indicators are within parameters. Abrupt recovery from slow airspeeds may result in rapid loss of altitude when forward cyclic is applied. Coordinated control movement of both cyclic and collective is required to fly the helicopter throughout the approach.

Ground Lighting Aids

4-150. A field lighting system provides fewer visual cues than a lighting system for a fixed landing site. Approaches to a field LZ normally are made without a landing light. The type and arrangement of lighting may vary considerably. Regardless of the lighting device, at least two lights should be used, separated by at least 15 feet, to identify the touchdown point. An illusion of movement (autokinesis) may occur when a single light source is viewed. When more than two lights are used to mark the LZ, spacing between the lights can be reduced.

4-151. Two tactical field lighting configurations are used as landing aids for aircrews—the inverted Y and the T. When operating with NATO aviation forces, aircrews should anticipate use of the T.

Inverted Y

4-152. The inverted Y system is best used for an approach initiated from terrain flight altitudes. Figure 4-17, page 4-33 shows light cues for six different approach alignments. Part A of this figure depicts the proper setup of the light system. Before the aircraft reaches the entry point for approach, lights in the stem will appear as a single light. This sight picture will also indicate the helicopter is on approach and below the desired approach angle (neither of these situations is depicted in figure 4-17, page 4-33). When the normal approach angle is maintained, the Y appears normal (part B). If the distance between the lights appears to increase, approach is too steep and the helicopter is above the desired approach angle (part C). If the distance between the lights appears to decrease, approach is too shallow and the helicopter is below the desired approach angle (part D). If the spacing between the front lights is uneven and the stem is shifted right of the centerline, the aircraft is too far right and should correct left (part E). If the spacing between the front lights is uneven and the stem is shifted left of the centerline, the aircraft is too far left and should correct right (part F). The desired touchdown point is inside the Y with the fuselage aligned with the stem lights. During the last 25 feet of the approach to a Y, aviators should divert their FOV away from the lights and concentrate on acquiring ground references.

Rotary-Wing Night Flight

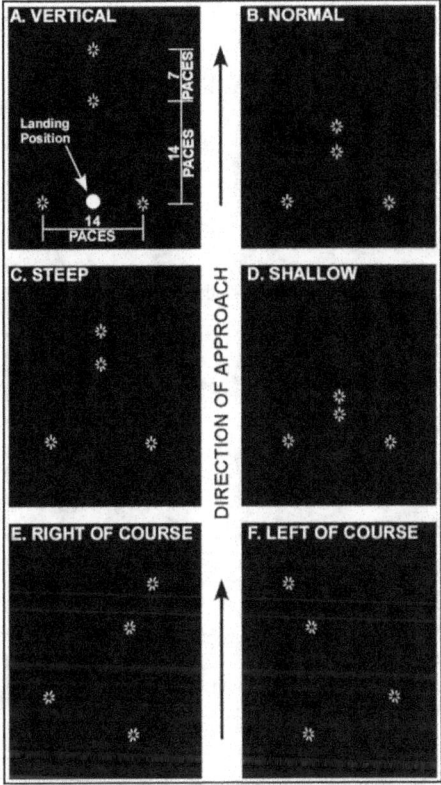

Figure 4-17. Approach to a lighted inverted Y

T Shape

4-153. The T, while seldom used by U.S. forces, may be encountered when working with allied forces. The T is best used for approaches initiated from an altitude above 500 feet AGL. Figure 4-18, page 4-34, shows light cues for six different approach alignments. Part A depicts the proper setup of the light system. The apparent distance between the lights in the stem of the T can be used as a reference for maintaining a constant approach angle. A change in the spacing of the lights will occur as the approach angle changes. Before the aircraft reaches the entry point to begin approach, the lights in the stem will appear as a single light. This sight picture may also indicate the aircraft is below the desired approach angle (neither of these situations is depicted in figure 4-18). After an approach angle is intercepted, the stem of the T appears similar to part B. If the distance between the lights appears to increase, the approach angle is becoming too steep and the helicopter is above the desired angle of descent (part C). If the distance between the lights appears to decrease, approach is becoming too shallow and the helicopter is below the desired approach angle (part D). If the stem of the T points left of the helicopter, the aircraft is too far right of the course and should correct left (part E). If the stem points right of the helicopter, the aircraft is too far left of the course and should correct right (part F). During the last 25 feet of approach to a T, aviators should divert their FOV away from the lights and concentrate on acquiring ground references.

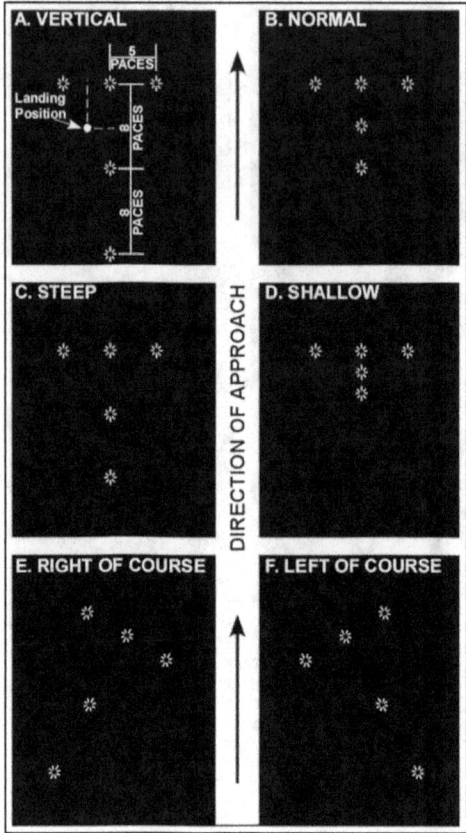

Figure 4-18. Approach to a lighted T

LIMITATIONS

4-154. Visual reference outside the aircraft is limited at night. Movement of the helicopter is difficult to detect as night terrain features often blend into one solid background. Hover altitude and ground track are also difficult to estimate. The degree of difficulty depends on ambient light level and aircraft altitude. This is also true when flying with NVG, although to a lesser degree.

4-155. Equipment, instruments, and control switches are easily located in a lighted cockpit; however, aviators should be able to locate and use cockpit equipment associated with immediate action emergency procedures without cockpit lighting ensuring the proper control switch can be identified. The use of artificial lighting, flashlight, map light, NVG supplemental lighting, or even chemical lights is recommended, time permitting. The location of items such as radios, mission equipment, and switches, must be standardized to ease this process and provide a consistent reference base.

4-156. Visual references providing positive identification during the day may be difficult to see at night. Objects that illuminate, such as airport beacons and towers with obstacle lights, are prominent night visual

NAVAIDs which are used with discretion. Visual flight is more demanding over sparsely inhabited areas with few ground lights. As the altitude AGL increases, visual references are less effective and aircrew members must rely more on instruments as a primary aid. Reduced visual references may cause crewmembers to focus on a single light or a group of lights in a concentrated area. This can induce illusions, most notably autokinesis. Even with NVG, aircrew members will find it difficult to estimate distance or altitude based on visual references.

4-157. When completely night adapted (if flying unaided), the eyes become extremely sensitive to light; light exposure causes a partial or complete, but temporary, loss of night vision. Aircrew members should avoid exposure to light sources, both outside and inside the aircraft, especially looking directly at or into a high-intensity light. This same type of exposure during aided flight may have similar effects especially when viewing unaided around the periphery of NVG.

4-158. Night flight is inherently more stressful than day flight. Therefore, crew mental and physical fatigue experienced while flying occurs sooner during night flight. Deteriorating performance and efficiency causes poor coordination and slowed reaction time while reducing the ability to see. When designing or using a training program, individual crewmember experience levels must be considered. AR 95-1 provides guidance regarding crew rest or fighter management.

EMERGENCY AND SAFETY PROCEDURES

BASIC CONSIDERATIONS

4-159. Emergency procedures for day and night flight are the same. Responding to an emergency situation usually takes longer at night. To minimize time delays in executing emergency procedures at night, the aviator must be familiar with the location of all controls and switches and know all immediate action emergency steps from the appropriate operator's manual. Established safety procedures prevent emergencies.

ELECTRICAL FAILURE

4-160. If a total or partial electrical failure occurs, the aircrew must execute appropriate emergency procedures for the aircraft being flown. The aircraft may be difficult for other aircrews to see, so the aircrew must avoid other aircraft, such as those in the same traffic pattern or formation. When on final approach (if at an airfield), the aircrew must decide if approach can be continued without creating an unsafe condition for other aircraft. During approach, the aircrew should watch the tower for light gun signals. The Department of Defense (DOD) flight information publication (FLIP), Flight Information Handbook, or the aeronautical information manual (AIM) contain information on ATC light signals.

EMERGENCY LANDING

With or Without Power

4-161. With power, descent to a lower altitude may aid the aircrew in locating and identifying a suitable landing area. Depending in part on atmospheric conditions, when landing with or without power the aircrew may turn on the landing light to assist in locating an LZ or identifying obstacles. An aviator must exercise caution when fog, haze, or other obscurations are present as the landing light tends to degrade night vision. Before landing, an aviator should attempt to advise the controlling agency of the situation and location. The crew should, depending on the enemy threat, remain with the helicopter after landing and identify its position by using appropriate signals.

GROUND SAFETY

4-162. During night operations, the number of support personnel, vehicles, and use of ground-handling equipment on the flight line should be limited to minimum essential for mission accomplishment. Aircrew members, support personnel, and other personnel should use lights when walking on the flight line to identify obstacles, locate nearby aircraft, and to be seen especially by taxiing aircraft. The lights should be equipped with an appropriate filter. During preflight inspection, aircrew members should pay particular attention to the structural components of the aircraft. Before moving on or around the aircraft, flashlights should be used to identify any obstacles or hazards. Serious injury can easily result if caution is not exercised or proper procedures are not used. Climbing onto the aircraft at night can be especially hazardous. Any surfaces should be checked for substances such as oil, hydraulic fluid, water, and frost.

AIR SAFETY

4-163. Aircrew members must be aware of the limitations of night vision and not overestimate their ability to perform duties. After initial qualification or refresher training, aircrew members must train continuously to remain proficient. If a long period has elapsed since the aviator's last aided or unaided night flight, a thorough review of basic tasks may be appropriate. Because flight attitude references are limited at night, visual illusions, disorientation, or vertigo may be induced. Aviators must be aware of these conditions and use flight instruments to assist in maintaining a normal flight attitude. When an unsafe condition develops while hovering, the aviator may turn on the landing light if environmental conditions permit. Continuous observation outside the helicopter is required so obstructions and other aircraft are avoided and SA is maintained. This is particularly true during multihelicopter and terrain flight operations.

AIRSPACE MANAGEMENT

4-164. Increased emphasis on night operations has resulted in a greater number of missions being flown at night. Aviators flying at night unaided have difficulty in detecting other aircraft, especially when those aircraft are "blacked out." Aviators wearing NVG also have difficulty in detecting other aircraft involved in different, unrelated missions. This problem is compounded during field training exercises when multiple aircraft or multihelicopter flights share the same airspace. The following help minimize the problem:

- Sound operational planning and operating procedures must be developed and practiced to prevent most airspace conflicts.
- The area commander should establish priorities and guidelines for airspace usage; these should be published in SOPs and operation orders (OPORDs) with subsequent changes disseminated as necessary.
- The commander or his or her designated representative should approve the use of airspace for preplanned operations.

Chapter 5

Rotary-Wing Terrain Flight

To survive and accomplish the mission, combat aviation units must use tactics that degrade an enemy's capability to detect aircraft. Darkness protects aircrews from visual and optical acquisition by the enemy. However, darkness will not protect aviation elements from electronic detection. Terrain flight is a tactic that uses terrain, vegetation, and manmade objects to mask aircraft from visual, optical, thermal, and electronic detection systems. This tactic involves a constant awareness of capabilities and positions of enemy weapon systems and detection means in relation to masking terrain features and flight routes. The most effective combination for detection avoidance is flying terrain flight altitudes at night. The ability to perform night terrain flight depends on ambient light level, flight proficiency, terrain familiarity, and effective use of various NVDs. This chapter provides a description of tactics and terrain flight planning considerations, flight techniques, training program guidance, and the environment in which aviation units will be required to operate.

SECTION I – TERRAIN FLIGHT OPERATIONS

MISSION PLANNING AND PREPARATION

5-1. Using elements of METT-TC is essential to safe and successful accomplishment of missions at terrain flight altitudes. As discussed earlier, various factors dictate most decisions made in terrain flight planning and preparation. Consistent with commander's intent, flight routes, LZs, PZs, and BPs will be determined and planned accordingly.

5-2. Contingency planning is also a critical element during this stage of the operation, including alternate flight routes, alternate LZs and PZs, and suspected enemy positions. The entire planning sequence must be a methodical and thorough effort, eliminating confusion and clarifying each step in the planned execution phase. This intensive level of preparation also better prepares each aircrew to react to changes, unexpected events, and emergencies. This planning phase must include appropriate personnel from the next lower level of operation ensuring adequate dissemination of information and mission accomplishment. Historically, Army aviation has witnessed many failures due to inappropriate exclusion of operations personnel and aircrews actually flying the mission from the planning process.

5-3. Another key element is rapid dissemination of information allowing maximum planning and familiarization time by aircrews, which also permits maximum time to brief the mission and addresses the body of questions and inquiries that inevitably result. There must be a sense of urgency in expediting flow of information to aircrews as quickly as possible. History reveals too many instances in which, to the detriment of mission accomplishment, critical information has been unnecessarily delayed at a higher operational level. Terrain flight planning and preparation also includes aircraft preparation ensuring aircraft are configured, preflighted, and readied for the ensuing mission. This is most effectively accomplished with a timely and continuous information flow from higher headquarters, such as through the battalion S-3, to lower units.

AVIATION MISSION PLANNING SYSTEM

5-4. The aviation mission planning system (AMPS) is an automated mission planning and battle synchronization tool designed specifically for aviation commanders. AMPS functions include tactical planning, mission management, and mission rehearsal capabilities. The tactical planning function includes all planning tasks performed, while the mission management function can be associated with actions taking place during mission execution. The system is also capable of mission briefing and rehearsal providing aircrews with the best possible preparation before mission execution.

5-5. The main element of the system's hardware is a lightweight computer unit (LCU) employed at each aviation unit headquarters. Brigade and battalion headquarters have two LCU systems each. Companies may also have one or more such units, depending on mission requirements. Additional peripherals include a CD-ROM drive, magneto-optical drive, data transfer receptacle (loads data transfer cartridges), and an uninterruptible power supply. The AMPS employs a menu-driven graphical user interface, allowing the operator to enter and view critical mission planning data. The AMPS is subordinate to the maneuver control system, with which it shares mission data and gains access to the joint common data base (JCDB). It also provides the means to generate mission data for use in either hard copy or electronic format.

5-6. Information generated on the AMPS can be distributed in electronic format to other systems, which rapidly reduces dissemination time and leaves aviation units with more time for mission planning and preparation. It also transfers mission data directly to aircraft by means of the data transfer system (DTS).

TERRAIN FLIGHT LIMITATIONS

5-7. Terrain flight imposes additional factors on aircrews and units not encountered on missions flown at higher altitudes. The following are considerations for missions at terrain flight altitudes:
- Mountainous or uneven terrain that restricts use of LOS radios, making it difficult or sometimes impossible to conduct normal communications.
- Aircrews should predict and plan limits on communications when operating near enemy forces.
- In terrain flight operations, control may be delegated to a lower level due to inherent problems. Aircrews and platoon, section, or team leaders must be knowledgeable enough to execute the mission using sound tactical judgment. This is a result of training and experience.

5-8. Such missions should be coordinated with higher headquarters ensuring appropriate airspace management and acquiring the latest intelligence updates. Even in a training scenario, the plan to conduct terrain flight operations must be disseminated ensuring safe use of the training area. The unit anticipates increased maintenance as a result of increased demands placed on aircraft and components.

5-9. Demands on aircrews increase dramatically when terrain flight operations increase, especially NVD terrain flight. Specifically, fighter management becomes a larger issue with an increase in psychological and physiological stress. The factors increasing stress include—
- Increased workloads (physical dexterity and mental processes).
- Limited FOV when using NVDs.
- Reduced visual acuity, viewing distances, and depth perception.
- More complex aircrew coordination.

Frequent training, physical fitness, thorough flight planning, and preparation can minimize these factors.

TERRAIN FLIGHT MODES

5-10. Terrain flight includes appropriate tactical application of low-level, contour, and NOE flight techniques (figure 5-1, page 5-3), as appropriate, diminishing the enemy's capability to acquire, track, and engage aircraft. For NVD training, terrain flight is conducted at 200 feet or less above the highest obstacle. Altitude and airspeed restrictions—for NVD flight training—are listed with the description of each mode. Terrain flight requires aircrew proficiency in map reading, preparation, and terrain interpretation. It also requires constant vigilance in identifying terrain features and hazards, and understanding effects of surrounding terrain, ambient light, and seasonal changes in vegetation. Continuous NOE or contour flight is unusual as terrain and vegetation vary. Normally, there is a transition from one mode to the other as the situation dictates. Modes of terrain flight are defined below.

Rotary-Wing Terrain Flight

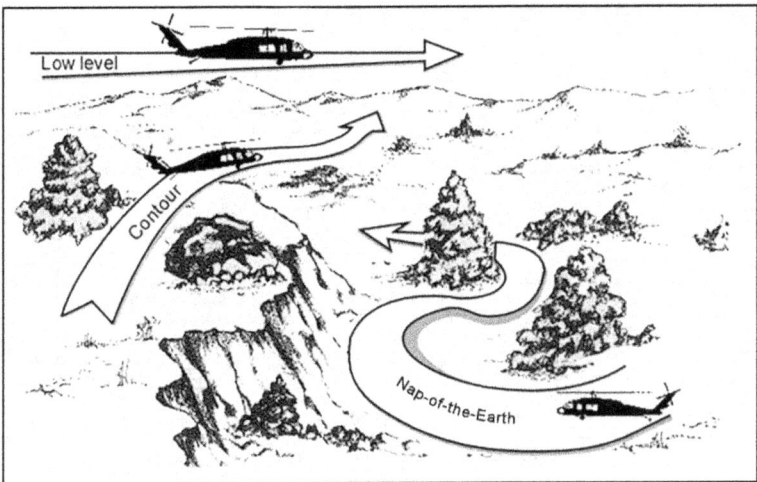

Figure 5-1. Modes of flight

> **WARNING**
>
> While unaided night flight at terrain flight altitudes is not prohibited, it is unwise and would usually fall into the extremely high risk category of the risk assessment process.

NAP-OF-THE EARTH FLIGHT

5-11. NOE flight is conducted at varying airspeeds as close to the earth's surface as vegetation and obstacles permit. For NVG training, NOE flight is further defined as operating with the skids or wheels up to 25 feet above trees and vegetation in the flight path. (For training, a safe airspeed is used based on ambient light, flight visibility, terrain, winds, turbulence, obstacles, and crew proficiency.) Aviators should decrease airspeed if weather and/or ambient light restrict visibility.

CONTOUR FLIGHT

5-12. Contour flight is conducted at low altitudes conforming to the earth's contours. It is characterized by relatively constant airspeeds and varying altitude as dictated by terrain and obstacles. For NVG training, contour flight is further defined as operating with the skids or wheels between 25 and 80 feet above highest obstacle (AHO). (For training, a safe airspeed is used based on ambient light, flight visibility, terrain, winds, turbulence, obstacles, and crew proficiency.) Aviators should decrease airspeed if weather and/or ambient light restrict visibility.

LOW-LEVEL FLIGHT

5-13. Aviators perform low-level flight at constant altitude and airspeed, dictated by threat avoidance. For NVG training, low-level flight is further defined as operating with the skids or wheels between 80 and 200 feet AHO at an airspeed commensurate with operational requirements and aircrew limitations. Aviators should decrease airspeed if weather and/or ambient light restrict visibility.

Chapter 5

SELECTION OF TERRAIN FLIGHT MODES

5-14. Aviators must determine which terrain flight mode to use in each segment of the planned route during the mission planning sequence. This determination is based on METT-TC (table 5-1).

Table 5-1. Mission, enemy, terrain and weather, troops and support available, time available, civil considerations and terrain flight modes

Mission	Influences selection of terrain flight techniques (especially if mission is performed at night). Factors such as light levels and moon illumination complicate night vision device (NVD) flight at terrain flight altitudes. The lack of visual acuity may demand a lower airspeed &/or higher altitude.
Enemy	Threat weapons can detect & engage aircraft at low altitudes. Select the appropriate terrain flight mode to avoid or minimize detection.
Terrain and Weather*	Vegetation and terrain features masking an aircraft from visual & electronic detection significantly degrade the capability of threat weapons to detect an aircraft. Determine a maximum safe flight altitude by availability of terrain features and vegetation. Use the highest terrain flight altitude for a specific condition. A higher flight altitude reduces difficulty in navigation, permits a higher airspeed, reduces hazards to terrain flight, and minimizes fatigue.
	Periods of deteriorating weather with low ceilings/restricted visibility may make all terrain flight modes extremely difficult or impossible. It also makes navigation more difficult & increases potential for (IIMC), especially when flying in formation or operating in an unfamiliar environment.
Troop	Factors, such as aircrew availability, experience level, effects of the fighter management program, and mission-oriented protective posture, may affect selection of terrain flight techniques.
Time	Influences selection of the terrain flight mode. Whenever possible, the route should be flown at the highest flight mode to permit the shortest completion time.
Civil Considerations	The selection of a particular mode must consider the safety of and potential threat from any civilian sector.

*See chapter 3 for more information

PICKUP ZONE/LANDING ZONE SELECTION

5-15. PZ/LZ selection is extremely important. Technical and tactical considerations must be analyzed ensuring the best choice for mission success is made. A poor LZ can jeopardize the entire mission.

PICKUP ZONE SELECTION

5-16. The first step in the loading plan is selection of a suitable PZ or PZs. Primary and alternate PZs should also be selected during this process. Multiple primary PZs may be necessary to facilitate a smooth flow of personnel and equipment. The mission may require separate PZs for troops and equipment (heavy and light PZs). The heavy PZ contains any external loads used for air assault, and the light PZ is where troops will be lifted from. Selection of PZs is based on METT-TC, commander's intent, location of assault forces in relation to the PZ, and size and capability of available PZs. PZ selection should be based on the considerations noted in table 5-2, page 5-5.

Table 5-2. Pickup zone selection considerations

Number	Multiple pick-up zones (PZs) may have an advantage over single PZs as they avoid concentrating the force in one location. Multiple PZ operations require detailed & precise planning by the supported & supporting units.
Size	Each PZ should accommodate all supporting aircraft at one time. Points to consider include— • Number and type aircraft. Minimum recommended landing point separation— CH-47: 80 meters UH-60/AH-64: 50 meters • Unit proficiency. • Nature of loads. • Climatic conditions. • Power management. • Day or night operations.
Obstacles	Plan for a 10 to 1 ratio for arrival and departure ends of PZ.
Location	PZs should be selected close to the troops being lifted (so they do not have to travel long distances) and accessible to vehicles moving support assets & infantry. However, locate PZs in an area limiting traffic from vehicles or personnel not directly involved. Mask PZs by terrain from enemy observation.

Conditions	Consider area surface conditions. Excessive slope, blowing dust or sand, blowing snow, & natural (tree stumps, rocks) and manmade (wires, foxholes) obstacles create potential hazards to PZ operations. Weather vulnerable, a perfect PZ could become unusable after a hard rain or fog from a nearby river. Other considerations are—
	• Blowing dust/sand/snow: increase separation between aircraft (as a general rule, by 50%).
	• Ground slope: Should be level terrain. As a guide:
	    Land upslope for 0 to 6 degree slope
	    Land side slope for 7 to 15 degree slope

Wind	Orient into the wind especially if aircraft are operating near maximum capacity or if the PZ is hazardous due to sand, dust, or snow.
Approach/ Departure Routes	Analyze terrain surrounding a possible PZ for air traffic patterns. In a tactical situation, avoid constantly approaching the PZ over the same ground track. Still, there are only so many ways to approach an area. Ideally, there should be an obstruction-free approach and exit path into the wind using the long axis. If required, mask routes from enemy detection.

LANDING ZONE SELECTION

5-17. Considerations for PZs apply to LZ selection. In coordination with the air mission commander (AMC) and liaison officer, the air assault task force commander (AATFC) selects primary and alternate LZs. The number and location of selected LZs is based upon ground scheme of maneuver and LZ availability. Aviation planners advise the AATFC on LZ suitability. Table 5-3, page 5-6, provides additional considerations for selecting a suitable LZ.

Table 5-3. Pickup zone selection considerations

Location	Locate the landing zone (LZ) in an area supporting the ground tactical plan of the air assault task force commander (AATFC). It may be located on the objective, close by, or at a distance. Consider mission, enemy, terrain and weather, troops and support available, time available, and civil considerations (METT-TC) factors when selecting LZs. Select LZs within range of supporting fires (artillery, close air support [CAS], naval gunfire) if required.
Capacity	LZ size determines how much combat power can be landed at one time. The selected LZ must be large enough to support the number of aircraft required by the AATFC. Squads must land intact in the LZ, and platoons must land in the same serial to ensure fighting unit integrity during air assault. This consideration also determines the need for additional LZs or separation between serials.
Alternates	An alternate LZ should be planned for each primary LZ to ensure flexibility.
Threat	The air mission commander (AMC) considers enemy troop concentrations, air defense artillery (ADA) locations, weapons ranges, and the enemy's ability to reposition ground forces to react to the air assault. LZ selection involves the air assault task force (AATF) staff operations officer (S-3), AMC, and staff intelligence officers (S-2s) from the AATF and aviation task force. S-2s provide intelligence affecting selection of LZs.
Obstacles	LZ selection must include existing as well as reinforcing obstacles on the LZ. Which side of the obstacles (away or same side as enemy) to use is determined by AATFC.
Identification	LZs should be easily identifiable from the air, if possible (more critical for the first lift).
Approach/ Departure Routes	Approach and departure flight routes should avoid continued flank and visual exposure of aircraft to the enemy.
Number	The decision to use a single or multiple LZs is based upon the ground tactical plan & AATFC intent. Advantages to using a single LZ are— • Make controlling operations easier. • Require less planning and rehearsal time. • Centralize any required resupply operations. • Concentrate supporting fires on one location. • Provide better security on subsequent lifts. • Amass more combat power in a single location. • Make detection of the air assault by enemy units more difficult as the air assault operation is confined to a smaller area of the battlefield and there are less flight routes Advantages for multiple LZs include the following: • Do not group the entire force in one location. • Force the enemy to fight in multiple directions. • Allow rapid dispersal of ground elements to accomplish tasks in separate areas. • Make determining size of the assault force difficult for the enemy. • Reduces troop/aircraft congestion.

ROUTE-PLANNING CONSIDERATIONS

CRITERIA

5-18. The route to and from the objective area must be tactically sound and conducive to successful navigation. Select routes with the final objective in mind. An aviator should base route selection, primarily, on the enemy tactical situation and, secondarily, on ease of navigation. Before route selection, an aviator should mark all known threat sites with weapons systems on the map. Criteria considerations using METT-TC are found in table 5-4, page 5-7.

Table 5-4. Route planning considerations

Mission
Supports the ground tactical plan. Does not hinder fire support plan. Avoids airspace control order (ACO) and special use airspace. Whenever routes coexist, avoid flying on or designating choke points; these are potentially midair collision points. When establishing ingress and egress routes, try to make separate routes or, at a minimum, establish different altitudes.
Enemy
Avoid being silhouetted when crossing ridge lines or by the moon on approach to objective areas. Cross major hydrographic features, major roads, and railroads at wide angles (90 degrees) reducing exposure time. Plan alternate routes in case the primary route is blocked due to weather or the enemy. Bypass known or anticipated enemy positions keeping a terrain mass or vegetation between the enemy and aircraft. When forced to plan a route near known or anticipated enemy positions, plan route at the edge of their weapons maximum effective range.
Terrain and Weather
Provide cover when terrain permits, placing terrain and/or vegetation between enemy and the aircraft. Negotiate large north-south valleys on the lighted side with respect to the position of the moon. This will avoid shadows cast by the moon and will silhouette terrain features for navigation. When the route direction is east to west (or west to east) in mountainous terrain, use narrow valleys or passes to cross north/south ridge lines so flight in shadows is avoided and terrain is generally silhouetted. Shadows do not aid in concealing aircraft but do make hazard identification and navigation more difficult. Avoid routes directly toward a low angle rising or setting moon. Alter the course, to include establishing a zigzag course, if no other options exist. Avoid paralleling linear features—roads or railways—associated with populated areas. Anticipate wires associated with all roads, towers, & buildings in open fields. When possible, avoid planning routes over large areas of low contrast such as large bodies of water, large fields, and snow-covered terrain. Evaluate potential weather problems in all areas (fog in river valleys, cloud covered ridge lines).
Troops
Extra troops ease navigation (day/night). For contingency planning, select air control points (ACPs) that can be used for day/night. Avoid planning route segments requiring heading changes of more than 60 degrees (especially critical during formation flight). Whenever possible, aviators plan en route altitudes at 200-500 feet above ground level (AGL) to reduce risk & avoid terrain flight hazards.
Time
Make routes as short as possible to amass fire power to allow greater on station time and flexibility for contingencies.
Civil
Avoid brightly lit areas and population centers. If not possible, reduce exposure time by maintaining cruise airspeed. Avoid navigational aids (NAVAIDs) and airports due to hazards associated with other aviation operations and to prevent detection by radar associated with these facilities.

AERIAL CHECKPOINT SELECTION

5-19. The two main types of aerial checkpoints are—
- **Air control point (ACP).** An air reference measure which is an easily identifiable point on the terrain or an electronic NAVAID used for navigation, mission command, and communication. ACPs are generally designated at each point where the flight route or air corridor makes a definite change in any direction and at any other point deemed necessary for timing or control of the operation.
- **Communications point (CP).** An air reference measure requiring serial leaders report either to the aviation mission commander or terminal control facility.

5-20. After selecting flight routes, an aviator selects ACPs using the following considerations:
- Select points controlling movement along the route, after determining general routing.
- Be detectable at a distance and not only visible when flying directly overhead.
- Contrast with surrounding terrain; for example, paved roads are poor choices in heavily vegetated terrain but are excellent in desert terrain. Another example is a small body of water, which provides little contrast in vegetated terrain but contrasts well in desert terrain.
- Avoid selecting points near towns that may have grown in size and can make detection difficult.
- Avoid points near bright lights.
- Avoid using manmade objects as primary points.
- Confirm selections with prominent adjacent features.
- Consider moon angle and effective illumination. Avoid selecting points within shadows cast by other features.
- ACPs should be 5 to 20 kilometers or nautical miles (NMs) apart. As a general rule, select ACPs 5 to 20 kilometers apart when utilizing map scales of 1:100,000 and below, and use NM when utilizing map scales of 1:250,000 and above. ACPs should be progressively closer as an aircraft nears the objective, facilitating timing and navigation. Type of terrain, illumination, total route distance, and accuracy of onboard navigation systems may allow selection of ACPs much further apart.
- Select prominent barriers near ACPs, particularly when planning significant turns. Use barriers to alert navigators an ACP has been overflown or bypassed and to cue for planned turns. As an ACP is passed, note actual time of arrival (ATA) and make necessary adjustment to the time and/or speed. A more difficult ACP with an excellent barrier is a better choice than a good ACP without a barrier.
- The start point (SP) and release point (RP) are important ACPs. Aviators use easily identifiable terrain features even if they must alter their route slightly. These points should be 3 to 8 kilometers or NM from the PZ/LZ to aid timing and navigation and should not involve significant turning. Avoid final legs between ACPs not having significant terrain features. The lack of significant terrain features precludes correct positioning and time management.
- Make note of MSL altitude of ground track and ACPs to aid in selecting an appropriate en route altitude.
- Select reference points between ACPs to ensure on-course navigation and time management. Use more reference points in low ambient light.
- The flight lead crew selects the final route and ACPs.

MAP SELECTION AND PREPARATION

5-21. While most of the following are techniques and suggestions for preparing maps used with NVDs, these same techniques apply to daytime operations.

SELECTION

5-22. Aviators assemble as many different types of maps as possible of the AO. Imagery is recommended, if available. Joint operations graphic (JOG) 1:250,000, tactical 1:100,000, 1:50,000 (or 25,000), and a

1:500,000 scale (tactical pilotage chart [TPC] or VFR sectional) are the primary maps used. When using a larger scale map, it generally requires an aircrew to fly at a higher altitude. For example, altitudes from surface to 200 feet AHO are easily navigated with a 1:50,000 map, while a 1:100,000 map works best from 200 feet AHO to 1,000 feet AHO.

5-23. The JOG should be the primary map for planning and flying the en route portion of a mission. The map scale covers a large area which permits a relatively small map uncluttered with extraneous information. It has latitude/longitude and universal transverse Mercator (UTM) features and is NVD compatible when properly prepared. The VFR sectional/TPC may be more appropriate for long-range navigation to the target area.

5-24. The tactical map should be used to accurately locate and confirm unique map features for transfer to the JOG. It displays more detail in areas absent or difficult to interpret on the JOG. Because en route landing and holding areas can be accurately plotted and studied on this map, aviators should use it during objective phases of operations. An aviator uses the tactical map for all flights flown at tactical altitudes and operations not less than 5 NM from the objective. Use caution when making the transition to different scale maps in flight as aircraft movement relative to the map scale may change radically. Aviators update tactical maps by using the chart update manual (CHUM) and current VFR sectional.

5-25. Aviators should consult the U.S. VFR sectional map which provides accurate information on major towers, airports, beacons, power lines, and magnetic variation. It is updated frequently and includes military training routes allowing aviators to bypass them.

PREPARATION

5-26. The following techniques prepare aviators to read a map in a near-dark cockpit with minimal lighting and simplify the task of map reading.

> **CAUTION**
> Maps marked with classified information become classified and must be handled and stored according to security regulations.

- Use permanent ink pens or markers only. When preparing maps by hand, use fine or medium black markers for routes, ACPs, and time distance heading data. If printing maps, utilize a line color that contrasts with the background. Use red markers for hazards. Iridescent fine red/orange hue markers may be substituted to highlight wires and towers. An iridescent yellow marker may be used to highlight hydrographic features. Do not use blue markers as they cannot be seen under the blue filters used in the cockpit. Recommended map preparation are—
    - Routes will be marked on the map with a solid line.
    - Corridor boundaries of a route will be marked on the map with alternating dashes and periods (- . - . - . -).
    - Alternate routes will be marked with dashed lines (- - - - - - -).
    - NOE routes will be marked on the maps with periods (. . . . . .).
- The map symbols used should include those indicated within this text, ADRP 1-02, JP 1-02, and designated by unit SOP. The symbology for common features—such as railroads and power lines—should replicate the legend information available on the map sheet or exaggerate existing information printed on the map itself such as bodies of water. The concern is clarity, simplicity, and immediate comprehension by any crewmember. Figure 5-2, page 5-10, depicts some typical route planning map symbols.

Chapter 5

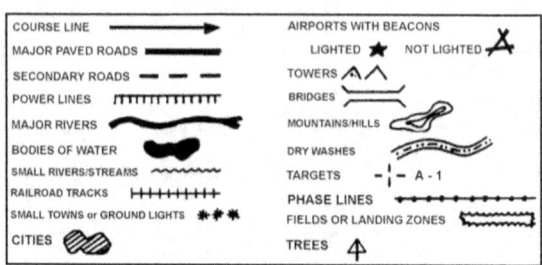

Figure 5-2. Route planning map symbols

- Do not over exaggerate map features.
- Orient notes and writing to direction of flight.
- Do not over prepare maps. Aviators highlight those features they expect to see and ensure they do not miss something important. Heavy vegetation or snow may prevent dirt roads, trails, and creeks from being seen. Highlighting these clutters the map. In a desert environment, however, if something shows on the map, it must be marked due to the lack of other cues in the desert.
- Place airspace control orders (ACOs), threat sites, and weapon system overlays on the map first. Then select route and ACPs.
- Place a large north (N) symbol on each fold of the map for rapid orientation in flight.
- Post all hazards not less than 10 NM on either side of the course line for safety during intentional or unintentional deviation.
- Highlight significant light sources, such as beacons and cities, out to a distance of at least 15 NM. Map sheets should not be trimmed until information, such as hazards and light sources, is posted.
- Transfer key features and hazards from VFR sectionals, tactical maps, and CHUMS to the maps, as necessary.
- Identify ACPs with a circle centered on a dot placed on the feature. Name/number the point and post planned arrival time to the side of the circle oriented in the direction of arrival.
- Identify SP/departure point/RP/initial points with graphics found in ADRP 1-02, JP 1-02, or unit SOP.
- Mark course lines with tick marks on either side to indicate elapsed time and distance. The information presented should always be in the same scale of measurement–for example, NM or kilometers–to prevent confusion. Time and distances should have a set side of the course line for standardization in a unit (for example time marks on the right side of course with distance marked on the left).
- Navigation information blocks (doghouses) provide crews with required navigational data from present waypoint to the next. When they are used, the following order of information within the block is suggested:
  - Designator of next waypoint.
  - Magnetic heading to next waypoint.
  - The distance to next waypoint identified with NM or kilometers.
  - Estimated time en route (ETE) to next waypoint.
- PZ and LZ will be identified with a triangle centered over the objective area.
- Individual aircraft/serial touchdown point will be identified by a "+" symbol.
- Hard times will indicated by "00:00:00." Used for time driven missions (H-hour).
- Soft time will be indicated by "0000." Use for missions that are not time sensitive.
- Elapsed times will be indicated by "00+00+00." Used for event driven missions.
- Intermediate times may be used at ACPs as a tool to ensure the aircraft arrive on time. Will be indicated by "00:00:00".

5-27. Figure 5-3 depicts a map preparation sample.

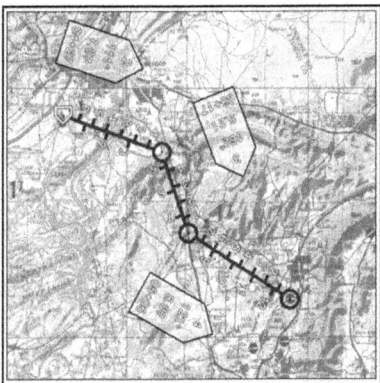

Figure 5-3. Sample–joint operations graphic map preparation

CHARTS, PHOTOGRAPHS, AND OBJECTIVE CARDS

5-28. The following tips are used for preparing charts, photographs, and objective cards:
- Post charts, photographs, and objective cards (not kneeboard) in the sequence of discussion during the OPORD brief. Accurately construct charts and objective cards. Label charts and objective cards not reproduced to scale as "not to scale." Place required items on charts, and refer to them, as necessary, during the mission brief.
- Prepare and orient charts and objective cards in the direction of approach, relating to magnetic north or in the direction of landing/takeoff. Photos and overhead imagery should be oriented as if viewed from the direction from which they were obtained.
- Include the following information on the objective card diagram:
    - Name of objective area.
    - Grid: military grid reference system or latitude/longitude.
    - Landing direction.
    - Landing formation.
    - Frequency and call sign.
    - Passenger entry/exit.
    - Go around direction.
    - Weapons control status/measures.
    - Fields of fire.
    - Hazards and markings.
    - Key terrain.
    - Alternate (if required).

ROUTE PLANNING CARD PREPARATION

5-29. Route planning (kneeboard) cards consist of navigation, en route, and objective cards. While used mostly in an NVD cockpit, route planning cards are also useful during daylight. They are intended to be easily viewed and lend organization to navigating and executing a mission. The following information applies to all three card types.

Chapter 5

- Aviators should write data in black ink contrasting with the card background. They should use letters and numerals at least ¼ inch in size and headings in degrees and nautical mile or kilometer abbreviations to preclude confusion.
- Time, distance, heading, and coordinates (UTM or latitude/longitude) will be triple-checked by members of the planning cell/aircrews before posting cards for briefings. Aviators should accomplish this check procedure using computers and manual measurements. Lead aircraft/AMC/lead navigators should resolve discrepancies.
- A completed card set is generated for each aviator and placed in plastic, transparent checklist pages. The card set is then fixed to the kneeboard, preventing loss during flight.

NAVIGATION CARDS

5-30. Navigation cards come in a variety of styles with different configurations tailored to suit a unit's needs (table 5-5). This format can be modified for specific needs.

Table 5-5. Example of a navigation card

Location	Heading	Destination	Distance nautical mile /kilometer	Ground Speed	ETE/ETA	Altitude AGL/MSL	Remarks
ACP 1	280°	CP 3	6 NM	100 kts	3+36	50 AGL	Bridge
Above ground level (AGL) Air control point (ACP) Check point (CP) Estimated time of arrival (ETA) Estimated time en route (ETE) Knots (kts) Mean sea level (MSL) Nautical mile (nm)							

EN ROUTE CARDS

5-31. En route cards reinforce map reconnaissance and display essential information for each phase/leg of flight (figure 5-4).

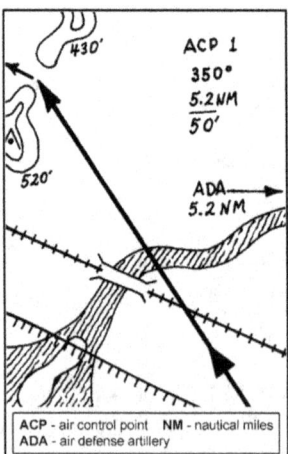

Figure 5-4. Example of an en route card

5-32. En route cards should be prepared with the following considerations:
- An aircrew may prepare en route cards for each leg and area of intended landing (objective, holding, or forward arming and refueling point [FARP]). An en route card may also highlight or detail an ACP position or turning point. En route cards are made with various elements close to scale giving attention to accuracy and detail. An aviator may refer to these cards instead of the map for quick orientation and reference.
- Xerographic copies of the JOG map cut to appropriate size, with route posted, serve as excellent en route cards once details have been highlighted according to map preparation guidelines for NVD use. This provides an en route card with an obvious accuracy advantage but must be studied in detail ensuring familiarity with the information.

OBJECTIVE CARDS

5-33. Objective cards reinforce map reconnaissance and provide a graphic picture of the LZ, PZ, and/or objective (figure 5-5). This card must be as accurate in detail as possible. It is important all crewmembers have the same understanding of where hazards, landing points, and loading points are located. Supported units may also receive a copy of the objective card so there is no doubt in positioning equipment and troops. When preparing an objective card, an aircrew depicts map elements to scale, as much as possible, reflecting relative sizes of each element.

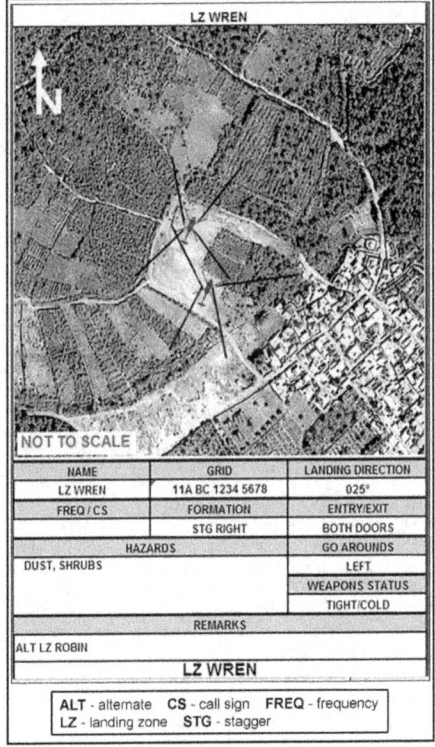

Figure 5-5. Example of an objective card

Chapter 5

HAZARDS TO TERRAIN FLIGHT

5-34. Specific hazards to terrain flight safety include physical, weather, and human factors.

PHYSICAL HAZARDS

5-35. Physical hazards are objects that the aircraft can actually contact during flight. Physical hazards are divided into two categories; manmade and natural.

Manmade Hazards

5-36. Manmade hazards are things made by man that pose a hazard to the aircraft. The list includes things such as buildings, bridges, towers, other aircraft, and wires. Manmade hazards are sometimes identified on maps but should be searched for continuously.

Wire Hazards

5-37. During terrain flight, regardless of location, aircrews continuously search for and expect wires. Throughout the world, wires are common at all altitudes and are found in the most unlikely places. Wire hazards consist of power lines, guy wires, communications wire, fences, missile-guidance wire, and wire barriers erected by the enemy. Under flight of wires could be hazardous due to attacks on the electrical power infrastructure. To minimize danger of wire strikes, aviators thoroughly review the AO before flight. Reference and update the operations map with any new information as part of a flight debriefing. In an unknown area, it may be necessary to reduce airspeed/increase altitude to provide increased reaction time. Two specific cues for locating wires include a swath cut through vegetation and the presence of supporting poles. Aviators may also detect these cues on aerial photos or a map—an essential reason for updating maps from the CHUM. Always expect wires along roads and waterways and near towers or buildings. If an aviator encounters wires, the safest way to cross them is overflying them at or near a pole. The aviator can see the pole more easily than the wires. The pole also provides a visual cue for estimating height above the wires. If forced to cross wires between poles, the aviator judges the necessary height by observing poles on either side of the aircraft and ensuring the aircraft is flown at an altitude at least as high as the poles.

> **CAUTION**
> Wires are nearly impossible to see with night vision devices (NVDs). They are also difficult to see during certain times of the day; for example, when an aviator is flying into a setting sun. Wires and poles can become coated in dust and snow making them even more difficult to identify. During any mission briefing, an aviator must review the presence of wires and identify the location and status of the wire-hazards map.

Natural Hazards

5-38. Natural hazards include trees, birds, and ambient light. Helicopters are particularly vulnerable to blade strikes during terrain flight especially when flying contour or NOE, or during masking/unmasking maneuvers. Trees are a problem during months when deciduous trees lose leaves or a tree is dead and branches are difficult to see. When flying NVDs, exercise caution when transitioning from high ambient light conditions to low ambient light conditions. In low ambient light, NVDs lose some resolution and reduce sharpness and definition of terrain features. Bird strikes are common and cause significant damage including penetrating the cockpit through the windscreen. Aviators should not try to avoid birds unless they are in a large flock as birds generally dive as an aviator flies toward them. Aviators should maintain a straight-ahead climb to clear birds. The best way to avoid either of these situations is through vigilance— looking outside the cockpit and maintaining a continuous visual scan. Flying terrain flight with the helmet visor lowered reduces potential eye damage resulting from tree or bird strikes.

Rotary-Wing Terrain Flight

WEATHER HAZARDS

Restricted Visibility

5-39. Weather can be a hazard if aviators do not exercise proper precautions. With reduced visibility, airspeed may have to be reduced or altitude (above the obstacles) increased to provide additional reaction time. When flying into a rising or setting sun, it is very difficult to detect obstacles ahead of the aircraft.

Wind Conditions

5-40. Strong wind conditions may create unsafe operating conditions for terrain flight. Gusting winds may create handling difficulties especially when using NVDs and with fewer visual cues present. Turbulence and thermals can be extremely dangerous especially at terrain flight altitudes. Terrain flight with external loads is especially dangerous under strong wind conditions.

HUMAN FACTORS

5-41. These factors include effects of fatigue and lack of ability to detect obstacles. The ability to maneuver and handle the aircraft effectively and safely is paramount to mission accomplishment at terrain flight altitudes. Each aircrew member must acknowledge his or her limitations and fly accordingly. These limitations may be based on such factors as lack of experience/proficiency or lack of familiarization with a particular environment. This information must be addressed during the crew brief to heighten crewmember awareness and ensure maximum aircrew coordination.

Fatigue

5-42. Terrain flight places unusual demands on the aircrew and is an extremely unforgiving environment. Fatigue is a difficult problem because it cannot be measured. Therefore, fatigue often goes unrecognized. The most common sign of fatigue is deterioration of performance and judgment, which slows reaction time and causes poor coordination and object fixation. The best way to combat fatigue is to establish and adhere to a fighter management program maximizing aircrew effectiveness.

Obstacle Detection Ability

5-43. This learned ability allows aircrew members to fully use peripheral vision and learn how closely they can maneuver an aircraft to an obstacle. Aviators use scanning techniques for accurate navigation and object recognition during terrain flight. In addition, with this acquired ability, the aircrew member must also understand how light, shadows, and seasons alter the appearance of terrain.

TERRAIN FLIGHT PERFORMANCE

5-44. The following considerations are important during any flight, especially during night flight–even with advances in NVDs.

AIRCREW COORDINATION

5-45. Aircrew teamwork is an essential element for mission accomplishment especially at terrain flight altitudes. One of the most important factors is crew station organization by each aircrew member. All necessary equipment must be readily available—including maps, DOD FLIPs, and flashlights (including NVD supplementary lighting). Aviators should secure this equipment preventing it from sliding down to the pedal area or blowing out of a window. There is little margin for carelessness or complacency. In this demanding environment, each aircrew member must be continuously vigilant in searching for potential obstacles and dangers threatening the safety of the aircraft. Regardless of duty position and rank, all aircrew members must contribute to safe flight and be heard and responded to. Each crewmember has a variety of duties. The demands of terrain flight complicate the normal performance of each crewmember's responsibilities. Every crew briefing must include assignment of duties, including scanning sectors. All crewmembers must completely understand the extent of their duties and mission intent. Whenever performance of such duties is impaired, aircrew members are obligated to inform other members. This allows adjustments to be made or changes implemented to compensate for shortcomings. Failure to work together as a team is a major contributor to aircraft mishaps and catastrophes.

NAVIGATION

5-46. Terrain flight navigation is difficult as the near-flat visual angle (low aircraft altitude) distorts shapes compared to those depicted on a map. Vertical relief (such as mountains or tall structures) is used as the primary means of identifying CPs. Accurate navigation requires proficiency in map interpretation and terrain analysis. Aviators must visualize how terrain will appear from information provided on a map. This ability to visualize three dimensionally, what appears two dimensionally on a map, and accurately identify the position of the aircraft is an acquired skill requiring continuous practice. This is more difficult at night as nearly all visual cues are less prominent making potential dangers harder to detect.

5-47. Navigation, conducted by an aviator, is augmented through information exchanged between aviators and often assisted by nonrated crewmembers (NCM). Rally terms, such as "turn left, stop turn, increase airspeed," and the use of clock positions to identify directions, are typical terms used to guide an aviator on the controls and aid in keeping his or her vision out of the cockpit. Aircrew members should agree on standardized terms identifying terrain features and eliminating regional language variations. This will help eliminate confusion and reduce unnecessary cockpit conversation. The navigating aviator must be able to project far enough ahead of the aircraft to facilitate timely information flow to the flying aviator, specifically, upcoming turns, airspeed and altitude changes, or expected terrain features he or she can assist in identifying. When an aviator becomes disoriented, it should be immediately acknowledged, and the aviator should start the reorientation process. The first step is to locate and identify a prominent feature in the immediate area. If this is not possible or practical, the aviator should attempt to return to the last known position. In a formation flight, if the lead aircraft becomes disoriented, the remaining aircraft should provide assistance. This assistance may be in the form of code words to guide the aircraft back onto course or, if necessary, by assuming the position and duties of lead aircraft. Aircrew members can use an established set of code words to guide the lead aircraft before it becomes disoriented or appears to be deviating off course. Chapter 4, section III contains additional information regarding navigation cues in terrain flight.

DETECTING AND AVOIDING THREAT

5-48. Rules for detection avoidance and use of operation security (OPSEC) measures will aid the aircrew in moving about the battlefield undetected, especially when searching for the enemy or if threat location is unknown. The following are guidelines for detection avoidance:

- Keep low and vary airspeed, altitude, and course to remain masked.
- When crossing an unavoidable ridgeline exposing the aircraft, select the lowest crossing point and move quickly down the forward slope to the nearest available concealment area.
- When crossing open/flat areas, cross at the narrowest point and move quickly across the area; try to use any available vegetation to mask the aircraft while following the lowest terrain.
- When flying parallel to a vegetated area, fly below and near vegetation.
- Fly as close to the ground as vegetation and manmade features will permit.
- When flying over dense vegetation, follow the lowest contours of the vegetation rather than the lowest contours of the earth.
- Do not fly into a situation in which there is no maneuver room, in case of attack.
- Always have an evasive maneuver planned in case of attack.
- Use communications equipment only when necessary, and limit transmission time.

SECTION II - TRAINING

5-49. Terrain flight is an essential element of an Army aviation unit's ability to accomplish the mission and is the fundamental element for mission success in a high threat environment. A unit must achieve maximum proficiency in executing missions at terrain flight altitudes, for day and NVD flight. The only way to achieve such proficiency is through training. Each unit must be committed to training for this demanding flight environment, maximizing every opportunity to practice terrain flight skills that can degrade quickly.

COMMAND RESPONSIBILITY

5-50. The commander has final authority for his or her unit's capability to perform properly on a high threat battlefield. The commander cannot delegate this authority and responsibility; rather, it must be exercised by personal participation in training. Commanders must use a hands-on approach including full qualification for all unit missions. This facilitates a depth of understanding and affords credibility with members of the unit. The commander ensures the unit proceeds with training and mission accomplishment using crawl-walk-run methodology to reach the needed level of proficiency. In addition, the Army's risk-management process must be consistently exercised along with elements of aircrew coordination.

IDENTIFICATION OF UNIT/INDIVIDUAL NEEDS

5-51. This is an ongoing assessment process with assigned personnel continuously changing and deploying to unfamiliar terrain or areas. There must be ongoing communication between major elements of a company/battalion; for example, commander, operations, standardization, safety, and maintenance. Each of these elements has important information to convey including status, trends, historical perspective, individual assessments, and progress towards proficiency in mission essential task list capability. The effective unit regularly discusses this information to identify strengths and weaknesses ensuring SA and continual progression.

TRAINING CONSIDERATIONS

5-52. The following references should be used to establish and maintain a training program:
- TC 3-04.11.
- The appropriate ATM for aircraft type.
- Any appropriate exportable training package; for example, the NVD operations exportable training package (NVD Branch, Fort Rucker, Alabama).

TRAINING SAFETY

5-53. Commanders may believe individual and unit terrain flying programs expose units to unacceptable training risks. While terrain flying is inherently riskier than other modes of flight, it is critical to train now prior to hostilities. Terrain flight training can be accomplished without neglecting controls and safeguards needed to help prevent accidents. Commanders must strictly supervise and control training. Thus, they ensure flight standardization procedures are strictly followed and training risks reduced.

This page intentionally left blank.

Chapter 6
Multi-Aircraft Operations

Multi-aircraft operations involve two or more aircraft flying together in briefed formations or while performing combat maneuvers. This chapter provides the basis for flight techniques which allows multiple aircraft, including dissimilar aircraft, to safely fly in close proximity to each other while providing aircrews with SA and the ability to mass combat power. These techniques support the successful accomplishment assigned missions such as air assaults or attack/reconnaissance operations.

SECTION I – FORMATION FLIGHT

6-1. Formation flight allows effective employment and control of two or more aircraft to accomplish a mission. The strengths of formations include control, predictability, flexibility, mutual support, and threat detection. These basic maneuvers and formations work well for team operations and can be enlarged to accommodate platoon size and larger formations. The following formations and maneuvers are building blocks that can be modified to support unit specific missions. Joint terminology has been used to facilitate joint operations on today's battlefield.

FORMATION DISCIPLINE

6-2. Discipline is the most important element for successful formations. On an individual basis, it consists of self-control, maturity, and judgment in a high-stress, emotionally-charged environment. Teamwork is an integral part of discipline; each individual must evaluate their own actions and how these actions will affect the flight and mission accomplishment. Discipline within a flight has a synergistic effect. If the flight lead and wingmen know their respective duties, they will work together as a team. Experience and realistic training leads to solid and professional air discipline.

CREW COORDINATION

6-3. The success and safety of multi-aircraft operations require all crew members in the flight to understand and utilize approved crew coordination techniques and terminology. Positive communication in and with each aircraft is necessary to maintain SA throughout the flight. Crew members should routinely update each other, highlight and acknowledge changes, and announce any hazards.

CREW RESPONSIBILITIES

AIR MISSION COMMANDER

6-4. The AMC is responsible for planning, organizing, and briefing the mission; delegating tasks within the flight; and ensuring flight integrity, flight discipline, and mission accomplishment. The AMC is in charge of all flight resources and should be aware of the capabilities and limitations of each crewmember. The AMC develops mission objectives to the lowest common denominator and provides correction to wingmen that are not performing their briefed responsibilities. Additionally, the AMC ensures all members are challenged and provided an opportunity to learn and grow. An effective AMC must maintain a high level of SA and ensure the information is provided to flight members. A good AMC must be able to control the aircraft, monitor the environment, observe the performance of wingmen, and control flight execution. Upon mission completion, the AMC must be able to mentally reconstruct the mission and make an accurate evaluation during the debriefing.

6-5. Under normal operations, the AMC should never relinquish the responsibility of ensuring mission accomplishment, flight safety, or air discipline. However, should the AMC be forced to leave a flight due to an in-flight emergency or the situation requires his or her aircraft to return to base, the designated alternate AMC assumes responsibilities.

6-6. An effective AMC is a leader and manager who conducts the mission in a decisive and highly professional manner. He or she begins by establishing a logical order of priorities and formulating a plan. The AMC will also—

- Use all available resources to gather pertinent data for the mission.
- Be assertive and communicate the plan and intentions.
- Encourage open communication so each crewmember is comfortable expressing their views.
- Listen carefully to inputs provided and consider them individually.
- Make sound decisions based on all factors; however, be willing to modify his or her position if someone advocates a better plan of action.
- Resolve conflicts as they arise within the crew or flight, and seek mission accomplishment through harmonious relations within the flight.

6-7. The AMC always evaluates and seeks information to ensure early detection of possible problems and reduce the potential for mishaps. He or she continuously challenges information and beliefs, including his or her own, with a firm leadership style.

FLIGHT LEAD (TEAM LEAD)

6-8. Flight (team) lead and wingman are roles flight members fulfill based upon their positions within the flight. Team lead is used to denote the flight lead for two aircraft operating in teams. Flight lead is the formation leader designated by the AMC and is generally the most proficient PC. Flight leads are selected based on ability and demonstrated knowledge of missions and tactics, and local SOPs. The flight lead's responsibilities include navigation, en route communication (between flight members, ATC, and supported units), obstacle and threat avoidance, wingmen position awareness, and the energy states of all aircraft. The AMC may delegate some of these duties throughout the flight. Chalk 2 should always be prepared to lead the flight.

TACTICAL LEAD

6-9. The tactical lead is a role of the first crew member to identify a threat or obstacle regardless of their position within the flight or aircraft. The tactical lead announces and selects an appropriate maneuver to engage, suppress, or bypass the threat. For example, Chalk 2 might become tactical lead when directing a break turn in response to enemy antiaircraft artillery (AAA) fire. tactical lead may change several times during the conduct of a mission. Flight lead will assume control of the flight when the situation permits and the threat is bypassed or neutralized.

WINGMEN

6-10. Wingmen (chalk 2, 3...) are assigned supporting roles in the flight. They help plan and organize the mission. Formation wingmen fly their aircraft in positions relative to lead. Their responsibilities include maintaining the desired formation and providing mutual support to the flight through lookout, navigation, and firepower. They also focus on collision avoidance as well as obstacle and threat avoidance. Wingmen are also responsible for accomplishing additional tasks assigned by the AMC and questioning lead any time a significant deviation occurs that jeopardizes mission accomplishment. Other duties include performing communications and backup navigation. Wingmen must always be prepared to assume lead if needed. Wingmen engage threats as briefed (or when directed by lead) and provide support during engagements. It is essential for wingmen to understand their briefed responsibilities and execute them in a disciplined manner.

INDIVIDUAL CREWMEMBERS

6-11. Each crewmember has the responsibility to provide security and mutual coverage for other aircraft in the formation. Mutual coverage is especially important in a combat environment where the flight is susceptible to an attack from enemy ground and airborne weapon systems.

Pilot on the Controls

6-12. The pilot on the controls (P*) has the primary responsibility of safely flying the aircraft to avoid all hazards through correct power/energy management and by scanning. The P* must also fly the aircraft in such a manner as to deny or minimize engagement by threats while maintaining a safe flight profile. The P* coordinates maneuvers with the flight. The P* also communicates to the crew intended plans of action to accomplish the mission or defend against a threat.

Pilot not on the Controls

6-13. The pilot not on the controls (P) monitors the flight profile of the aircraft, providing the P* with information regarding altitude, power requirements, terrain avoidance, and airspeed. The P should accomplish all tasks inside the cockpit such as changing radios or switch positions. The P is normally tasked to navigate, communicate, and copy all clearances and reports. The P must be able to immediately assume control of the aircraft any time the P* becomes incapacitated. He or she also must keep all crewmembers updated on the progress of the mission to enhance their SA whenever possible.

Nonrated Crewmember

6-14. The NCM must maintain SA relative to the terrain, threats, and other formation members. NCMs are also responsible for notifying the pilot of all changes in the relative position of other aircraft in the formation. This can be extremely demanding in a combat environment, especially during defensive maneuvering, where the crew is often required to direct the actions of the formation.

CONSIDERATIONS

6-15. In terrain flight, a greater number of aircraft can be more easily detected than a lesser number. In addition, a larger group requires more terrain relief to remain concealed. If a large group is necessary for the mission, dispersion can be achieved by using numerous routes with small flights instead of one large flight. The enemy situation, however, may mandate the use of one route and mass concentration of troops, which would require the larger flight. In a well-planned tactical formation flight, at terrain-flight altitudes, individual aircraft within the flight move like individual infantrymen in a squad. Flight lead selects the general direction of travel, but within those boundaries, each aircraft picks the exact piece of terrain to fly over. The aviator of each aircraft must be careful not to maintain equal distances from preceding aircraft or fly over exactly the same terrain as preceding aircraft, as this will aid enemy air defense artillery (ADA) or small-arms fire.

TECHNIQUES OF MOVEMENT

6-16. Multi-aircraft operations in a high-threat environment may require greater flexibility than is possible with basic flight formations. The flexibility required to conduct multi-aircraft operations at lower terrain-flight altitudes is best achieved by employing maneuvering formations in conjunction with techniques of movement. The three methods of movement used when conducting multi-aircraft operations are traveling, traveling overwatch, and bounding overwatch.

Traveling

6-17. Traveling is used to move rapidly over the battlefield when enemy contact is unlikely, or the situation requires speed for evading the enemy. All aircraft move at the same speed. This technique is the fastest method for aircraft formation movement but provides the least amount of security. Units often employ low-level and contour flight at high airspeeds using the traveling movement technique.

Traveling Overwatch

6-18. Traveling overwatch is used when speed is essential and enemy contact is possible. This technique is normally associated with reconnaissance, security, and attack missions when threat and/or environmental conditions preclude use of bounding overwatch. Lead aircraft or teams move constantly, and trail aircraft or teams move as necessary maintaining overwatch of lead. Overwatching aircraft key their movement to terrain and distance from the main element. It also remains ready to fire or maneuver, or both, providing support to main elements. Units often employ contour or NOE flight with the traveling overwatch technique using high and varying airspeeds depending on weather, ambient light, and threat.

Bounding Overwatch

6-19. Bounding overwatch is used when enemy contact is expected and the greatest degree of concealment is required. It is the slowest movement technique; too slow for high-tempo operations and too vulnerable for non-linear and/or urban operations. Individual aircraft or aircraft teams employ alternate or successive bounds.

6-20. One element remains in position to observe, fire, or maneuver before the other element moves. Overwatching elements cover the progress of bounding elements from a covered and concealed position, which offers observation and fields of fire against potential enemy positions.

6-21. The length of the bound depends on terrain, visibility, and effective range of the overwatching weapon system. Units normally employ contour and NOE flight with the bounding overwatch technique. Airspeed during each bound is varied depending on availability of vegetation and terrain for concealment.

Sight Picture

6-22. Sight picture is a particular angle, based on particular components a trailing aircraft sees or cues on when flying in formation on another aircraft. This is based on aircraft type and may cue on formation lights—especially at night or with NVDs. An aviator must become proficient and comfortable with this sight picture as it allows an aviator to judge attitude changes and relative position to the preceding aircraft.

Formation Angle

6-23. This is the angle relative to the aircraft being followed in formation flight. Zero degrees would be directly behind and ninety degrees would be abeam. While the angle is traditionally 30 or 45 degrees (figure 6-1, page 6-5), it may have to be different due to aircraft limitations. For example, at a 45-degree viewing angle between aircraft, the UH-60 helicopter has windshield posts that obstruct the aviator's ability to see, mandating a slightly different angle to accommodate this design flaw.

Formation Separation

6-24. The space between aircraft in any given formation represents a tradeoff between the previously mentioned formation characteristics. The capability of all members of the flight to navigate and avoid obstacles without the excessive concern of colliding with other flight members is a primary factor in determining formation spacing. METT-TC considerations drive spacing between aircraft. For example, low illumination nights usually require close spacing, while day flights can assume large separations, enhancing lead's ability to maneuver. In choosing a sound tactical formation, lead should consider the following factors and how they affect the formation:

- Threat.
- Terrain.
- Illumination.
- Time of day.
- Visibility.
- Communications environment.
- Capabilities of the crews and aircraft in the flight.

The wingman is ultimately responsible for maintaining adequate separation to prevent collision by anticipating (and providing clearance for) maneuvering by lead.

HORIZONTAL DISTANCE

6-25. Formations are defined and expressed in rotor diameters (based on type of aircraft being flown) between tip-path planes or the rearward edge of the disk on the leading aircraft and the forward edge of the disk on the trailing aircraft. This distance is usually predetermined during the mission brief and established by the chalk 2 aircraft in the flight. Aircraft after chalk 2 should follow the established pattern. Horizontal distance is defined as (figure 6-1)—

- **Tight.** The horizontal distance for tight is approximately two rotor disks.
- **Close.** The horizontal distance for close is three to five rotor disks.
- **Loose.** The horizontal distance for loose is six to ten rotor disks.
- **Extended.** The horizontal distance for extended distance is more than ten rotor disks, as dictated by tactical requirements.

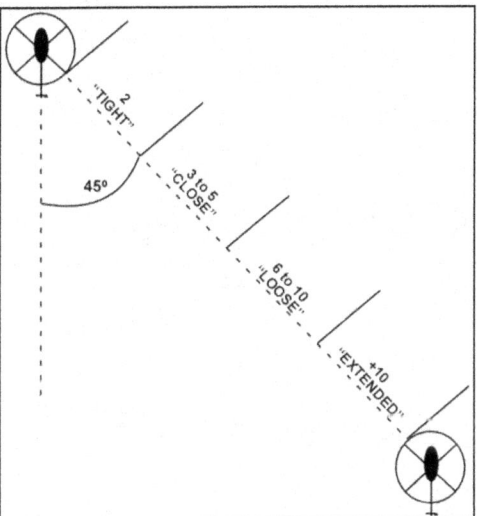

Figure 6-1. Horizontal distance

Vertical Separation

6-26. Flat, stepped-up (figure 6-2, page 6-6), and stepped down are vertical separations.

- **Flat.** All aircraft are flown at the same altitude.
- **Stepped-up.** Vertical separation of 1 to 10 feet higher between lead, chalk 2, and each successive aircraft.
- **Stepped-down.** Vertical separation of 1 to 10 feet lower between lead, chalk 2, and each successive aircraft.

Chapter 6

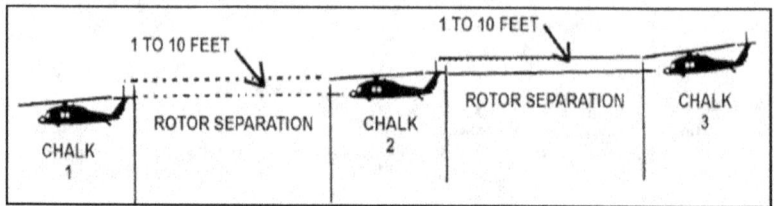

Figure 6-2. Stepped-up vertical separation

Note. In stepped-down formation, trailing aircraft may experience wake turbulence. To avoid this turbulence, they will need to adjust their relative position. Trailing aircraft require more power to fly in this formation.

Flat Terrain

6-27. Generally, in flat terrain, formation separation should increase as such formations are more difficult to detect. If the enemy detects the formation, it must choose one helicopter and potentially lose SA on the second. This aircraft may pass completely unnoticed and provide mutual support. This is true for both air and surface threats.

Rough Terrain

6-28. Rough terrain may require closer formation spacing. The tactical advantages of wide formations must be balanced with the difficulty of controlling those formations in rough terrain. The formation selected should enhance cover and concealment of all aircraft in the flight and the ability for each member of the flight to select terrain and seek concealment while still maintaining SA on lead (visual contact is desired but not required at all times).

MULTI-AIRCRAFT OPERATIONS BRIEFING

6-29. Regardless of the number of aircraft in the formation, the lead/wing concept should be applied. During multi-aircraft operations, additional crew actions must be considered. All multi-aircraft operations are briefed using a unit approved multi-aircraft/mission briefing checklist and should include the following:

- Formation type(s).
- Altitudes.
- Airspeeds.
- Aircraft lighting.
- Lead change procedures.
- Loss of visual contact/in-flight link-up.
- Loss communications procedures.
- IIMC procedures.
- Actions on contact.
- Downed aircraft procedures.
- Separation.

LIGHTING FOR MULTI-AIRCRAFT OPERATIONS

6-30. AR 95-1, AR 95-2, aircraft design limitations, local regulations, and SOPs govern lighting for multi-aircraft operations (table 6-1, page 6-7).

Table 6-1. Sample lighting conditions

LITECON	Description	Anticollision	Visible Position NAV Lights	Formation Lights	Covert IR Position Lights
1	FAA-Day	White	OFF	OFF	OFF
2	FAA-Night	Red	Bright	Bright	ON
3	Night Form	OFF	Dim	Bright	ON
4	Covert	OFF/IR	OFF	Covert	ON
5	Total Blackout	OFF	OFF	OFF	OFF

Federal Aviation Agency (FAA)
Infrared (IR)
Lighting condition (LITECON)
Navigation (NAV)

FORMATION TAKEOFF

6-31. A formation takeoff is two or more aircraft leaving the ground simultaneously and then maintaining a predesignated relative position during the takeoff. Most formation takeoffs are made from the ground and liftoff simultaneously at a prearranged signal from the lead aircraft. The leading aircraft should accelerate slightly faster than a VMC takeoff, allowing the following aircraft to gain translational lift; care must be taken, however, to not accelerate too quickly and leave the flight scrambling to catch up. The initial rate of climb must be enough to clear barriers with a safety margin. Trailing aircraft maneuver into the en route formation and attain a stepped-up vertical separation as soon as possible permitting acceleration and climb to undisturbed air. Once the flight is airborne and established, the lead aircraft can slowly and smoothly accelerate to normal climb or cruise airspeed. Takeoffs should only be into the wind, especially for dust/sand/snow conditions. For moderate to heavy dust/sand/snow conditions, aircraft should take-off separately in chalk order and then conduct an in-flight join-up.

FORMATION FLIGHT-EN ROUTE

6-32. Formation flying is the maneuvering of aircraft according to established TTP. It includes rapid, but controlled, change from a specific formation suitable for one set of conditions to another formation meeting requirements of an entirely different set of conditions. Safe and orderly formation flight is the result of extensive training, continuous practice, and a high degree of discipline.

6-33. The aviator flying each aircraft maneuvers with primary reference to only one other aircraft. The constant vigilance necessary to fly, reference the other aircraft, avoid obstacles, and incorporate an instrument scan precludes the P* from observing other aircraft. However, P can observe aircraft other than the primary reference aircraft. In formation types requiring observation of two aircraft such as diamond or staggered, the P* must do so with great care and precision while mainly viewing the primary aircraft.

6-34. Aviators must anticipate aerodynamic interference between aircraft during formation flight. Aviators flying trailing aircraft may encounter wake turbulence (section V) if they permit their aircraft to go below leading aircraft. Flight in turbulence may result in rapid attitude (pitch, roll, and yaw) changes.

6-35. Distance between aircraft can be increased or decreased to fit the tactical situation. At terrain flight altitudes, aircraft may spread out to take advantage of the terrain/tactical situation. In addition, it is less fatiguing to fly loose or extended formations as opposed to tight or close formations.

6-36. All aircraft should have the P navigating in the event they must take over the lead position and assist the flight with ensuring navigational accuracy to complete the mission.

6-37. Altitude and airspeed changes should be smooth and gradual especially during tight and close formations. This allows all aircraft in the formation to act in unison. Abrupt changes in altitude and airspeed by the lead aircraft may cause an "accordion" effect. This results when all remaining aircraft in the formation make correspondingly abrupt altitude and airspeed changes to maintain their relative position, and the effects

Formation Turns

6-38. The lead aircraft should make smooth constant rate turns and avoid angles of bank greater than 30 degrees. Turns at reduced bank angles require larger turning radiuses, particularly in the landing pattern, and must be considered in planning. If a large turn is required, flight lead enters the turn as early as possible to avoid excessive bank angles and subsequent recovery. This allows the flight to react in a timely manner. During a turn, the inside aircraft must decelerate slightly and drop slightly lower than the leading aircraft, while the outside aircraft must accelerate slightly and climb slightly to maintain its position in the formation. Whenever possible, the aviator avoids turns in which aircraft are forced inside the lead aircraft's turning arc. This is usually addressed during the planning process and briefed accordingly. Aircrews should avoid planning route segments requiring heading changes of more than 60 degrees.

Formation Changes During En Route Flight

6-39. Formation changes en route require a high degree of proficiency and therefore are executed with caution and only when necessary. Any changes to a formation are specifically briefed and understood by all aircrews involved. As a technique, trail formation could be used as a transitional formation before executing the next briefed formation.

Lead Changes

6-40. Lead changes are inherently difficult, potentially dangerous, and should be executed on the ground, whenever possible. A lead change is never initiated, day or night, by accelerating to overtake the lead aircraft. Only the lead aircraft may give the signal to initiate lead changes. Flight lead initiates by a prearranged signal, and the flight acknowledges beginning with chalk 2. The lead aircraft then makes a 30- to 90-degree heading change in the prebriefed direction to depart the formation and establish separation space. Lead maneuvers a minimum of eight rotor disks to the announced side and begins to parallel the formation. When chalk 2 (the new lead) confirms and announces the former lead is clear of the flight, the former lead will slow to 10 KIAS less than the en route airspeed. The former lead visually (and possibly verbally) confirms each aircraft in the flight as it passes to prevent rejoining the flight prematurely causing a midair collision. After the last aircraft (former trail) has passed by, the former lead aircraft will rejoin the flight and assume the duties of the trail aircraft to include displaying appropriate lighting. The former trail aircraft then reconfigures its lighting to conform to the rest of the formation.

Formation Landing

6-41. All aircraft touch down at the same time while maintaining their relative positions within the flight. The rate of closure throughout approach and landing is somewhat slower at night than during the day. Flight lead should maintain straight-and-level flight until the desired approach angle is intercepted. Lead then maintains a constant approach angle and, where terrain and obstacles permit, makes the approach to the ground avoiding hovering turbulence and brownout or whiteout conditions. If the rate of closure is too fast, the aviator should avoid S-turns to lose airspeed. Instead, execute a go-around if unable to slow to the appropriate airspeed, especially with heavily loaded aircraft.

6-42. Lead must plan to touch down far enough forward in the PZ/LZ to provide sufficient landing space for the entire flight. When planning the touchdown, consideration should be given to obstacles and power availability on the departure. If potential whiteout or brownout conditions exist, the flight may have to spread out to the briefed landing disk separation before the approach is established facilitating safe landing conditions. The AMC should consider, based on aviator experience and the environment, stacking down and landing in reverse chalk order once flight lead initiates an approach. This reduces the possibility of being caught in the cloud from the preceding aircraft and is especially true with CH-47s when executing formation flight approaches to a snow field where potential exists for sliding after touchdown. Finally, if safety is in doubt regarding landing or landing conditions, the flight lead should execute a go-around. The go-around should be executed prior to descending below any obstacles or losing ETL to prevent sudden high power demands on the other aircraft.

FORMATION BREAKUP

6-43. The following are four examples of formation breakup. These techniques can be adapted for use with other formation types. In addition, a flight can be disbanded by simply landing somewhere and departing separately, beginning with lead or trail aircraft and continuing in an orderly fashion.

BREAKUP INTO SINGLE AIRCRAFT

Method 1

6-44. Aviators may use this maneuver when an LZ is large enough for only one aircraft at a time. Figure 6-3 shows an echelon formation before breakup. Lead aircraft designates the interval (determined by required ground time, 10 seconds in this example) between breaks. Lead issues the command to execute, then turns 90 degrees away from the formation. Lead is followed 10 seconds later by chalk 2, then chalk 3, and so on. When aviators use this maneuver for landing in a single-ship LZ, the formation ideally approaches the LZ on the landing heading and starts the breakup over the LZ as shown in figure 6-4.

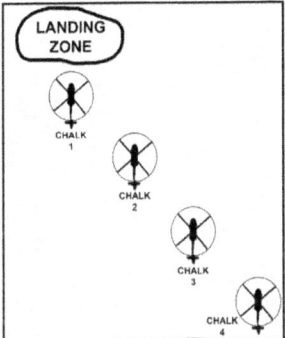

Figure 6-3. Echelon formation before breakup

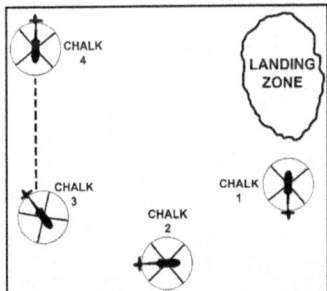

Figure 6-4. Left break with 10-second interval for landing

Method 2

6-45. This maneuver may be initiated anytime multihelicopter operations are terminated. Lead issues the command to break up, and then the trail aircraft turns 30 to 90 degrees away from the formation to the clear

Chapter 6

side. Once the trail aircraft is visually confirmed clear, the remaining aircraft, in reverse chalk order, turn 30 to 90 degrees away from the formation to the clear side. Once clear of the formation, each aircraft must adjust lighting and avionics as appropriate.

BREAKUP INTO ELEMENTS

6-46. Aviators execute this maneuver from the staggered formation and breakup into elements of two or more aircraft as required. Lead announces the time interval between elements and receives an acknowledgment from each aircraft if not briefed. After lead has issued the command to execute, the first element aircraft continues on course. The remaining aircraft slow or turn by elements until each attains the desired separation. Aviators adjust exterior lighting and avionics as necessary. Figure 6-5 shows a flight of five becoming an element of two and three.

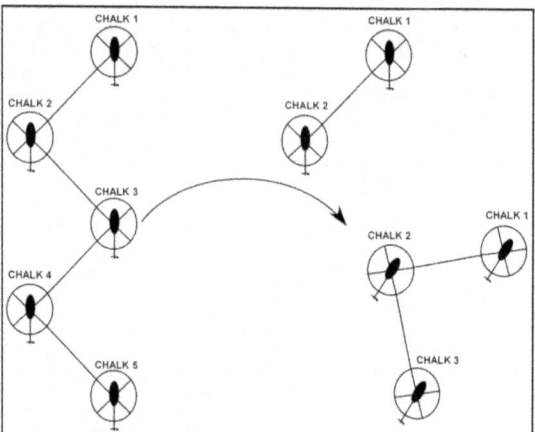

Figure 6-5. Breakup into two elements

INADVERTENT INSTRUMENT METEOROLOGICAL CONDITIONS BREAKUP PLANNING

6-47. Helicopter flight crews must be trained to cope with marginal weather conditions they may encounter during formation flight. All multihelicopter operation mission briefs must include a planned response for encountering IIMC. As well as being an established part of an SOP, IIMC must be planned and briefed for all phases of the mission. During the breakup procedure, all aircraft should remain in contact with the lead aircraft and also contact ATC in chalk order for further guidance. Communication is key to a safe execution of this procedure. Aviators should perform all turns, airspeeds, and climbs at a predetermined standard rate. They should maintain prescribed headings and altitudes for each aircraft at least 30 seconds after breakup to gain separation before executing any additional procedures. The following procedures are guidelines for units to further develop their own procedures, based on mission, terrain, weather, and enemy situation.

Breakup Procedure

6-48. It is unlikely more than two or three aircraft will enter IIMC before the situation is recognized and remaining aircraft take prebriefed evasive action. Vigilance, communication, and SA are important factors in avoiding IIMC. If any aircraft encounters IIMC, they will notify the rest of the flight via the radio using a prebriefed code word or plain language. An example call would be "Lead is IMC, executing breakup procedure, heading 090". The lead aircraft heading is important as the other aircraft will plan their headings accordingly. A good heading choice is 10 degrees times chalk position from lead's announced heading to the clear side of the formation. Upon hearing this message, the formation begins the breakup procedure (if unable to remain VMC) according to the prearranged plan. When aviators initiate IIMC recovery, the following procedures–for a staggered formation–are suggested. The following information relates to figure 6-6.

- Flight lead continues straight ahead and reports the magnetic heading and altitude he or she will climb to and maintain.
- Chalk 2 executes a 20-degree turn away from the flight (if staggered left, chalk 2 would turn left) and climbs 500 feet higher than the lead aircraft.
- Chalk 3 executes a 30-degree turn away from the flight (if staggered left, chalk 3 would turn right) and climbs 500 feet higher than chalk 2 (1,000 feet higher than lead).
- Chalk 4 executes a 40-degree turn away from the flight (if staggered left, chalk 4 would turn left) and climbs 500 feet higher than chalk 3 (1,500 feet higher than lead).
- Chalk 5 executes a 50-degree turn away from the flight (if staggered left, chalk 5 would turn right) and climbs 500 feet higher than chalk 4 (2,000 feet higher than lead).

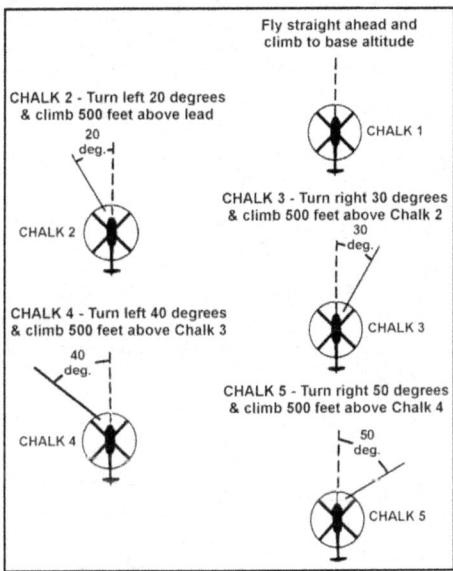

Figure 6-6. Formation breakup–inadvertent instrument meteorological conditions

6-49. There are many variations to this technique (lead climbs to highest and others stack down 500 feet); however it offers the simplicity of correlating chalk number to the number of degrees turning. In addition, the direction of turn is simplified by stating, in staggered left formation as an example, even-numbered chalk positions turn left and odd-numbered chalk positions turn right. While an additional 500 feet might seem excessive for each chalk number to climb above the previous chalk number, this technique offers an additional safety margin. Considerations for IIMC procedures include the following:

- Enemy ADA capabilities.
- Terrain elevation and relief.
- Emergency minimum safe operating altitude.
- Availability of location of recovery airfields.
- Fuel considerations.
- ACO requirements.
- Turns should not exceed standard rate.

Chapter 6

- When flying near hostile borders and prohibited or restricted areas, consideration must be given to avoid flying into these areas.
- IIMC should be briefed when forecasted weather conditions are less than 1,000/3.
- Mountainous terrain requires detailed IIMC and innovative planning.

THREAT BREAKUP

6-50. Threat breakup is executed to evade an observed enemy engagement threatening the flight. Since combat cruise uses the two-ship section as its basic building block, large formations can easily be broken down and dispersed if attacked. Premission planning should include an evasive action plan and procedures for rejoining the formation and continuing the mission.

6-51. Standard threat terms listed in the appropriate ATMs should be used to identify threats. Codes such as "bandit break" for an air threat or "enemy break" for a ground threat are used to execute a threat breakup procedure. This breakup should be a last response to the enemy taking action against the formation. Formations with an odd number of aircraft could have lead, chalk 2, and chalk 3 break to the clear side of the formation and remaining pairs break in opposite directions. Aircrews must remain oriented with the other aircraft executing a threat turn in the same direction. Aircrews should descend to cover and dispense chaff or flares if equipped. PCs determine if external loads are to be jettisoned.

RENDEZVOUS AND JOIN-UP PROCEDURES

6-52. Rendezvous and join-up procedures are inherently difficult and dangerous maneuvers whether executed at day or night. The difficulty comes with identifying joining aircraft and judging airspeeds and rates of closure. When the tactical situation permits, rendezvous and join-up should be executed on the ground to reduce hazards.

GROUND

6-53. Aircraft conducting rendezvous and join-up should arrive at the rally point as briefed. Once all aircraft are on the ground, they are organized into formation to continue the mission.

IN FLIGHT

Rendezvous

6-54. Rendezvous is definitely the more dangerous maneuver, especially at night with multiple aircraft joining within minutes. Vigilance is key with aircrew coordination both within the aircraft and the flight. If an airborne rendezvous is necessary, the flight lead approaches the rendezvous point at the preplanned time and altitude. After arrival at the rendezvous point, lead enters an orbit in the prebriefed direction using a standard rate (or less) turn and airspeed of 70 KIAS or as briefed. Aircraft joining the flight should approach the lead aircraft by entering its orbit at the assigned airspeed. Aviators adjust airspeed and heading to enter the formation in the prebriefed position. For join-up, a safe rate of closure is essential as it is easy to overrun the aircraft ahead. It is important to brief and maintain planned airspeeds for both the flight and closing aircraft. Each aircrew must exercise extreme caution to avoid overrunning the aircraft directly to the front as aircrew members cannot see the silhouette of an aircraft at night except at a close distance.

Join-Up

6-55. This procedure is normally required when aircraft are not in a position to observe other aircraft departing individually or during brownout/whiteout conditions. Aircraft will depart in chalk order when ready and will display appropriate single ship lighting until established in the formation. Lead will accelerate to the appropriate briefed airspeed, normally 70 KIAS. Subsequent chalks will accelerate to no more than 10 knots greater than the briefed airspeed. Once all aircraft have completed the in-flight join-up, lead will accelerate to the en route airspeed.

LOST VISUAL CONTACT PROCEDURES

6-56. In the event an aircraft in the flight loses visual contact with the aircraft they are following, they will immediately make a radio call to lead. Lead will announce heading, altitude, airspeed, and distance to next waypoint if available. The aircraft that has lost visual contact with the flight will immediately assume flight lead's heading and airspeed and attempt to regain visual contact. Lead must maintain this heading, altitude, and airspeed until all aircraft have rejoined the flight. The flight will begin reorientation procedures. The most important consideration when an aircraft has lost visual contact with the flight is reorientation. Except for enemy contact, all mission requirements are subordinate to this action.

6-57. Unit SOPs should provide procedures for reestablishing contact with the flight. Considerations should include, but are not limited to, rallying to a known point, use of covert/overt lighting, and ground rally. METT-TC, power available, and ambient light will influence how contact is reestablished. When a flight rallies to a known point, the point may be an ACP along the route, a present position report or waypoint sent by lead, or a terrain feature. Situations may occur when an aircraft rejoins the flight in another position than briefed. Mission commanders may use altitude, a target reference point/priority fire zone, cardinal direction, or other method to maintain separation. Only after the entire flight is formed can the mission commander proceed with the mission.

COMMUNICATION DURING FORMATION FLIGHT

6-58. Radio communications during formation flight must be efficient and brief. The need for radio communications can be greatly reduced through use of visual signals, established procedures in the unit SOP, and a thorough mission brief covering all contingencies. The ability to execute multi-aircraft radio-silence missions requires proficiency aircrew members achieve only through training and practice. Radio-silence missions should be used with discretion, with safety being the priority. The following situations are examples of formation flight without radio communication.

- **Forming of flight.** Aviators maneuver the helicopters into position for the formation takeoff. At this point, the anti-collision light should be on. The pilots will then turn off the anti-collision light when their aircraft is ready for takeoff and after the preceding aircraft's anti-collision light is off. When the trail aircraft is ready for takeoff and the preceding aircraft's anti-collision light is off, it will announce to lead the flight is ready using a codeword or plain language. Trail will leave its anti-collision light on for the flight. The flight will then depart after the codeword, ATC call, and/or on time as per the mission briefing.
- **Formation landing.** Upon landing, all aircraft will immediately turn their anti-collision light back on. Whenever their aircraft is ready for takeoff and the preceding aircraft's anti-collision light is off, the aviator will turn off the anti-collision light. When trail aircraft is ready for takeoff and the preceding aircraft's anti-collision light is off, they will announce to lead the flight is ready using the code word or plain language. Again, the trail aircraft will leave its anti-collision light on for the flight. The flight will then depart as briefed.

SECTION II – FORMATION TYPES

6-59. Common formations used during multi-aircraft operations include fixed formations (such as echelon, staggered, or trail) and maneuvering formations (including combat cruise and combat spread). Army aviators should be familiar with basic formations and maneuvers described in the following paragraphs. All angles and distances can be modified based on aircraft and mission. The two helicopter team/section is the building block for all formations from which can be built upon to create platoon- and company-sized formations (figure 6-7, page 6-14). The intent is to allow aircraft to be able to fly together using common terminology and techniques. The only authorized formations for night/NVG flight at 80 feet AHO and below are combat cruise formations in conjunction with techniques of movement according to TC 3-04.11.

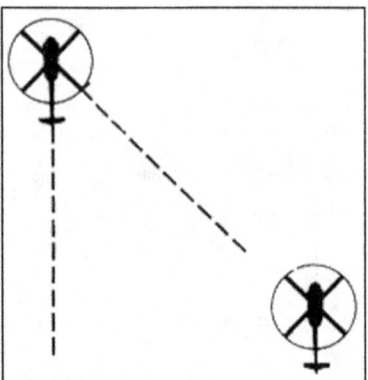

Figure 6-7. Two-helicopter section/element

TWO-HELICOPTER TEAM

6-60. A team usually consists of two helicopters flying as lead and wingman. The wingman may fly to the left or right rear of the lead aircraft. When flying to lead's left rear, the wingman is flying in echelon or staggered left. When flying to the lead's right rear, the wingman is flying in echelon or staggered right. The correct angular location is approximately 45 degrees with consideration given to aircraft limitations.

FIXED FORMATIONS

6-61. These formations are used when more control is required. The flight acts as one aircraft regardless of the number of aircraft in the flight, and the movements of lead are mirrored throughout the flight. Fixed formations are useful for departure and arrival at LZs, airfields, administratively transiting airspace, deployment, and when environmental conditions do not allow or require tactical separation. When lead locks the wingman into these fixed formations, lead must consider the wingman's obstacle clearance and provide appropriate horizontal and vertical clearance. Wingmen, as well as lead, must consider the reduction in altitude wingmen have when flying on the inside of turns and ensure adequate obstacle/terrain clearance. Spacing and separation must be considered during changes in altitude and headings.

STAGGERED

6-62. This is one of the most commonly used formations in Army aviation and is flown as a staggered right or staggered left (figure 6-8, page 6-15). Each aircraft of the formation holds a position approximately 45 degrees astern of the aircraft to its front, alternating left and right. Chalk 2's position determines if the formation is staggered right or staggered left. Chalk 3 (and any other odd-numbered wingmen) flies in trail directly behind lead. A staggered formation is essentially a continuous, alternating series of the basic two-helicopter section/element. This formation is not limited to any prescribed number of aircraft. The mission requirement dictates its size. This formation gives wingmen the ability to estimate distance and rates of closure and allows some flexibility in relation to adjacent aircraft while affording lead control of the flight. Staggered formations are common formations used through congested airspace, for large formations in a low-threat area, for air assault approaches and takeoffs, or for traveling through narrow canyons. Formation changes between a left and right staggered formation are directed by lead. During the crossover, wingmen maintain appropriate clearance. Chalk 2 should use a heading change of approximately 5 to 10 degrees to cross from one side to the other. Chalk 3 maintains position behind lead. A slight vertical stacking is recommended during the crossover to avoid rotor wash. Staggered formation has the following advantages and disadvantages:

- Advantages:

- Fixes position of wingmen.
- Allows lead maneuverability.
- Simplifies prepositioning of loads.
- Allows rapid deployment of troops for all-round security.

● Disadvantages:
- Increases pilot workload to maintain relative position to the aircraft in front of it when flying tight or close.
- Requires a relatively long and wide landing area.
- Places some restriction on suppressive fire by door gunners.

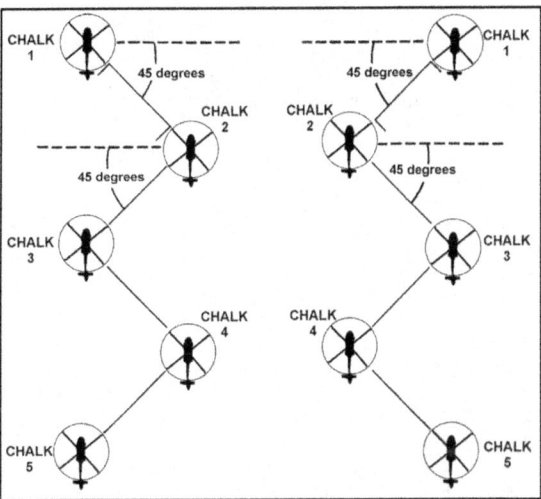

Figure 6-8. Staggered right and left formation

ECHELON

6-63. This formation (figure 6-9, page 6-16) is flown as either echelon right or echelon left. Wingmen fly a fixed position on an approximate 30- to 45-degree offset from lead's 6 o'clock position. All formation aircraft are positioned on the same side of lead at briefed horizontal and vertical distances. There is no requirement for wingmen to maintain a level plane when turning. This is especially true for turns toward the wingmen; wingmen may stack slightly low as required to keep the preceding aircraft in sight. Echelon formation has the following advantages and disadvantages:

● Advantages:
- Provides ease in maintaining view of the entire formation.
- Allows rapid deployment of troops to the flank.
- Allows nearly unrestricted suppressive fire by door gunners.
- Provides excellent formation for dust/sand/snow takeoffs and landings.

● Disadvantages:
- Severely limits flight maneuverability of the flight. The lack of maneuvering room makes aircraft more vulnerable during a threat engagement.
- Requires a relatively long and wide landing area.
- Presents some difficulty in prepositioning loads.

Chapter 6

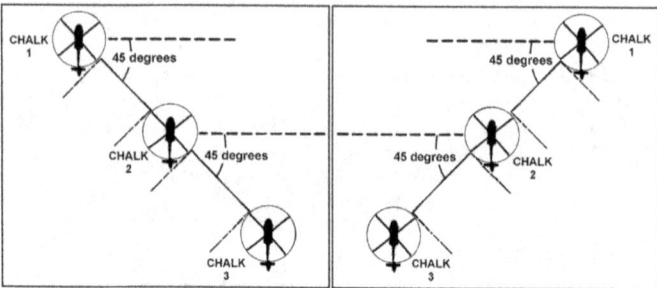

Figure 6-9. Echelon right and left formation

TRAIL FORMATION

6-64. The trail formation is the most difficult of the fixed formations (figure 6-10, page 6-18). Each wingman/chalk follows leads movement within 10 degrees of the preceding aircraft. Trail formation can be used for landings and takeoffs and as a transition during formation changes. Trail formations should not be flown for extended periods of time as distances and rates of closure between aircraft are difficult to determine. It is important to note flight at the 6 o'clock position makes it very difficult for the preceding aircraft to scan for wingmen and can degrade SA in the flight. Trail formation has the following advantages and disadvantages:

- Advantages.
  - Simplifies prepositioning of loads.
  - Allows nearly unrestricted suppressive fire by door gunners.
  - Allows rapid deployment of troops to the flanks.
- Disadvantages.
  - Creates difficulty in interpreting aircraft spacing and relative motion while in flight, especially during night flight–aided or unaided.
  - Presents a poor choice during dust/sand/snow takeoffs and landings. Aircraft can be engulfed by the cloud of the preceding aircraft.
  - Requires a relatively long landing area.

Rotary-Wing Terrain Flight

Figure 6-10. Trail formation

V-Formation

6-65. This formation consists of a leader and two wingmen, each in echelon, left and right (figure 6-11, page 6-18). The wingmen hold a position approximately 45 degrees astern of the leader, both left and right. Aviators must scan for both aircraft to maintain proper position in the formation. V-formation has the following advantages and disadvantages:
- Advantages:
    - Allows rapid deployment of troops for all-around security.
    - Requires a relatively small landing area.
    - For dust/sand/snow condition takeoffs and landings, small V-formations can be used with light wind conditions. Increased rotor disk separation prevents being engulfed in the cloud from the preceding aircraft.
- Disadvantages: Restricts suppressive fire from inboard door gunners.

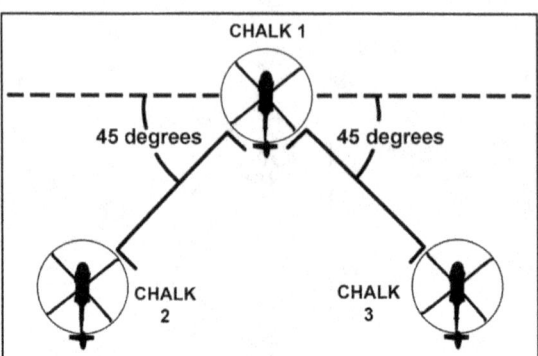

Figure 6-11. V-formation

MANEUVERING FORMATIONS

6-66. Maneuverability is the prime consideration for formations flying in tactical situations. The following formations provide the basis for team maneuvering flight and are used to provide maximum maneuverability, flexibility, and survivability due to greater separation between aircraft. They also promote security by providing overlapping fields of view and fields of fire. These formations allow lead to maintain formation integrity, yet maneuver the formation with few restrictions. Wingmen must maintain a position that will not hamper the preceding aircraft's ability to maneuver. Wingmen must also understand that due to their authority to maneuver, lead is free to maneuver near terrain, expecting wingmen to provide their own horizontal and vertical clearance.

6-67. The positions and distances described in this document are guidelines and can be modified as the situation dictates. Over open terrain or during high illumination, greater spacing is used to increase survivability and flexibility. Formation spacing should be tighter in rough terrain or reduced illumination/visibility. Formation positions nearer the abeam make scanning more difficult in keeping the preceding aircraft, as well as approaching terrain, in sight. Many wingmen move to the outside of turns to more easily keep lead and approaching terrain in sight, while maintaining altitude (or stacking high). Conversely, wingmen must be extremely vigilant if assuming a position on the inside of turns, as a rapid scan is required to maintain SA on lead and approaching terrain. This is especially critical when stacking low in the turn. It is important to avoid flying the entire formation over the same spot on the ground. Variations in flight path between teams should be the rule.

6-68. The mission will dictate aircraft separation and team separation. Aircraft and team separation may range from three to five rotor disks to 1 kilometer or more. Primary concern when establishing separation is METT-TC and the ability to provide mutual support. Basic team formations are combat cruise, combat cruise left/right, combat trail, and combat spread. They can be enlarged to accommodate multiple teams, platoon size, and larger formations.

COMBAT CRUISE OR COMBAT CRUISE TEAMS IN TRAIL

6-69. Combat cruise replaces the term free cruise to incorporate joint terminology. Combat cruise is the basic formation utilized by a team and provides maximum flexibility and adequate mutual support. Lead retains the freedom to maneuver and engage targets without affecting his or her wingman's flight path unless aircraft are flying in tight formation. Observation sectors must be divided between lead and wing to provide overlapping observation and fire. The wingman should inform lead when changing from one side to the other if this information is required for SA. The wingman is allowed to vary separation and angle anywhere in the maneuver area from approximately 3 to 9 o'clock (figure 6-12, page 6-19). Since the formation does not require an absolute position, flight crews can concentrate on navigation, terrain masking, and enemy detection/avoidance. Wingmen position themselves where they can best visually cover lead (optimum position is 45 degrees) and should be prepared to deliver ordinance in support of lead.

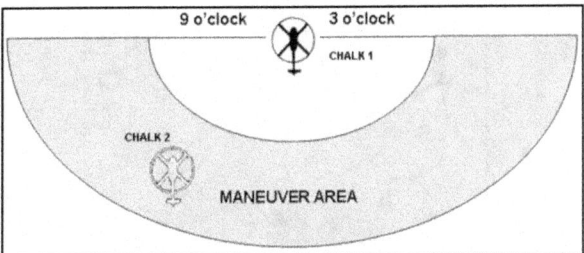

Figure 6-12. Team combat cruise

6-70. In rough terrain, the formation is normally tighter than in open terrain. When lead initiates a turn, wingmen maintain longitudinal clearance on the aircraft directly ahead by sliding and utilizing the radius of the turn created by lead. As soon as lead rolls level, positions are resumed. Since the position in combat cruise varies, the wingman should avoid presenting a linear target during break turns. Extended flight in lead's 6 o'clock is not recommended.

6-71. Formations of more than two aircraft can utilize combat cruise. Figure 6-13 shows a flight of four in combat cruise with the maneuver area limited to 45 degrees. Each subsequent aircraft flies a relative position off the preceding aircraft. To maintain team integrity for attack/recon scout weapons teams, the term combat cruise teams in trail can be used and spacing between teams are extended slightly.

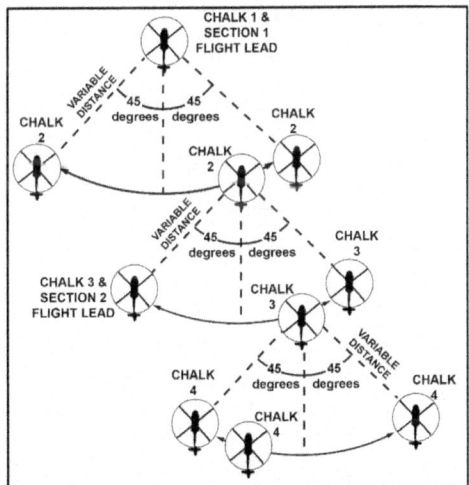

Figure 6-13. Flight combat cruise

COMBAT CRUISE LEFT/RIGHT

6-72. Another formation used by flight lead to limit maneuverability is combat cruise left/right (figure 6-14, page 6-20). Combat cruise left/right is a modified staggered formation which allows for tactical maneuverability and spacing yet maintains some predictability. Subsequent aircraft will remain in either right or left cruise and change sides only after briefed by flight lead. Using combat cruise left/right, the wingman remains in an arc 0 degrees aft to 90 degrees abeam of lead to the left or right side. Optimum position is 45 degrees. Observation sectors are divided between lead and wing providing overlapping observation and fire. Figure 6-15 illustrates combat cruise left for more than two aircraft.

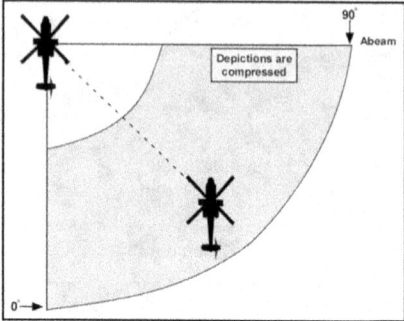

Figure 6-14. Combat cruise right

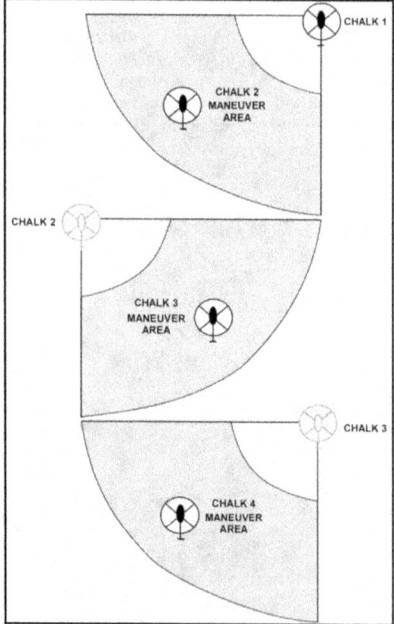

Figure 6-15. Combat cruise left

COMBAT TRAIL

6-73. While combat cruise allows wingmen maximum flexibility, there may be instances where flight lead requires more control of the flight and must restrict some maneuverability. Combat trail can be used to limit wingmen's movement to plus or minus 30 degrees from the preceding aircraft (figure 6-16). This formation is useful for negotiating narrow terrain or landing in narrow LZs. It should not be flown for extended periods of time or at night due to the difficulty of determining rates of closure for preceding aircraft.

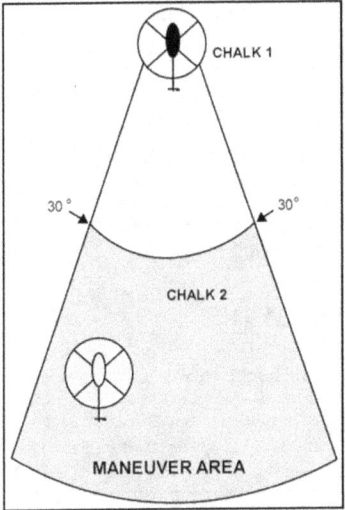

Figure 6-16. Combat trail

COMBAT SPREAD

6-74. Combat spread is a formation used when maximum observation to the front is desired or an attempt to limit package exposure time over open areas is made. When flight lead announces combat spread, he or she includes the command right or left. Wingmen should move toward that abeam position, either lead's 3 or 9 o'clock position (figure 6-17). Flying in combat spread requires a rapid scan to maintain SA on the other aircraft and approaching terrain. This requires even more vigilance at night.

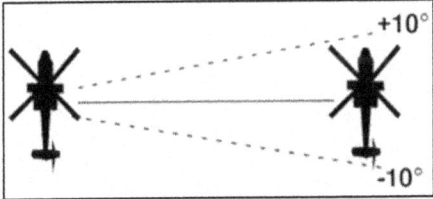

Figure 6-17. Combat spread

SECTION III – BASIC COMBAT MANEUVERS

6-75. Basic combat maneuvers (BCMs) are essential elements for successful multi-aircraft operations. Team maneuvering flight relies on standardized maneuvers and terminology to defend against deliberate or chance encounters with enemy forces occurring throughout the battlefield. Each team member must be able to communicate and understand each maneuver to enhance mutual support within the team and flight while performing various missions. These missions include attack, reconnaissance, assault, and lift conducted by

Chapter 6

one or more teams or during missions involving dissimilar aircraft such as rescue escort, casualty evacuation (CASEVAC), and personnel recovery.

MANEUVERING FLIGHT COMMUNICATIONS

6-76. It is essential every crewmember understand the maneuver to be performed. Communication is an integral part of training for lead and wing. It provides a basis and the control measures required to practice maneuvering team flight. As a team gains proficiency, the communication between lead and wing may evolve to a more abbreviated form but the basics should remain. The more abbreviated form of communication does not constrain the use of these maneuvers in application. During training, each pilot should acknowledge the maneuver and respond with the command of execution (for example, Lead: Team one, Break left, Ready; Wing: Team one, Break left, Go). Several maneuvers have standard turn changes. This may be modified in the communication prior to executing the turn (for example, Team one, Break left 270, Ready; Wing: Team one, Break left 270, Go). Engagement criteria and target identification may also be added for clarity (for example, Team one, Cross turn and cover high engage; Wing: Team one, Cross turn and cover low, tally target).

BASIC COMBAT MANEUVERS

6-77. BCMs provide the team with a "toolbox" of maneuvers to choose from when encountering threats during tactical flight (figure 6-18). These maneuvers facilitate the suppression of enemy fire destruction of targets, Mission command, and the reorganization of the flight following the encounter. Maneuvers are divided into two categories to include maneuvers required to engage—

- Close in threats, less than 1.5 kilometers inside area weapons system (AWS) ranges.
- And/or bypass threats outside weapons ranges, 1.5 to 5 kilometers outside area weapons ranges.

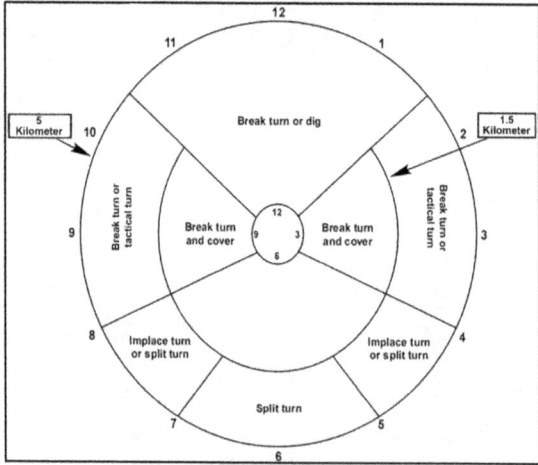

Figure 6-18. Basic combat maneuver circle

Note. Unless an engagement forces the tactical lead to change from one aircraft to another, flight lead will not change during any of these maneuvers. Rather, the wingman should delay initiating the turn or vary the angle of bank or airspeed to assume an appropriate position once the maneuver is completed. For example, a split turn initiated while a team is in combat cruise formation will require the wingman delay initiating his or her turn or vary the angle of bank so not to roll out of the turn in front of the tactical lead.

TACTICAL TURNS

6-78. The tactical turn is used to maneuver the flight, maintain observation sectors, and allow mutual support. These maneuvers are used to change the direction of the formation (usually approximately 60 to 120 degrees) and change wingman side. tactical turns also enable aircrews to turn the formation in a smaller area by eliminating the need for the wingman to fly the outside arc of lead's turn. All tactical turns follow three basic principles—

- The aircraft on the outside of the turn always turns first.
- The wingman always changes sides in the formation.
- The wingman is always responsible for separation.

A turn of 90 degrees is understood, if not stated. If a smaller or larger heading change is desired, lead may specify a magnitude of heading change (for example, Team 1, tactical left to heading 270).

Tactical Turn (Away from Wingman)

6-79. From combat cruise or combat spread, lead maintains heading, and wing turns immediately to the new heading. When wing passes the 5 or 7 o'clock position, lead turns to the new heading and formation change is understood. A vertical component (cover) may be added by stating "Cover high" or "Cover low". Figure 6-19 depicts a tactical turn away.

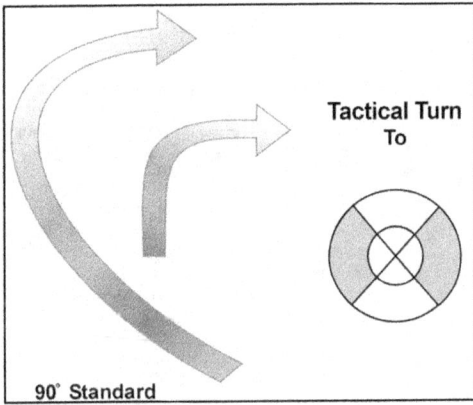

Figure 6-19. Tactical turn away

Tactical Turn (Toward the Wingman)

6-80. From combat cruise or combat spread, on acknowledgement, lead immediately turns to the new heading and passes in front of the wingman. The wingman maintains heading (or alters slightly to lead's tail) until lead passes 2 o'clock. Wing then turns to the new heading. If separation is not adequate for lead to cross the wing position, wing may initiate a turn in the opposite direction to facilitate lead's turn. Maneuver reverses each aircraft's relative position (combat cruise right will now be combat cruise left). Formation change is understood. A vertical component (cover) may be added by stating "Cover high" or "Cover low". Figure 6-20 shows a tactical turn to wingman.

Chapter 6

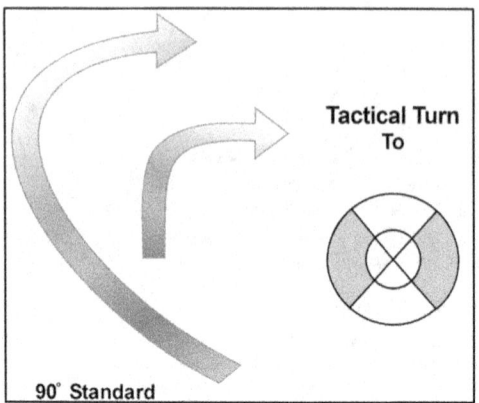

Figure 6-20. Tactical turn to

DIG, RESUME, AND PINCH

6-81. Dig and pinch may be used in combination or separately to react to a threat or hazard in the forward quadrant. These maneuvers allows for rapid dispersion and incremental control of the formation with short, precise commands from lead. If a threat is discovered in the forward quadrant, little or no time may be available to engage the target. In formation, a dig will split the lead team to enable a follow-on team to engage a forward quadrant threat.

6-82. From combat cruise or combat spread, aircraft simultaneously turn 30 to 45 degrees away from each other. When desired lateral separation is attained, the tactical leader calls "resume". To decrease lateral separation, the tactical leader calls "pinch". Both aircraft simultaneously turn 30 to 45 degrees toward the inside. Aircraft will decrease lateral separation until they return to the previous formation and separation or until a "resume" call is made. Resume is defined as a "return to mission heading and maintain current separation". Figure 6-21 illustrates the dig and pinch maneuvers.

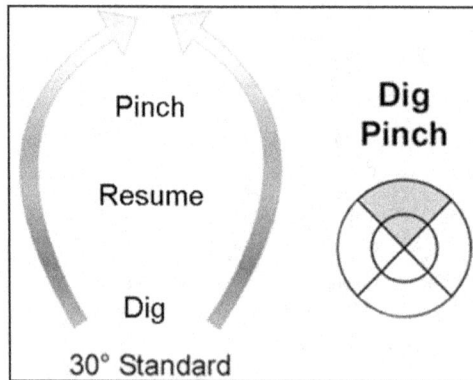

Figure 6-21. Dig and pinch maneuvers

SPLIT TURN

6-83. This maneuver rapidly reverses team direction to engage or bypass an enemy located at 5 and 7 o'clock, outside weapons range. When aircraft within the formation are in combat cruise and rapid dispersion of the team is desired, a split turn is the preferred method. This maneuver forces the threat to either bypass or commit earlier to the lead or wingman.

6-84. A split turn changes a formation's heading from 120 to 240 degrees. Both aircraft will execute a left and right turn to the new heading. If no heading is given, a turn of 180 degrees is understood. Angle of bank and power must be maintained so the aircraft should be tail to tail at the apex of the turn. When the maneuver is complete, lead and wingman will have reversed relative positions. Figure 6-22 depicts a split turn.

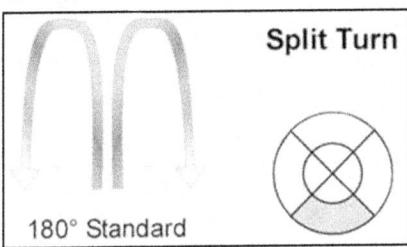

Figure 6-22. Split turn maneuver

IN-PLACE TURN

6-85. An in-place turn (reversal) (figure 6-23) rapidly reverses team direction to engage or bypass an enemy located at 5 and 7 o'clock, outside weapons range. This maneuver allows the wingman to keep the lead aircraft in sight at all times. An in-place turn may also be used to egress a static BP.

6-86. On acknowledgement, both aircraft will execute a left or right turn to new heading. An in-place turn may be used for both small (30 degrees or less) and large heading changes (120 to 240 degrees). If a specific heading is not given, a heading change of 180 degrees is understood. To initiate small heading changes, both aircraft turn to the new heading and relative position is maintained. To initiate large heading changes both aircraft turn in the specified direction. Angle of bank and power must be maintained so aircraft are in trail at the apex of the turn. As the team continues its turn to the new heading, the wingman switches relative position (combat cruise right is now combat cruise left).

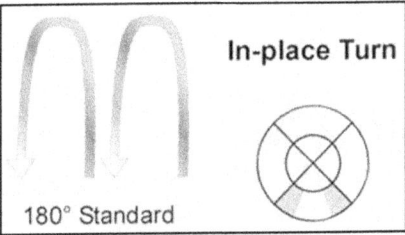

Figure 6-23. In-place turn

CROSS TURN

6-87. A cross turn (figure 6-24, page 6-26) is used to rapidly orient the team on an engaging threat from the rear quadrant. This turn may also be used to reverse the flight's heading in channelized terrain. Cross turns change a formation's heading approximately 120 to 240 degrees. If a specific heading is not given, a heading change of 180 degrees is understood. Cross turns may be performed from either combat cruise or combat spread. Unless specified, lead should fly the outside turn allowing wing to turn inside. The aircrew initiating the turn may specify whether they will be flying the outside turn or the inside turn by stating "cross turn inside" or "cross turn outside", especially during execution from a combat spread formation. Initial separation determines the angle of bank needed. Angle of bank must be adjusted to maintain position in the flight. The cross turn should not be used in situations where an enemy might deliver ordnance at the apex of the turn, since both helicopters are closely aligned at this point.

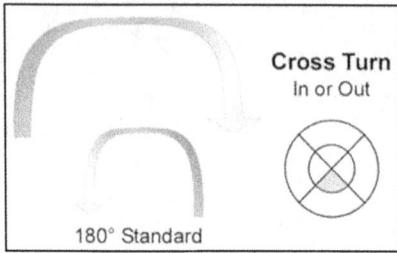

Figure 6-24. Cross turn in or out

CROSS TURN AND COVER

6-88. The cross turn and cover maneuver (figure 6-25) provides for vertical separation as well as lateral separation between lead and wing. It is also a variation of the cross turn. This maneuver is initially used as a defensive maneuver to orient the team on the enemy while confusing the enemy. Once oriented, the team then engages targets within weapons range.

6-89. This maneuver reverses the flight's direction and provides split-phase, split-plane engagement of targets. It may be initiated from combat cruise or combat spread. The aircrew sighting the enemy first initiates the maneuver, calls "cross turn cover high" or "cross turn cover low", and executes the "high" or "low" altitude. The high aircraft maneuvers to a lookdown position on the enemy. The low aircraft turns to face the enemy and maneuvers to provide mutual support.

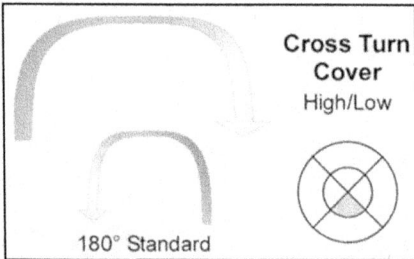

Figure 6-25. Cross turn cover (high/low)

BREAK TURN

6-90. Break turns are maximum aircraft performance maneuvers that may be used to orient the flight toward an enemy aircraft that has penetrated within weapons engagement parameters, break away from hostile ground fire or bring weapons to bear immediately on a target.

6-91. Break turns are used as an initial formation response when a member of the formation has spotted a threat outside the AWS range. Flight lead further develops the situation into a drill once the flight is properly oriented to/from the threat.

6-92. On acknowledgment, both aircraft will execute a left or right turn to the new heading (figure 6-26). If a specific heading is not given, a heading change of 90 degrees is understood. To maintain the same relative position, an adjustment of speed may be required to compensate for steeper turns. Break turns may be executed to the left or right.

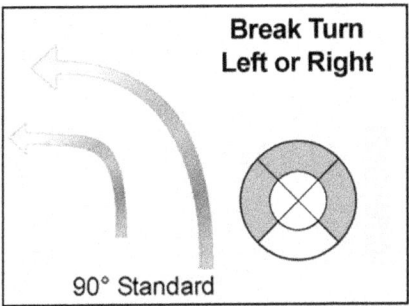

Figure 6-26. Break turn left/right

BREAK TURN AND COVER

6-93. The break turn and cover provides an immediate break with vertical separation of aircraft to engage a target. This is an immediate action maneuver used when enemy is spotted abeam (2 to 4 o'clock or 8 to 10 o'clock) within weapons range. Described simply, it is a break turn with vertical separation to engage a common threat.

6-94. After the execution call and acknowledgment, the aircraft closest to the enemy initiates the maneuver. The aircraft closest to the enemy begins an immediate climb while simultaneously turning to confront the enemy. As aircraft closest to the enemy begins to climb, the wingman turns to an angle-off flight path and maneuvers to provide mutual support. This maneuver achieves maximum vertical separation and should force the enemy to choose between two possible targets maneuvering to engage. Figure 6-27 depicts a break turn and cover.

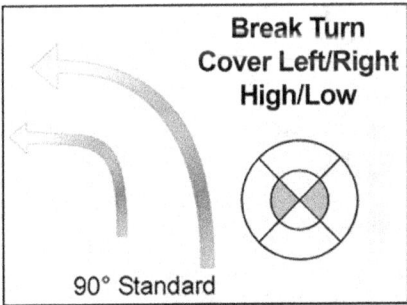

Figure 6-27. Break turn left/right (high/low)

SHACKLE TURN

6-95. This maneuver allows aircraft to thoroughly observe the 6 o'clock position of the flight. If the section consists of three aircraft, only the last two aircraft should perform the maneuver. If a flight consists of two or more sections, the same applies. Only the last two aircraft will check rear security for the flight.

6-96. Shackle turns (figure 6-28) can be executed from both combat cruise and combat spread. The command for the shackle turn is "shackle turn" followed by the execution command "go" and acknowledgement. Lead maintains his or her heading, while wingman initiates a 30-degree turn toward the lead. Lead verifies the wingman has initiated a turn. Lead then initiates a 30-degree turn in the opposite direction. As wing passes the 6 o'clock position, lead returns to original heading. If performed at night, a greater off-angle may be required based on sensor limits. The off-angle continues until lead calls "resume". At the completion of this maneuver, lead and wing have changed relative position.

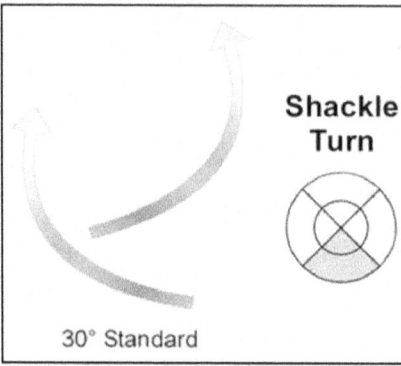

Figure 6-28. Shackle turn

SECTION IV – PLANNING CONSIDERATIONS AND RESPONSIBILITIES

PLANNING CONSIDERATIONS

6-97. The factors considered in determining the best formation, or sequence of formations, are as follows:
- Mission requirements include the mission of the supported unit and aviation unit.
- Enemy considerations include current enemy situation, enemy ADA capability and placement, and accessibility to enemy visual/electronic surveillance.
- Fire support plan considerations include artillery support available, LZ preparation planning, air support availability and requirements, and naval gunfire—including planned types of ordnance and any en route suppression of enemy air defense.
- Terrain and weather considerations include configuration of en route obstacles and/or corridors, LZ characteristics, obstacles in/or affecting approaches to the LZ, ceiling and visibility, wind and turbulence, and ambient light levels throughout the mission.
- Formation maneuver and flexibility considerations include possible changes in the mission or situation and evasive tactics to be used.
- Armed aerial escort considerations include the number and type of armed escort aircraft required and available.
- Formation control considerations include the degree of control required and method of control such as radio, visual signals, and prearranged timing.
- Other considerations include type of aircraft, type of NVDs used, OPSEC and safety measures required, level of crew training and experience, and aircraft capabilities.
- When different types of aircraft operate in a formation, external lighting capability of the various aircraft types must be evaluated. In addition, when aircraft types are mixed at night, differences between NVG and FLIR must be identified and considered in planning.
- Installed aircraft survivability equipment (ASE) and the impact different formations have on it versus threat.

PLANNING RESPONSIBILITIES

6-98. Supported ground unit commanders should brief the supporting aviation unit on the following items:
- Fire and EW support plans.
- Frequencies and call signs.
- Details of friendly troops including location, numbers, and unit identification.
- Number of troops to be airlifted.
- Description, amount, size, and weight of cargo.
- Location, details, control provided, and specific landing points for primary and alternate PZs.
- Safe routes to and from the LZ based on available intelligence.
- Desired arrival time in the LZ.
- Location, details, control provided, and specific landing points for primary and alternate LZs.
- Location of the ground unit commander, if airborne.

6-99. The air mission or flight commander is responsible for effecting liaison with the supported ground unit and supporting aviation units. The aviation brief to the supported ground unit should include the following:
- Safety requirements.
- Use of aircraft lights providing aircraft identification means to the supported unit.
- Frequencies, call signs, and troop commander seat assignment including availability of aircraft headset and communication capability.
- Probable en route and landing formations.
- Aircraft troop and cargo load capability and identification of aircraft carrying both.
- Downed crew pickup points and downed aircraft procedures.
- PZ/LZ lighting requirements and aircraft separation requirements.
- Thorough passenger briefing, including appropriate warnings regarding aircraft ingress and egress, and approach/departure paths to/from the aircraft. This briefing must include seat belt availability, placement of personal equipment, and emergency procedures.

6-100. The mission brief to the aviation unit should include the following:
- Route of flight.
- Rules of engagement.
- Time schedule.
- Details of the PZ and LZ.
- Number of aircraft required for the mission.
- Troop load (including aircraft ACL) and cargo load.
- Formations to be used.
- Numbering system (identification) for aircraft; for example, in case of a formation change, starting with lead or chalk 1 continuing backwards through the flight; lead departs the flight, then chalk 2 becomes lead, chalk 3 becomes chalk 2, and so on.
- Assigned duties for each chalk number.
- Horizontal distance and vertical separation.
- Use of aircraft lights.
- Signal requirements including lights and communication.
- IIMC procedures.
- Emergency breakup procedures (threat response).
- Method of changing formation.
- Rendezvous and join-up procedures.
- Available intelligence regarding routes and PZs/LZs.
- Lost communication procedures.
- Downed aircrew pickup points and downed aircraft procedures.

Chapter 6

- Status of armed escort aircraft.
- Refueling and rearming instructions, including FARP locations.
- Emergency medical procedures.
- Location of AMC and aviation unit commander.

SECTION V – WAKE TURBULENCE

6-101. Information on wake turbulence is placed in this chapter as aviators are likely to experience turbulent conditions while operating around other aircraft. Successful aviators understand and recognize conditions conducive to wake turbulence and take appropriate countermeasures. Larger aircraft create more turbulence and are greater hazards.

IN-FLIGHT HAZARD

6-102. Every aircraft in flight generates wake turbulence. This disturbance is caused by a pair of counter-rotating vortices trailing from the wing tips. It is possible the wake of another aircraft can impose rolling moments exceeding the control authority of the aircraft. Additionally, if encountered at close range, wake turbulence can damage the aircraft and/or cause personal injury to the occupants. It is important to imagine the location of the vortex wake generated by other aircraft and adjust the flight path accordingly.

GROUND HAZARD

6-103. Hazardous turbulence is not only encountered in the air. During ground operations and takeoff, jet engine blast (thrust stream turbulence) and rotorwash can cause damage and disturbance if encountered at close range. Exhaust velocity versus distance studies at various thrust levels have shown a need for light aircraft and helicopters to maintain an adequate separation behind large turbojet aircraft.

VORTEX GENERATION

6-104. Lift is generated by the creation of a pressure differential over the wing surfaces. This pressure differential triggers rollup of airflow aft of the wing resulting in swirling air masses trailing downstream of the wingtips (figure 6-29). After rollup is completed, the wake consists of two counter-rotating cylindrical vortices. Most of the energy is within a few feet of the center of each vortex; however pilots must avoid the region within approximately 100 feet of the vortex core.

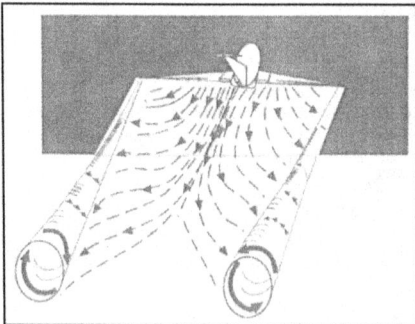

Figure 6-29. Wake vortex generation

STRENGTH

6-105. Vortex strength is governed by the weight, speed, and shape of the generating aircraft's wing. The basic factor is weight; vortex strength increases proportionately with an increase in aircraft operating weight.

Peak vortex speeds up to almost 300 feet per second have been recorded. The greatest vortex strength occurs when the generating aircraft is heavy, clean, and slow.

BEHAVIOR

6-106. Trailing vortices have certain behavioral characteristics which can help aviators visualize wake location and take avoidance precautions.

- Vortices are generated from the moment an aircraft leaves the ground because trailing vortices are a by-product of wing lift. Prior to takeoff or landing, pilots should note the rotation or touchdown point of the preceding aircraft.
- Vortex circulation is outward, upward, and around the wing tips when viewed either ahead or behind the aircraft. If persistent vortex turbulence is encountered, a slight change of altitude and lateral position (preferably upwind) should provide a flight path clear of the turbulence.
- Flight tests have shown vortices from aircraft sink at a rate of up to several hundred FPM, slowing their descent and diminishing in strength with time and distance behind the generating aircraft. Atmospheric turbulence hastens breakup. Aviators should fly at or above the preceding aircraft's flight path, altering their course as necessary to avoid the area behind and below the generating aircraft.
- A crosswind will decrease lateral movement of the upwind vortex and increase movement of the downwind vortex. This results in the upwind vortex remaining in the touchdown zone for a period of time and increases drift of the downwind vortex toward another runway.

INDUCED ROLL AND COUNTER CONTROL

INDUCED ROLL

6-107. In rare instances, a wake encounter can cause in-flight structural damage of catastrophic proportions. The most common hazard is associated with induced rolling moments which can exceed the roll control capability of the encountering aircraft. During flight tests, aircraft have been intentionally flown directly up trailing vortex cores of larger aircraft. These tests prove the capability of an aircraft to counteract the roll imposed by the wake vortex primarily depends on wing span and counter control responsiveness of the encountering aircraft.

COUNTER CONTROL

6-108. Counter control is usually effective and induced roll is minimal in cases where the encountering aircraft extends outside the affected area of the vortex. It is more difficult for aircraft smaller than the aircraft generating the vortex to counter the imposed roll induced by vortex flow. Although aviators of short span aircraft and helicopters must be especially alert to vortex encounters, the wake of larger aircraft requires the respect of all aviators.

OPERATIONAL PROBLEM AREAS

6-109. Wake turbulence encounters can be one or more jolts with varying severity depending upon direction of the encounter, weight of the generating aircraft, size of the encountering aircraft, distance from the generating aircraft, and point of vortex encounter. The probability of induced roll increases when the encountering aircraft's heading is generally aligned or parallel with the flight path of the generating aircraft.

- Avoid the area below and behind the preceding aircraft especially at low altitude where even a momentary wake encounter could be hazardous.
- Aviators must be particularly alert in calm wind conditions and maneuvering situations in the vicinity of the airfield where the vortices could—
  - Remain in touchdown area.
  - Drift from aircraft operating on a nearby runway.
  - Sink into takeoff or landing path from crossing runway.

- Sink into traffic patterns from other airport operations.
- Sink into flight path of aircraft operating VFR.

6-110. Aviators should visualize the location of the vortex trail behind a larger aircraft/helicopter and use proper vortex avoidance procedures to achieve safe operation. It is equally important aviators of a larger aircraft/helicopter plan or adjust their flight paths, whenever possible, minimizing vortex exposure to other aircraft.

HELICOPTERS

6-111. In a slow hover taxi or stationary hover near the surface, helicopter main rotor systems generate downwash producing high velocity outwash vortices to a distance approximately three times the diameter of the rotor. When rotor downwash hits the surface, the resulting outwash vortices have behavioral characteristics similar to wing tip vortices produced by FW aircraft. However, vortex circulation is outward, upward, around, and away from the main rotor(s) in all directions. Aviators should avoid operating within three rotor diameters of any helicopter in a slow hover taxi or stationary hover. In forward flight, departing or landing helicopters produce a pair of strong, high speed trailing vortices similar to wing tip vortices of larger FW aircraft. Aviators must use caution when operating or crossing behind landing and departing helicopters.

JET ENGINE EXHAUST

6-112. Engine exhaust velocities, generated by larger jet aircraft during ground operations and initial takeoff roll, dictate the desirability of lighter aircraft awaiting takeoff to hold well back of the runway edge at the taxiway hold line. It is also desirable to align the aircraft to face any possible jet engine blast effects.

6-113. The FAA has established standards for the location of runway hold lines. For example, runway intersection hold short lines are established 250 feet from the runway centerline for precision approach runways served by approach category C and D aircraft. For runways served by aircraft with wingspans over 171 feet, such as the C-5, taxiway hold lines are 280 feet from the centerline of precision approach runways. These hold line distances increase slightly with an increase in field elevation.

VORTEX AVOIDANCE TECHNIQUES

6-114. Under certain conditions, airport traffic controllers apply procedures for separating aircraft operating under instrument flight rules (IFR). The controllers also provide VFR aircraft with the position, altitude, and direction of larger aircraft followed by the phrase "caution–wake turbulence." Whether or not a warning has been given, aviators are expected to adjust their operations and flight path(s) as necessary to avoid serious wake encounters.

6-115. The following vortex avoidance procedures are recommended:
- Landing behind a larger aircraft on the same runway. Stay at or above the larger aircraft's final approach flight path, note the touchdown point, and land beyond it.
- Landing behind and offset from a larger aircraft landing on a parallel runway that is closer than 2,500 feet. Consider possible vortex drift onto your runway, stay at or above the larger aircraft's final approach flight path, note its touchdown point, and land beyond it.
- Landing behind a larger aircraft on a crossing runway. Cross above the larger aircraft's flight path.
- Landing behind a departing larger aircraft on the same runway. Note the larger aircraft's rotation point and land well prior to rotation point.
- Landing behind a departing larger aircraft on a crossing runway. Note the larger aircraft's rotation point. If past the intersection, continue the approach and land prior to the intersection. If prior to the intersection, abandon the approach unless a landing is assured well before reaching the intersection and avoid flight below the larger aircraft's flight path.
- If departing behind a larger aircraft, note the larger aircraft's rotation point and rotate prior, and then continue to climb above the larger aircraft's climb path until turning clear of its wake. Avoid subsequent headings crossing below and behind a larger aircraft. Be alert for any critical takeoff situation possibly leading to a vortex encounter.

- For intersection takeoffs on the same runway, remain alert for adjacent large aircraft operations particularly upwind of the runway. Avoid headings crossing below a larger aircraft's path.
- When departing or landing after a larger aircraft executing a low approach, missed approach or touch-and-go landing, because vortices settle and move laterally near the ground, the vortex hazard may exist along the runway and in the flight path after a larger aircraft has executed a low approach, missed approach or a touch-and-go landing, particularly in light quartering wind conditions. Ensure an interval of at least two minutes has elapsed before takeoff or landing.
- When en route VFR, avoid flight below and behind a larger aircraft's path. If a larger aircraft is observed above on the same track (meeting or overtaking), adjust position laterally, preferably upwind.

AVIATOR RESPONSIBILITY

6-116. Government and industry groups are making concerted efforts to minimize or eliminate hazards of trailing vortices; however, the flight discipline necessary to ensure vortex avoidance during VFR operations must be exercised by the aviator. Vortex visualization and avoidance procedures are exercised by the aviator using the same degree of concern as in collision avoidance.

6-117. Aviators are reminded in operations conducted behind all aircraft, acceptance of instructions from ATC in the following situations is an acknowledgement that the aviator ensures safe takeoff and landing intervals and accepts responsibility for providing wake turbulence separation:
- Traffic information.
- Instructions to follow an aircraft.
- Acceptance of a visual approach clearance.

For operations conducted behind heavy aircraft, ATC specifies the word "heavy" when this information is known.

6-118. For VFR departures behind heavy aircraft, air traffic controllers are required to use at least a 2-minute separation interval unless an aviator has initiated a request to deviate from the 2-minute interval and indicated acceptance of responsibility for maneuvering the aircraft, thereby avoiding wake turbulence hazard.

AIR TRAFFIC CONTROL WAKE TURBULENCE SEPARATION

Required Separation

Behind Heavy Jets

6-119. Due to possible effects of wake turbulence, controllers are required to apply no less than specified minimum separation for aircraft operating behind a heavy jet and, in certain instances, behind large non-heavy aircraft.

6-120. Separation is applied to aircraft operating directly behind a heavy jet at the same altitude or less than 1,000 feet below—
- Heavy jet behind heavy jet—4 miles.
- Small/large aircraft behind heavy jet—5 miles.

6-121. Also, separation, measured at the time the preceding aircraft is over the landing threshold, is provided to small aircraft—
- Small aircraft landing behind heavy jet—6 miles.
- Small aircraft landing behind large aircraft—4 miles.

6-122. Additionally, departing aircraft will be separated by either two minutes or the appropriate 4 or 5 mile radar separation when takeoff behind a heavy jet will be—
- From the same threshold.
- On a crossing runway and projected flight paths will cross.

- From the threshold of a parallel runway when staggered ahead of the adjacent runway by less than 500 feet and when runways are separated by less than 2,500 feet.

6-123. Aviators, after considering possible wake turbulence effects, may specifically request a waiver of the 2-minute interval. Controllers may acknowledge this statement as aviator acceptance of responsibility for wake turbulence separation and, if traffic permits, issue takeoff clearance.

Behind Larger Aircraft

6-124. A 3-minute interval will be provided when a small aircraft will takeoff—
- From an intersection on the same runway (same or opposite direction) behind a departing large aircraft.
- In the opposite direction on the same runway behind a large aircraft takeoff or low/missed approach.

This 3-minute interval may be waived upon specific aviator request.

6-125. Controllers may not reduce or waive the 3-minute interval if the preceding aircraft is a heavy jet and operations are on either the same runway or parallel runways separated by less than 2,500 feet.

6-126. Aviators may request additional separation, that is, two minutes instead of 4 or 5 miles for wake turbulence avoidance. This request is made as soon as practical on ground control and at least before taxiing onto the runway.

6-127. Controllers may anticipate separation and need not withhold a takeoff clearance for an aircraft departing behind a large/heavy aircraft if there is reasonable assurance the required separation will exist when the departing aircraft starts takeoff roll.

Chapter 7

Fixed-Wing Aerodynamics and Performance

This chapter presents aerodynamic fundamentals for FW flight and should be used with information found in chapter 1 that applies to FW flight.

SECTION I – FIXED-WING STABILITY

MOTION SIGN PRINCIPLES

7-1. An aircraft has six directions of motion around three mutually perpendicular axes (figure 7-1). These three axes are vertical, lateral, and longitudinal.
- Vertical axis about which the aircraft yaws.
- Lateral axis about which the aircraft pitches.
- Longitudinal axis about which the airframe rolls.

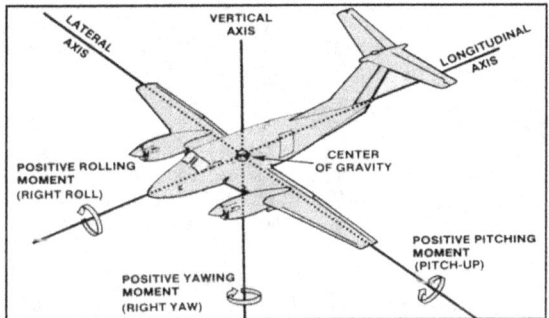

Figure 7-1. Stability nomenclature

7-2. Sign principles are assigned to each motion. The right-hand rule can be applied to remember the signs. A right roll, right yaw, and pitch-up are all positive. For example, a positive rolling moment rolls the aircraft right and a negative rolling moment rolls the aircraft left.

STATIC STABILITY

7-3. Static stability is the tendency an object possesses after it has been displaced from its equilibrium (figure 7-2, page 7-2). Newton's first law of motion implies if the sum of the forces and moments about the CG of an object are equal to zero, then no acceleration will take place. This state is called equilibrium.

Chapter 7

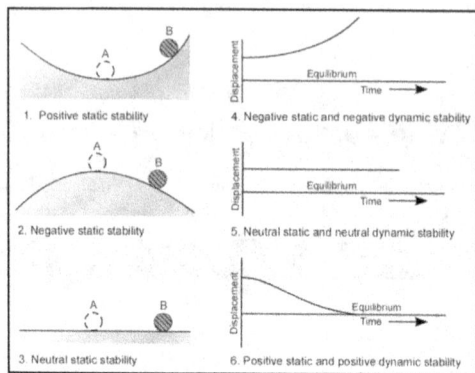

Figure 7-2. Nonoscillatory motion

POSITIVE STATIC STABILITY

7-4. If an object possesses positive static stability, it tends to return to its equilibrium position after having been moved. Part 1, point A, shows the ball in equilibrium. If the ball is moved to point B, it tends to roll back toward point A. This tendency demonstrates positive static stability. The ball may not actually return to point A, but it does tend to return. Therefore, it has positive static stability.

NEGATIVE STATIC STABILITY

7-5. In part 2, the bowl has been inverted. Point A is the equilibrium position. The ball has been moved to point B and tends to roll away from point A. This tendency toward movement away from the equilibrium position demonstrates negative static stability. The ball may or may not roll away from point A, but it tends to roll away. Therefore, it has negative static stability.

NEUTRAL STATIC STABILITY

7-6. In part 3, the ball has been placed on a flat surface. When the ball is moved to point B, it neither tends to return to nor roll away from point A. This demonstrates neutral static stability.

DYNAMIC STABILITY

7-7. The word dynamic implies motion, while dynamic stability refers to movement of an object with respect to time. When dynamic stability of an object is considered, static stability of the object must also be considered. Figure 7-2 also depicts the follow categories of dynamic stability.

NONOSCILLATORY MOTION

Negative Static and Negative Dynamic Stability

7-8. An object possessing negative static stability tends to move away from its equilibrium position. Part 4 reflects nonoscillatory dynamic stability.

Neutral Static and Neutral Dynamic Stability

7-9. If an object has nonoscillatory neutral dynamic stability if it has been displaced and does not move toward or away from its equilibrium position (part 5).

Positive Static and Positive Dynamic Stability

7-10. An object with a positive static stability and strong positive dynamic stability results in nonoscillatory positive dynamic stability (part 6). This is called deadbeat damping. Dynamic stability is particularly stressful on an aircraft structure and could eventually cause material failure.

OSCILLATORY MOTION

7-11. To have oscillatory motion, an object must possess positive static stability. Following is a discussion of various types of dynamic stability coupled with positive static stability.

Positive Static and Positive Dynamic Stability

7-12. An object has positive static stability if it is displaced and tends to return to its equilibrium position. As indicated by the elapsed time shown in figure 7-3, the object at part A moves toward its equilibrium position. This motion continues but diminishes until the object comes to rest at its equilibrium position. A decrease in the amplitude of the oscillations indicates the object has positive dynamic stability and will eventually come to rest in an equilibrium position.

Positive Static and Neutral Dynamic Stability

7-13. Figure 7-3 part B, shows that elapsed time indicates neutral dynamic stability. An object displaced moves toward equilibrium and overshoots it. Positive static stability makes the object move back toward the equilibrium position. Again, the object overshoots the equilibrium position with its oscillations equal to the oscillations in the first displacement. As time passes, the amplitude of the oscillations is the same on both sides of the equilibrium position and never comes to rest. Because the oscillation amplitude does not increase or decrease, the object has neutral dynamic stability.

Positive Static and Negative Dynamic Stability

7-14. In figure 7-3, part C, positive static stability makes the object oscillate, as in the first two examples. However, the increasing amplitude of the oscillations as time passes indicates negative dynamic stability. Lines drawn tangent to the top and bottom of each oscillation diverge.

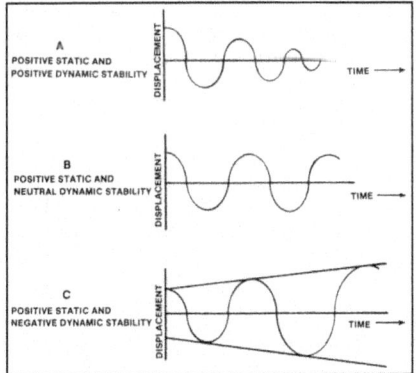

Figure 7-3. Oscillatory motion

PITCH STABILITY

7-15. Pitch, or longitudinal, stability is longitudinal axis stability about the lateral axis of the aircraft.

OSCILLATORY PITCHING MOTION

7-16. An aircraft is a well-designed, complex mass in motion. When moved from its equilibrium position, it develops a moment of inertia. If it has positive static stability, the aircraft becomes oscillatory unless a damping force prevents the motion. Moderate damping forces cause the oscillation to converge with equilibrium. The oscillation period is a function of the inertia moment and damping force.

7-17. The oscillation period varies depending on aircraft characteristics at a given airspeed. A long-term oscillation (more than 5 seconds) is called phugoid motion (pronounced foogoid). Because the pitch attitude change is slight, gain or loss in altitude and forward movement of the aircraft result in the motion occurring at a relatively constant AOA. An aviator makes subconscious pitch-attitude corrections. Because AOA is essentially constant, the phugoid motion is relatively unimportant for aircraft stability.

7-18. A buzz is a short-term oscillation (0.3 to 0.5 second) and occurs so quickly the stability of the aircraft dampens out the motion before the aviator can respond.

7-19. A medium-term oscillation (1.5 to 5 seconds) lasts about as long as the aviator's response time. This can lead to a sudden and violent divergence in pitch attitude, resulting in large positive and negative load factors aggravated when the aviator attempts to control the oscillation. This is called pilot-induced oscillation. During the landing sequence, it is called porpoising.

PITCHING MOMENTS ABOUT THE CENTER OF GRAVITY

7-20. A controllable aircraft has positive static longitudinal stability and positive dynamic longitudinal stability. Dynamic stability involving motion with respect to time is not fully covered in this manual. This section primarily covers static stability. By considering the pitching moments about the aircraft CG, the aviator can analyze any tendency the aircraft has when it is displaced from its equilibrium position.

7-21. The coefficient of pitching moment (C_M) comes from the pitching moment equation. The sign of the pitching moment coefficient indicates whether a pitching moment will pitch the nose of the aircraft up (+) or down (-). This section of the manual discusses direction of the pitching moments created by various components of the aircraft. It does not discuss magnitude of the pitching moment; therefore, the pitching moment equation will not be explained further in this section.

7-22. Pitching moments about the aircraft CG are caused by changes in total lift as it is distributed between the wings, fuselage, and tail surfaces. Total lift acts through the aerodynamic center of the entire aircraft; it is called the neutral point. To simplify discussion, drag changes and compressibility effects are omitted. If controls are fixed at the trim position, a constant AOA results and a zero pitching moment exists. Figure 7-4 shows this trim point.

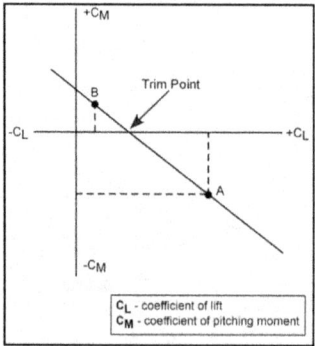

Figure 7-4. C_M versus C_L

7-23. Figure 7-4 also shows the variation of C_M with respect to changes in C_L. The C_L can be used instead of AOA as their relationship is linear, except as maximum value of the coefficient of lift ($C_L max$) is

Fixed-Wing Aerodynamics and Performance

approached. At any AOA above $C_L$max, the airfoil begins to stall. At the value of C_L where the curve crosses the horizontal axis, the pitching moments are zero. At this AOA (trim point), stability considerations are made.

7-24. If the AOA is increased to a higher value of C_L than indicated at the trim point, a negative pitching moment must be present to return the aircraft to the trim AOA. Figure 7-4, point A reflects this. The opposite is also true. If the AOA decreases from the trim point, a positive pitching moment must be created that tends to return the aircraft to the trim point. Figure 7-4, point B shows this. For an aircraft to exhibit positive longitudinal stability, the slope of C_M versus the C_L curve must be negative. The degree of the slope indicates the degree of stability. A steeper slope shows stronger pitching moments with changes in C_L so greater stability exists.

7-25. The trim point location is important in aircraft design. The trim point must occur at some usable AOA, between zero lift and the stalling AOA. To satisfy the preceding requirements, the neutral point, or aerodynamic center of the aircraft, must be aft of the CG of the aircraft (figure 7-5). If a sudden gust pitches the aircraft to a higher AOA, the increase in overall lift of the aircraft, acting through the aerodynamic center, creates a negative pitching moment. This tends to return the aircraft to its equilibrium position. A gust pitching the aircraft nose down causes a decrease in the AOA and an overall net decrease in lift forces. This results in a pitching moment about the aircraft CG and tends to return the aircraft to equilibrium.

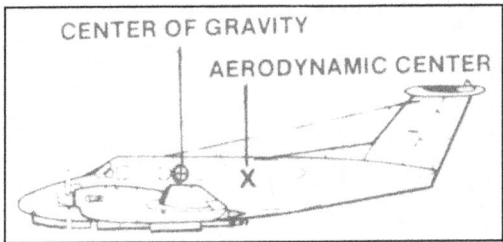

Figure 7-5. Fixed-wing aircraft center of gravity and aerodynamic center

WING CONTRIBUTION

7-26. Overall static stability of the aircraft depends on the CG's position in relation to the aerodynamic center of the aircraft. All parts of the aircraft contribute to its static stability. The moments contributed by the wing or other parts depend on the location of the wing's aerodynamic center or the other aircraft parts being considered in relation to the CG. Together, these moments determine where the neutral point is located. If airflow is incompressible, the wing's aerodynamic center is about the 25 percent chord of the wing. In figure 7-6, part A (page 7-6), the aircraft CG is behind the wing's aerodynamic center. An external disturbance pitching the wing to a higher AOA increases the pitching moment toward the stalling AOA. This additional increase in AOA also increases the lift and pitching moment. Unless another force counters the effect, the aircraft will pitch upward repeatedly. As figure 7-6, part B shows, when the wing's aerodynamic center is aft of the aircraft CG, longitudinal stability is enhanced. Likewise, if the aircraft CG and wing's aerodynamic center coincide, the wing contributes neutral stability to the aircraft.

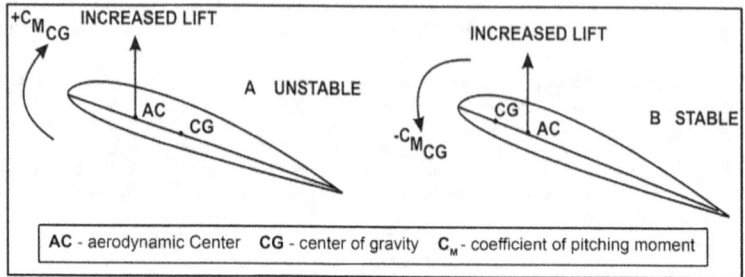

Figure 7-6. Wing contribution to longitudinal stability

7-27. It is accepted that lift normally acts through the aerodynamic center of an airfoil, when actually, the aerodynamic force, of which lift is a component, acts through the center of pressure. As angles of attack change, the center of pressure moves back and forth on the airfoil. Unequal pressure distribution on the wing creates a moment. The moment about the wing's aerodynamic center should not be confused with the moment about the aircraft CG (figure 7-7).

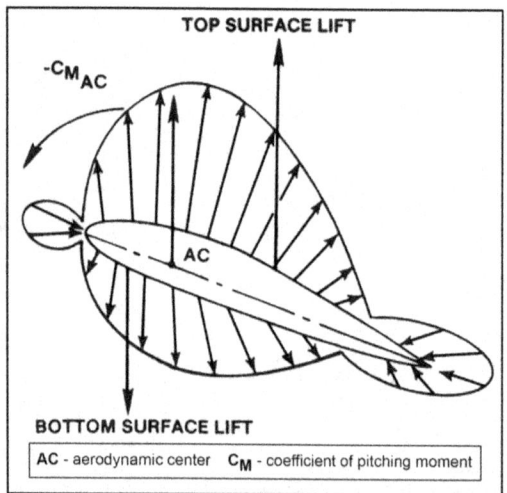

Figure 7-7. Negative pitching moment about the aerodynamic center of a positive-cambered airfoil

7-28. An aircraft with a positive-cambered airfoil and a CG forward of its aerodynamic center cannot be trimmed unless the negative moment is balanced by a positive moment about the CG. This balance can be accomplished only when the value of C_L is negative. At any greater value of C_L, the net result of the moments will be negative (figure 7-8, page 7-7). For a positive-cambered wing to contribute positive stability to the aircraft, it must be trimmed at an unusable AOA. The negative moment can be overcome by a positive moment from the horizontal tail.

Fixed-Wing Aerodynamics and Performance

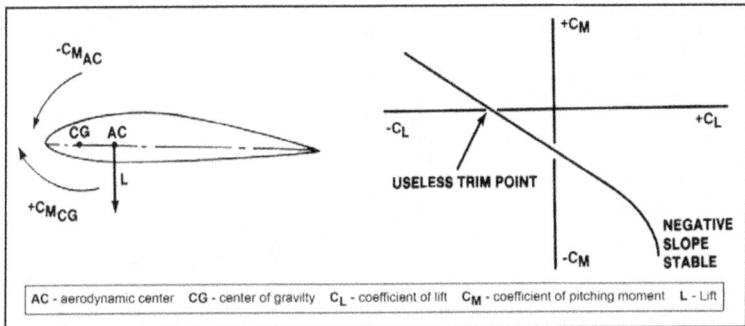

Figure 7-8. Positive longitudinal stability of a positive-cambered airfoil

7-29. If this same wing has the aircraft CG located behind its aerodynamic center, then a positive lift force at the aerodynamic center creates a positive pitching moment. This positive pitching moment balances the negative moment about the aerodynamic center present in cambered airfoils. The wing can be trimmed at an AOA above $C_L=0$ and below the stalling AOA (figure 7-9). The wing, however, contributes negatively to the stability of the aircraft. This can also be overcome by the horizontal tail.

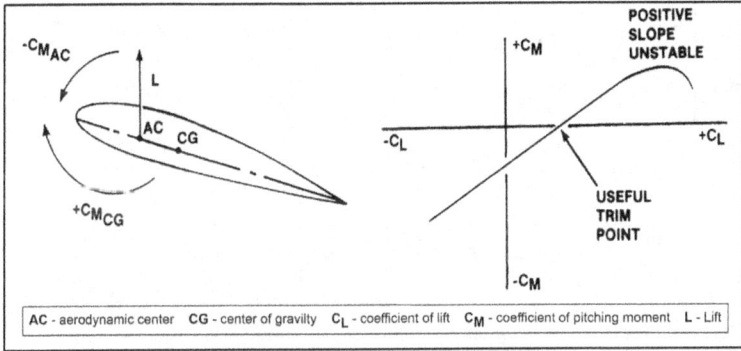

Figure 7-9. Negative longitudinal stability of a positive-cambered airfoil

7-30. When the aircraft CG is forward of the wing's aerodynamic center, a positive-cambered airfoil contributes positively to stability. The aircraft, however, can only be flown in the usable flight mode if a horizontal tail balances the moments. If the CG is aft of the wing's aerodynamic center, the CG contributes negatively to the stability of the aircraft. The aircraft, however, can be balanced by a horizontal tail.

FUSELAGE AND ENGINE NACELLE CONTRIBUTION

7-31. A symmetrical body in a perfect fluid at a positive angle develops pressure distributions, but no resultant force exists. A streamlined fuselage is a good example. The airstream is not a perfect fluid, and the fuselage is not perfectly symmetrical. The fuselage produces positive pitching moments at negative angles of attack. Induced flow from the wing (upwash ahead of the wing and downwash behind the wing) adds to unstable contributions of the fuselage. An engine nacelle located on the wing's leading edge is also influenced by wing upwash and adds to longitudinal instability. The fuselage-engine combination has an aerodynamic center at about 25 percent of the fuselage length rearward from the fuselage nose. Normally, no fuselage resultant force exists. The fuselage-engine aerodynamic center is placed in a position relative to the aircraft

Chapter 7

aerodynamic center so it makes a negative contribution to aircraft longitudinal stability. This negative contribution is corrected by using the horizontal stabilizer.

HORIZONTAL STABILIZER CONTRIBUTION

7-32. The horizontal stabilizer, which is usually a symmetrical airfoil, is located well aft of the aircraft CG. Because the airfoil is symmetrical, it can produce either positive or negative lift, depending on AOA. The entire horizontal stabilizer is located behind the CG so its aerodynamic center is aft of the CG, creating a stable relationship.

7-33. The stabilizing moment created by the horizontal stabilizer can be controlled in two ways. The distance (moment arm) can be increased between the CG and horizontal stabilizer's aerodynamic center, or the stabilizer's surface area can be increased. Because the stabilizer is an airfoil, it produces lift as the stabilizing force. An increase in the tail area or the distance between the CG and stabilizer's aerodynamic center increases the tail force and stabilizing moment.

7-34. If the aircraft pitches up to an AOA higher than the trim AOA, the increased AOA on the horizontal stabilizer increases the positive lift of the tail (figure 7-10). This produces a negative pitching moment and returns the aircraft toward the equilibrium trim point. If the aircraft AOA decreases, the horizontal stabilizer AOA also decreases and produces less, or even negative, lift. This creates a positive pitching moment about the CG and returns the aircraft to equilibrium.

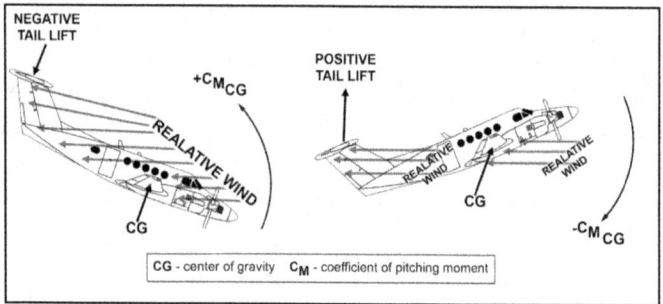

Figure 7-10. Lift as a stabilizing moment to the horizontal stabilizer

WING, FUSELAGE, AND HORIZONTAL STABILIZER COMBINATION

7-35. Depending on the location of the aircraft CG relative to the wing's aerodynamic center, the wing may stabilize or destabilize to the entire aircraft. The horizontal stabilizer is used to overcome the trim problem of a stabilizing wing (CG ahead of the wing's aerodynamic center). The horizontal stabilizer also provides the stability needed with an unstable wing (CG aft of the wing's aerodynamic center). Because the fuselage is destabilizing to the entire aircraft, the horizontal stabilizer is also used to solve this negative contribution. To create positive static stability, the aerodynamic center of the entire aircraft must be located aft of the aircraft CG.

THRUST AXIS CONTRIBUTION

7-36. The line along the thrust force vector is called the thrust axis. If the thrust axis is located above the aircraft CG, an increase in thrust creates a negative pitching moment (figure 7-11, page 7-9). The horizontal stabilizer must also balance this moment. The aviator must be able to trim the aircraft at any power setting. If thrust is located below the CG, opposite pitching moments are created when thrust is increased.

Fixed-Wing Aerodynamics and Performance

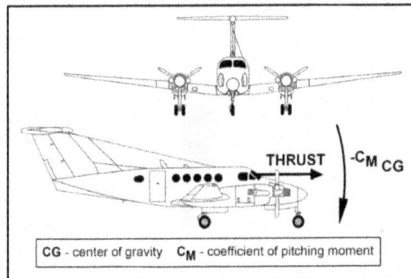

Figure 7-11. Thrust axis about center of gravity

DIRECTIONAL STABILITY

7-37. Directional stability involves motion of the aircraft about the vertical axis, or the yawing motion of the aircraft. Directional stability also involves sideslip and yawing moments produced about the CG due to sideslip.

SIDESLIP ANGLE

7-38. Sideslip angle, which is symbolized by using the Greek letter beta (β), is the angle between relative wind and the longitudinal axis of the aircraft. When relative wind is right of the aircraft's nose, the sideslip angle is positive. Figure 7-12 shows a positive sideslip angle.

YAWING MOMENTS VERSUS SIDESLIP ANGLE

7-39. This section covers the direction of the moments about the CG. The magnitude of these moments will not be developed. The coefficient of a yawing moment (C_N) denotes the direction of the yawing moments developed by the various components of the aircraft. The negative (-) sign is used for left yawing moment, and the positive (+) sign is used for right yawing moment.

7-40. The aircraft should be in directional equilibrium if the relative wind is parallel to the aircraft's nose or along the longitudinal axis. If no sideslip exists, no yawing moment exists. However, if the aircraft has a positive sideslip angle (figure 7-12), a positive yawing moment is required for static directional stability. If relative wind is coming from the right, the aircraft should tend to yaw toward the right. This yawing motion to the right will return the relative wind to the aircraft's nose and reestablish equilibrium.

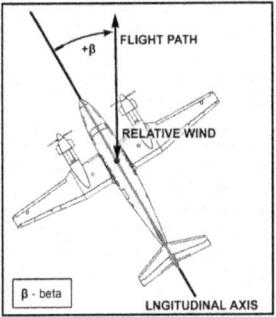

Figure 7-12. Positive sideslip angle

Aircraft Component Contributions

Fuselage and Engine Nacelle

7-41. In considering longitudinal stability, the fuselage and engine nacelles are influenced by upwash and downwash of the airstream as it passes over the wing. This adds to the instability created by the fuselage. In directional stability, wing influences do not affect the yawing moments created by the fuselage and engine nacelles. The only consideration is the side area of the fuselage and engine nacelles ahead of the CG when that area is compared to the side area behind the CG. Most aircraft have a larger side area ahead of the CG than behind it. Therefore, a relative wind striking the aircraft from either side causes a larger yawing moment ahead of the CG than behind it. This creates an unstable directional condition as the aircraft yaws away from the relative wind (figure 7-13). In other words, a positive (right) sideslip angle on the fuselage produces a negative (left) yawing moment.

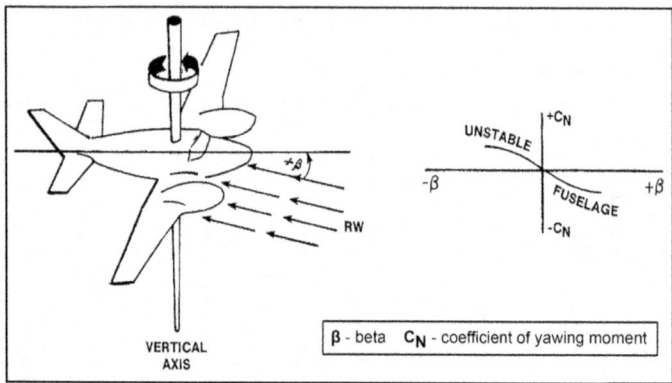

Figure 7-13. Directional stability (β versus C_N)

7-42. To achieve directional stability, the side area of the aircraft must be larger behind the CG. Therefore, a fin, or vertical stabilizer, must be added to the fuselage to increase the area, create a desirable yawing moment, and produce positive directional stability.

Vertical Stabilizer

7-43. The vertical stabilizer is a symmetrical airfoil. Like the horizontal stabilizer, it is located behind the aircraft CG. Therefore, the aerodynamic center of the vertical stabilizer is located in a position producing positive static directional stability.

7-44. As with the horizontal stabilizer, the vertical stabilizer area, or distance from the CG, can be varied to obtain desired stabilizing moments. However, increasing the vertical stabilizer area too much increases the height of the tail, which increases the frontal area and drag. To decrease drag, the height of the tail is often decreased and a dorsal fin added (figure 7-14, page 7-11). This also decreases the aspect ratio of the vertical stabilizer, which makes the tail effective at higher angles of sideslip. Because the vertical stabilizer is an airfoil, it is subject to aerodynamic stalls. Aerodynamic stalls can occur at high sideslip angles. Adding the dorsal fin, which decreases the aspect ratio of the tail, increases the stalling AOA. The tail is then effective at larger angles of sideslip. This is of particular importance to a multiengine aircraft subjected to large sideslip angles due to asymmetrical power conditions; for example, an inoperative engine on one wing.

Fixed-Wing Aerodynamics and Performance

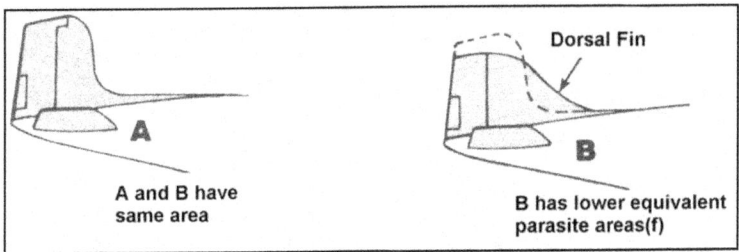

Figure 7-14. Dorsal fin decreases drag

Wing

7-45. Wing contribution to aircraft directional stability is small. However, it becomes greater with increases in swept-back wing design. To be effective, this swept-back angle must be at least 30 degrees.

Final Configuration

7-46. Figure 7-15 depicts aircraft final configuration. This aircraft configuration must have a positive slope of the C_N curve. The graphic also shows how the vertical stabilizer must produce a stabilizing moment strong enough to overcome the destabilizing moments generated by the fuselage and engine nacelles.

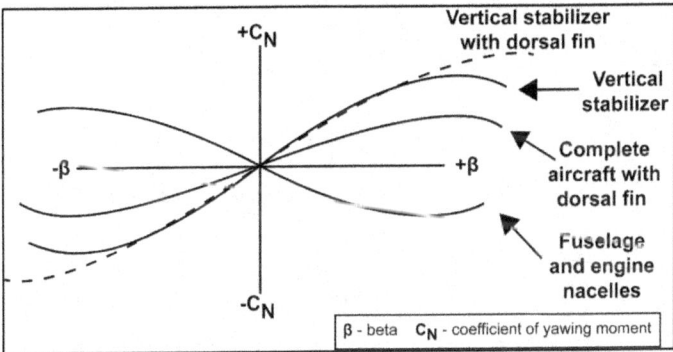

Figure 7-15. Fixed-wing aircraft configuration positive yawing moment

LATERAL STABILITY

7-47. Lateral, or roll, stability involves lateral axis stability about the longitudinal axis. Motion about the longitudinal axis is roll. A right roll is indicated with a positive sign; a left roll, with a negative sign. Wing design is important to aircraft stability. It is the primary lift-producing and roll-stabilizing surface. As with directional stability, the aircraft achieves its roll stability through the sideslip angle. With roll stability only, stabilizing rolling moments are created by the sideslip acting on the wing.

SIDESLIP CAUSED BY WING DOWN

7-48. In figure 7-16, page 7-12, the aircraft has its right wing down. This tilts the wing's lift vector to the right so a horizontal component of lift acts to the right. Because there is no opposing force, this horizontal force moves the aircraft to the right. This motion to the right, along with the forward motion of the aircraft, produces a positive sideslip angle. With the right wing down and a positive sideslip angle generated, the aircraft must develop a negative, or left, rolling moment for positive static lateral stability (the coefficient of the rolling moment [C_l] is not the coefficient of lift.). A curve of the rolling moment versus the sideslip angle must have a negative slope to indicate positive static lateral stability (figure 7-17). The curve must go through the origin as a rolling moment should not be generated when the aircraft wings are level and the relative wind is on the nose. The degree of slope, as with other stabilizing curves, indicates the degree of lateral static stability of the aircraft.

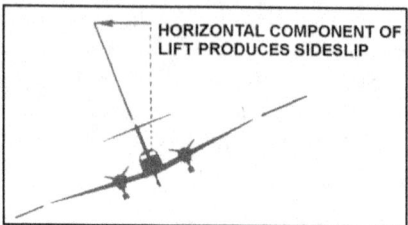

Figure 7-16. Horizontal lift component produces sideslip

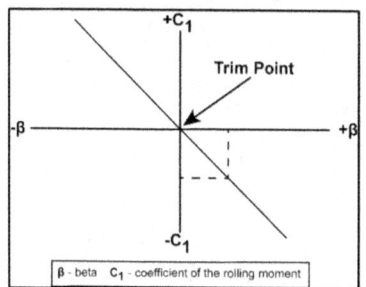

Figure 7-17. Positive static lateral stability

DIHEDRAL

7-49. Dihedral of the wing is the angle between the wing and a plane parallel to the lateral axis (figure 7-18), creating a stabilizing moment. When the wings droop, they have an anhedral, a negative dihedral, or cathedral angle. In other words, the wing produces a destabilizing rolling moment.

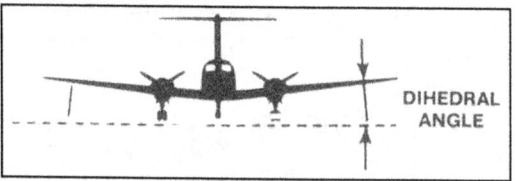

Figure 7-18. Dihedral angle

Stability

7-50. To understand how dihedral produces a stabilizing moment, a three-dimensional picture is required; however, figure 7-19, page 7-13, shows the relative wind approaching the airfoil from the left side of the figure. The aircraft is moving forward in a right sideslip. Therefore, the third dimension must be added to the figure. The third dimension is the relative wind caused by the forward velocity of the aircraft. The low wing has a higher AOA than the high wing. This higher AOA gives the lower wing a larger coefficient of lift than the higher wing. The lower wing now creates more lift than the high wing. Therefore, a negative rolling moment is created by the differential in lift forces between the two wings.

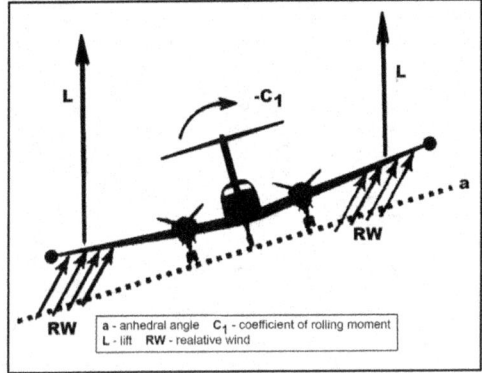

Figure 7-19. Dihedral stability

7-51. An anhedral angle produces the opposite reaction. The positive sideslip angle produces a positive rolling moment. Anhedral laterally destabilizes the aircraft.

Effects

7-52. Dihedral was the first method used in the construction of aircraft to gain lateral stability. Factors other than dihedral can contribute to lateral stability of the aircraft. Their contributions are called dihedral effects. They can either stabilize or destabilize; therefore, their contributions are classified as either positive or negative dihedral effects.

7-53. The vertical location of the wing in relation to the CG affects lateral stability. A high wing contributes a positive dihedral effect; a low wing contributes a negative dihedral effect. Wing position has a significant effect on lateral stability, causing the large dihedral angle normally used on aircraft with low wings. Wings mounted near the CG have essentially no effect on the lateral stability of the aircraft.

7-54. The vertical stabilizer makes a slight, positive contribution to the lateral stability of the aircraft. Because the vertical stabilizer is a large area above the aircraft CG, the sideward force caused by the sideslip angle produces a favorable rolling moment and helps stabilize the aircraft laterally.

TOTAL AIRCRAFT AND POSITIVE STATIC LATERAL STABILITY

7-55. The total aircraft must demonstrate a tendency toward positive static lateral stability. Some components might produce negative stabilizing moments. However, they must be overcome by stabilizing moments from some other component of the aircraft so the total aircraft is laterally stable.

CROSS-EFFECTS AND STABILITY

7-56. The sideslip angle is the main factor used to achieve directional and lateral stability. Because yaws and rolls both produce sideslips, cross-effects between directional and lateral stability exist. A sideslip angle produces a yawing and a rolling moment at the same time. The magnitude of the moments and inertia of the aircraft, or its resistance to react to the moments created, can produce certain cross-effects. Some of these effects are desirable; some are not.

Adverse Yaw

7-57. Adverse yaw is produced by rolling the aircraft with the ailerons, sometimes called adverse aileron yaw. The aircraft in figure 7-20 has its right aileron down and left aileron up. This generates a differential in the lift force acting on each wing and produces a left roll. The right wing has a higher coefficient of lift because the down aileron causes increased camber. Therefore, the induced drag is greater on the right wing than the left wing. This increased drag causes the aircraft to yaw toward the right about the CG. As the aircraft rolls, the relative wind resulting from the roll on the down-going wing is upward (opposite its direction of movement). This relative wind, when added vectorially to the free stream relative wind, resolves into an inclined relative-wind vector. Because the lift force produced by the down-going wing is perpendicular to its relative wind, the lift force acts forward. The opposite relative wind must occur on the up-going wing; therefore, its lift vector acts in a rearward direction. The different directions of the lift forces produce a condition adding to the adverse yaw caused by the drag differential.

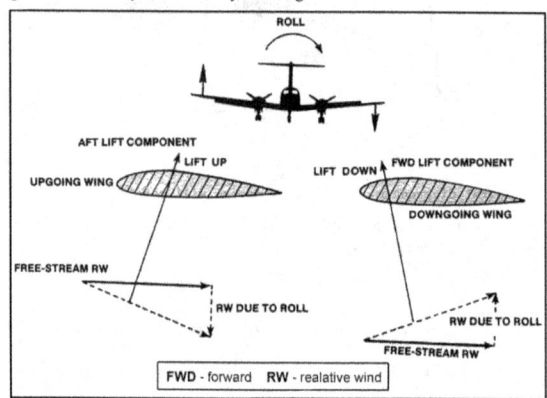

Figure 7-20. Adverse yaw

7-58. Modern aircraft subject to adverse yaw have controls to overcome this problem. They may be equipped with spoilers extending on the down-going wing. They spoil some lift and add drag to counter adverse yaw. Differential ailerons are also used to control adverse yaw. The down-going wing aileron extends up more than the up-going wing aileron extends down. This produces more drag on the down-going wing and counters effects of adverse yaw. Frise ailerons add drag to the down-going wing. They extend the part of the aileron forward of the hinge line and down into the airstream while the half of the aileron behind the hinge line extends up into the airstream. Without these or other similar devices, aircraft using only ailerons for lateral control tend to yaw to the right when they roll left and vice versa. Even with compensating devices, adverse yaw is present during flight and is usually corrected with a coordinated application of rudder.

Proverse Roll

7-59. Proverse roll is encountered when an aircraft yaws. In this case, an aircraft is put in a right yaw when the aviator applies right rudder. This creates a left sideslip angle. Because a negative sideslip angle produces a positive rolling moment, the aircraft rolls to the right. Because an aircraft wing cannot determine whether it is level, it responds to a negative sideslip with a positive roll. Another factor contributing to proverse roll is the difference in velocities of each wing. In the example above, the left wing has a greater velocity than the right wing due to the yawing motion about the aircraft CG. Increased velocity increases the lift force on the left wing, which causes a positive roll as long as the aircraft is yawing.

DIRECTIONAL DIVERGENT STABILITY

7-60. The degree of directional stability compared with degree of lateral stability of an aircraft can produce three conditions. These conditions are directional divergence, spiral divergence, and Dutch roll.

Directional Divergence

7-61. Directional divergence results from negative directional stability. This cannot be tolerated because directional divergence allows the aircraft to increase its yaw after only a slight yaw has occurred. This continues until the aircraft turns broadside to the flight path or until it breaks up from the high pressure load imposed on the side of the aircraft.

Spiral Divergence

7-62. Spiral divergence results if static directional stability is strong when compared with the dihedral effect. If an aircraft with strong directional stability has its right wing down, a positive sideslip angle is produced. As a result of strong directional stability, the aircraft tries to correct directionally before the dihedral effect can correct laterally. The aircraft chases the relative wind, and the resulting flight path is a descending spiral. To correct this condition, the aviator only needs to raise the wing with the lateral control surfaces, and the spiral stops immediately.

Dutch Roll

7-63. A Dutch roll results from a compromise of directional and spiral divergence, occurring somewhere between the two. In this case, lateral stability of the aircraft is stronger than directional stability. Directional tendencies of the aircraft have been reduced from the condition leading to spiral divergence.

7-64. If the aircraft has the right wing down, the positive sideslip angle corrects the wing position laterally before the nose of the aircraft tries to line up with the relative wind. As the wing corrects, a lateral directional oscillation starts. Therefore, the nose makes a figure-eight pattern on the horizon. The rolling and yawing oscillation frequencies are the same, but they are out of phase.

SLIPSTREAM ROTATION AND PROPELLER-FACTOR

7-65. Most aircraft engines rotate the propeller clockwise, as viewed from the cockpit. This induces a clockwise airflow about the fuselage striking the left side of the vertical stabilizer. Therefore, the vertical stabilizer is subjected to a negative sideslip angle, while the rest of the aircraft is not. As shown in figure 7-21, page 7-16, this negative sideslip produces a negative yawing moment and tends to move the nose of the aircraft to the left. The propeller can also develop a negative yawing moment when the aircraft is at a high AOA (figure 7-22, page 7-16). Because the propeller disk is inclined to the flight path, the down going blade (right side) has a greater AOA and velocity than the up going blade (left side). Therefore, the down going blade produces a larger thrust than the up-going blade, which produces a negative yawing moment added to the slipstream yawing moment. The propeller disk asymmetric loading is called the propeller-factor (P-factor).

Chapter 7

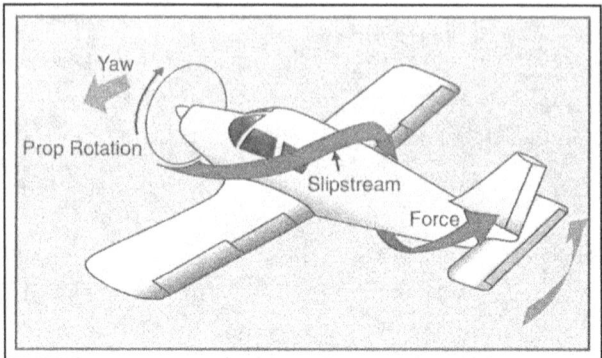

Figure 7-21. Slipstream and yaw

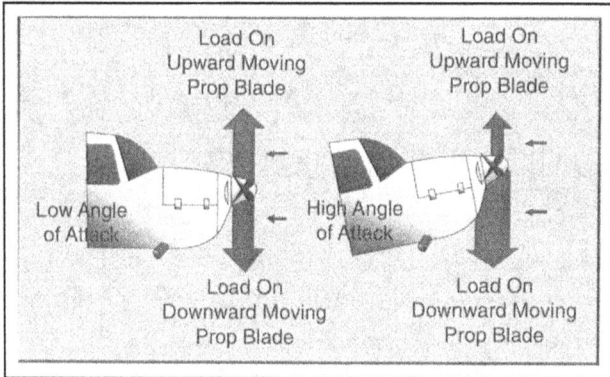

Figure 7-22. Asymmetric loading (propeller-factor)

7-66. The directional requirement resulting from slipstream rotation and P-factor is important to propeller-driven aircraft performance. The rudder must be able to overcome the negative yawing moments of both the P-factor and slipstream rotation. It must also be able to maintain directional control. The adverse moments created by slipstream rotation and the P-factor at high angles of attack are increased as the aircraft slows. However, the rudder moment used to counteract the adverse yawing moments decreases as velocity decreases. Therefore, the rudder must be deflected even more. This can also be a critical control requirement.

SECTION II – HIGH-LIFT DEVICES

PURPOSE

7-67. Low-speed characteristics of an aircraft can be as important as high-speed performance, if not more so. Army aviators spend much of their time in the air below 3,000 feet and at airspeeds of less than 150 knots as takeoffs and landings are made at relatively low altitudes and primarily involve low speeds. For this reason, aircraft designers must turn to high-lift devices, which increase the $C_L$max by various means.

Lift Force

7-68. A high-lift device is not used to increase lift but to obtain a required lift force at lower velocities. For example, an aircraft flying at 250 knots is developing 10,000 pounds of lift. When landing, the aircraft still requires 10,000 pounds of lift; however, it might now be flying at 100 knots.

Stall Speed

7-69. The slowest velocity an aircraft can fly depends on the maximum value of C_L attainable. This is shown in the stall-speed equation. The stall speed is inversely proportional to the square root of the value of $C_L$max. If this value is increased, then the stall speed is lowered or a greater weight can be supported with the same stall speed. Increasing the payload of an aircraft is another example of when high-lift devices are required. All high-lift devices increase the value of $C_L$max. The two most common ways to increase the value of $C_L$max are by increasing the camber of the airfoil or delaying the boundary-layer separation.

INCREASING THE COEFFICIENT OF LIFT

Increasing Camber

7-70. Of the two usual methods of increasing $C_L$max, increasing the camber of the airfoil is most often used.

7-71. A wing with more camber has a greater velocity differential between the wing's top and bottom surfaces. This greater velocity differential creates a large pressure differential across the wing. The pressure differential has been previously related to the value of C_L for a given AOA. Therefore, by increasing the camber of an airfoil, the value of C_L is increased.

7-72. Figure 7-23 shows use of trailing-edge flaps as the usual method of increasing the camber. This C_L curve is shown for flaps up and flaps down. The basic airfoil is a symmetrical airfoil; the wing has a zero-lift point at an AOA of zero degrees. With the flap extended, the airfoil now has a positive camber, and the zero-lift point has shifted to the left. The value of $C_L$max has increased, and the curve of the basic wing has shifted up and to the left as the flaps lowered. In this manner, all high-lift devices increasing camber affect an increase in the value of $C_L$max. The AOA at which the wing will stall has been decreased. The basic wing stalled at an AOA of about 18 degrees; with increased camber, it stalls at 15 degrees. However, the value of $C_L$max at 15 degrees (flaps down) is greater than at 18 degrees on the basic wing.

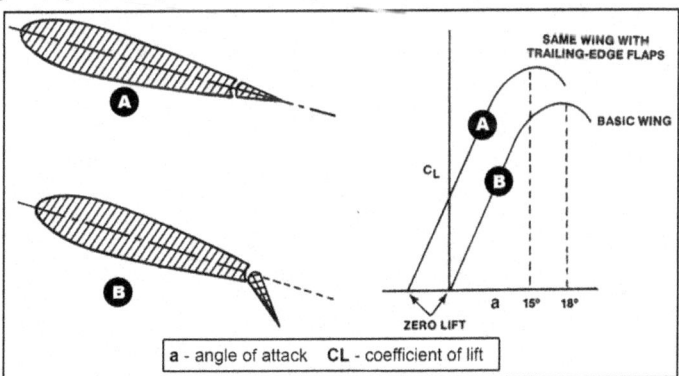

Figure 7-23. Increasing camber with trailing-edge flap

Delaying Boundary-Layer Separation

7-73. The other common method of increasing the $C_L$max value is by delaying boundary-layer separation. The maximum value of CL is limited by boundary-layer separation. The basic wing mentioned before stalled at an 18-degree AOA. If the energy level of the boundary layer over the wing is increased, then the wing rotates to higher angles of attack before a stall occurs. This technique uses boundary-layer control (BLC) to increase the value of $C_L$max. The energy level of the boundary layer can be increased by suction or blowing BLCs or vortex generators. Figures 7-24 through 7-26 show these methods of boundary layer control. Figures 7-24 and 7-25 also include C_L curves with the results of BLC. This manual does not consider the slight difference between suction and blowing BLCs due to airflow changes on the wing's surface.

Suction Boundary-Layer Control

7-74. The suction BLC in figure 7-24 draws off the low-energy, aerodynamically dead and turbulent air below the boundary layer, causing the higher energy layers above to be lowered closer to the airfoil surface. This makes the airfoil effective at angles of attack where it previously stalled. Part A shows boundary-layer separation when the airfoil is stalled at an 18-degree AOA with BLC off. In part B, the airfoil is also at an 18-degree AOA. With suction BLC on, however, boundary-layer separation no longer occurs. Suction BLC is rather inefficient; it requires a heavy vacuum pump or turbine to handle the large volume of air being drawn off the airfoil. This increases the weight of the aircraft; the extra weight partially offsets the advantages gained by the increased value of $C_L$max.

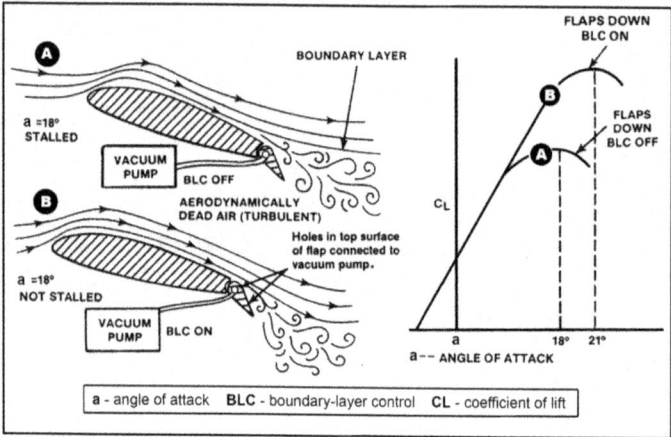

Figure 7-24. Suction boundary-layer control

Blowing Boundary-Layer Control

7-75. Blowing BLC (figure 7-25, page 7-19) increases the energy level of the boundary layer by introducing high-energy air through a nozzle, usually mounted ahead of the flap. This method can be thought of as blowing the turbulent air from the top surface of the airfoil. Blowing BLC is more commonly used than suction BLC because it is more efficient. The compressor section of a turbine engine can be used to supply the high energy air needed. Therefore, aircraft weight does not increase.

Fixed-Wing Aerodynamics and Performance

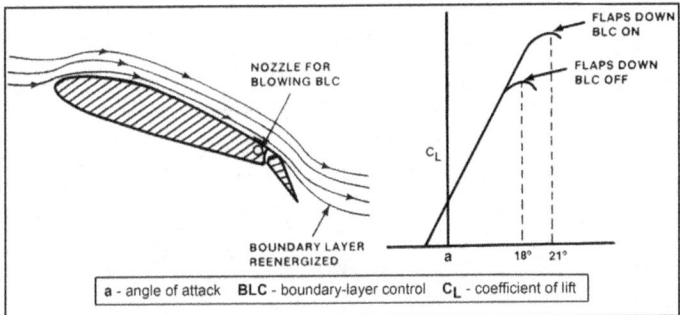

Figure 7-25. Blowing boundary-layer control

Boundary-Layer Control by Vortex Generators

7-76. Figure 7-26 shows another method of reenergizing the boundary layer using vortex generators. These small strips of metal are placed along the wing, usually in front of the control surfaces or near the wing tips. The turbulence caused by these strips mixes high-energy air from outside the boundary layer with boundary-layer air. The effect of the vortex generators on the lift curve is similar to other BLC devices.

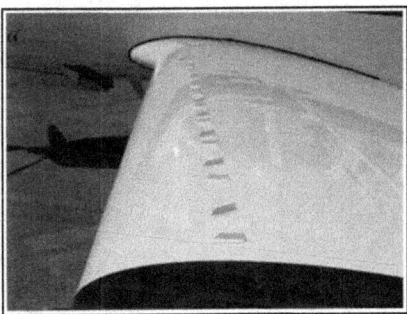

Figure 7-26. Vortex generators

TYPES OF HIGH-LIFT DEVICES

7-77. The following paragraphs cover various types of high-lift devices. These devices increase either the camber of the airfoil or the energy of the boundary layer.

TRAILING-EDGE FLAPS

7-78. Trailing-edge flaps are the most common type of high-lift device. These flaps have advantages and disadvantages. A trailing-edge flap increases the camber of the wing, thereby increasing the value of $C_L$max. However, in so doing, it moves the lift force toward the trailing edge of the wing, resulting in a negative, or nose-down, pitching moment. This moment limits the use of flaps to aircraft having horizontal stabilizers and elevators. When a trailing edge flap is extended, the angle of incidence is increased as the chord line of the airfoil changes. Figure 7-27, page 7-20, shows how the change in angle of incidence changes the zero lift line. The nose-down pitching moment on the fuselage results in better forward visibility during landings and takeoffs. Flaps also increase drag on the aircraft. This is useful in landing; the aircraft can make a steeper approach without increasing airspeed. However, this drag increase is not desired on takeoff. Most aircraft having large and effective trailing-edge flaps use only partial flaps on takeoff; thus they have the benefit of increased C_L without a large increase in drag. Some of the common types include the plain flap, split flap, Fowler flap, slotted flap, and slotted Fowler flap.

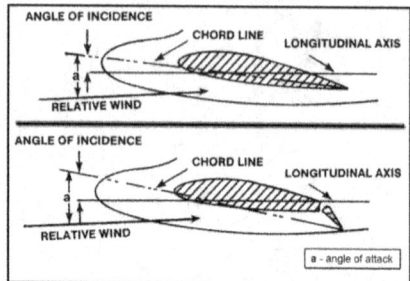

Figure 7-27. Angle of incidence change with flap deflection

Plain and Split Flaps

7-79. Figure 7-28, parts A and B, shows these two basic flaps. Both increase the camber of the airfoil. The split flap does not produce as large a nose-down pitching moment as the plain flap. The split flap also creates a greater drag force due to the low-pressure, high-turbulence area between the wing trailing edge and flap trailing edge.

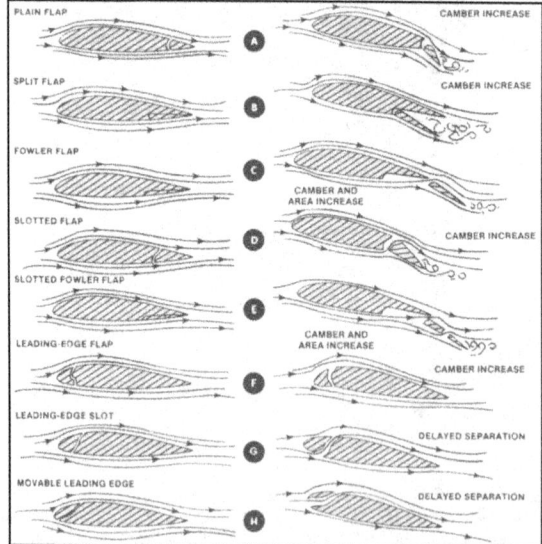

Figure 7-28. Types of high-lift devices

Fowler Flap

7-80. On aircraft that lift heavy loads from short fields, such as jet transports, the Fowler flap (figure 7-28, part C) is often used. When extended, this type of flap moves rearward as well as down. This increases $C_L$max because of an increase in camber and wing area. The Fowler flap then reduces stall speed (V_S) by increasing $C_L$max and wing area. Although aerodynamically the Fowler flap is the most efficient flap, it has disadvantages. With the huge surface extending so far behind the wing, a large twisting moment is set up in the wing. Therefore, the wing must be strong enough to withstand the load. The increased structural strength and more complicated actuating mechanisms account for large increases in weight and internal wing volume. The Fowler flap cannot be used on thin, high-speed airfoils.

Slotted Flap

7-81. To increase efficiency, most flaps can be slotted. Using slotted flaps combines the principle of BLC with a camber change. Together, the effects are cumulative. A plain flap curve is added before and after the slot (figure 7-28, part D). After adding the slot, separation over the flap area is delayed so the wing can be rotated to a higher AOA. The increased energy required to delay the boundary-layer separation comes from the low-velocity and high-pressure air under the flap. The air is directed through the slot over the top surface of the flap. This increases the energy of the boundary layer over the flap. A slotted Fowler flap is even more efficient. A slotted Fowler flap increases both camber and wing area. In fact, the $C_L$max value of a multiple-slotted Fowler flap may be twice that of the basic wing (figure 7-28, Part E).

LEADING-EDGE DEVICES

7-82. Figure 7-28, parts F through H, depicts leading-edge devices to include the leading-edge flap, leading-edge slot, and movable leading edge.

Leading-Edge Flap

7-83. Some aircraft use leading-edge flaps, (figure 7-28, part F). These increase the camber of the airfoil to increase $C_L$max. Unlike a trailing edge flap, the leading-edge flap does not produce a negative pitching moment. However, it may create a slight positive (nose-up) pitching moment, depending on its effectiveness.

Slots and Slats

7-84. BLC devices have both advantages and disadvantages. They are usually used with camber-changing devices because BLC alone is not as effective as camber change. Suction and blowing BLC devices and vortex generators have already been mentioned; however, the leading-edge and movable leading-edge slots are also forms of BLC (figure 7-28, parts G and H).

7-85. The slot through the wing (part G) will vent high-pressure air from the underside of the wing over the top surface. This delays a stall when the wing is at a high AOA. Because the slot is not exposed to the airstream, drag does not increase much.

7-86. Most modern carrier aircraft have movable leading-edge slats (figure 7-29, page 7-22). This is simply a slot that can be opened and closed. When opened at high angles of attack, the slat moves forward (some also move downward), increasing the camber and area. This occurs when the AOA is high, whether at low or high speeds, during high-G maneuvers. A BLC device does not create any pitching moment; therefore, aircraft without horizontal stabilizers use this type of device as it allows aircraft to rotate to higher angles of attack. However, aircraft can be rotated to such a high degree the aviator has difficulty seeing the landing area. This limits effective use of devices delaying boundary-layer separation.

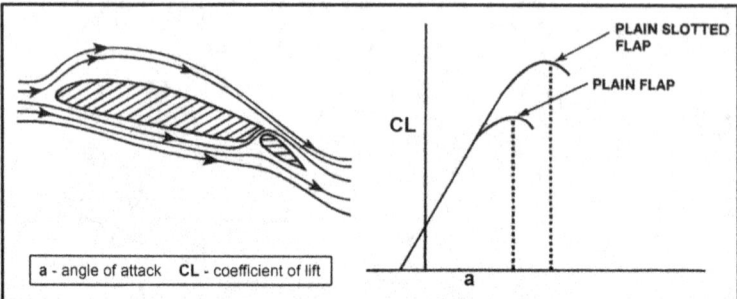

Figure 7-29. Coefficient of lift maximum increase with slotted flap

SECTION III – STALLS

7-87. In the early years of aviation, the advice was to fly low and slow. Because this condition affords a minimum distance to fall, it seemed to be sound reasoning. Actually, it is probably one of the most dangerous conditions of flight. To produce required lift at slow airspeeds, aviators must fly at a high AOA—near the AOA for the aerodynamic stall. When this stall occurs, lift decreases and drag increases, and there is usually a loss of altitude. In addition, there can also be a loss of control. Under these conditions, the aircraft can enter a spin. A considerable loss of altitude is possible before control of the aircraft can be regained.

7-88. Takeoffs and landings involve a combination of low airspeeds and altitudes making them hazardous phases of flight. One of the most frequent causes of takeoff and landing accidents is stall. When a stall occurs, there is usually insufficient altitude for recovery. During takeoff and landing, an aviator must be prepared to operate the aircraft at slow speeds and high angles of attack, a potentially hazardous configuration. Knowing this, aviators must thoroughly understand stall characteristics. This section defines stall and discusses its causes, warnings, and characteristics.

AERODYNAMIC STALL

7-89. An aerodynamic stall occurs when an increase in the AOA results in a decrease in lift coefficient. This is due to separation of the boundary layer (a thin layer of air near the surface of the wing) from the upper surface of the wing. When this boundary layer separates, turbulence occurs between the boundary layer and wing's surface. This causes static pressure on the upper surface of the wing to increase. The definition of stall does not refer to airspeed. The only condition that can cause a stall is an excessive AOA.

STALL ANGLE OF ATTACK

7-90. In figure 7-30, page 7-23, all angles of attack greater than the AOA for maximum lift coefficient fit the definition of stall. An increase in the AOA beyond the AOA for Coefficient of lift maximum ($C_L$max)(14 degrees) decreases the value of C_L. The crosshatched area is called stall region. When the aircraft operates at an AOA within this region, it is stalled, whether its airspeed is 60 or 160 knots.

Fixed-Wing Aerodynamics and Performance

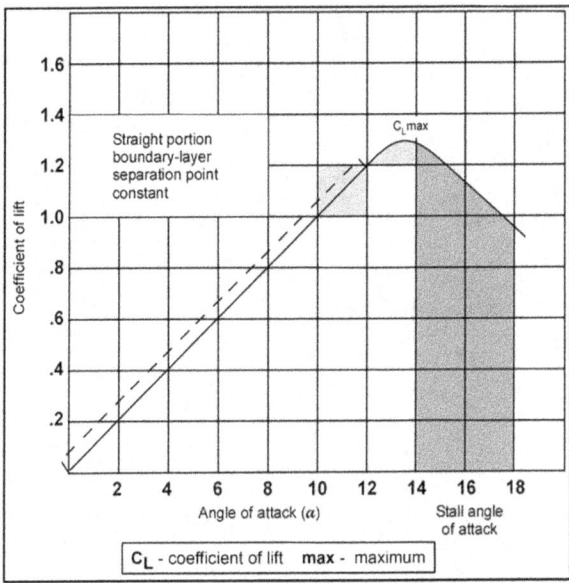

Figure 7-30. Coefficient of lift curve

CAUSES OF STALL

7-91. The cause of a stall is relatively easy to understand. The wing or airfoil is designed with a certain camber to give a definite pressure differential between the top and bottom surfaces. As the AOA increases, the C_L increases due to an increased pressure differential. At all angles of attack corresponding to the straight portion of the C_L curve to the left of the stall region, the airflow follows the curvature of the top surface until it almost reaches the trailing edge. At that location, the boundary layer breaks away and a small turbulent wake is formed (figure 7-31).

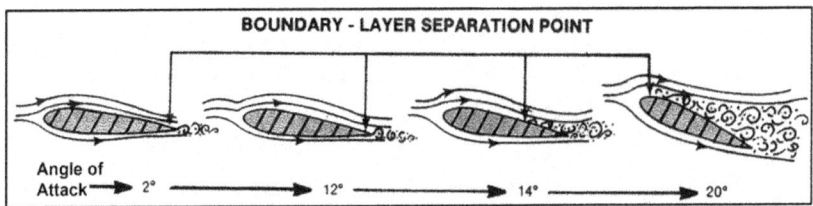

Figure 7-31. Various airfoil angles of attack

7-92. The point where the boundary layer separates from the airfoil stays essentially constant so long as the AOA is of a value where the C_L curve is a straight line. If the AOA increases beyond the straight portion of the C_L curve, the point of boundary-layer separation moves forward. This actually decreases the top surface area of the wing producing lift. The airflow under the boundary layer is turbulent; therefore, in that area, static pressure is increased, compared to the area where no separation occurs. The increase in the AOA increases the pressure differential on the portion of the wing where no separation exists. This increase in pressure differential is partially offset by the loss of some of the effective area of the wing. This results in a smaller increase in the C_L for each degree increase in AOA. The slope of the C_L curve decreases and continues to decrease. As the AOA increases, the separation point of the boundary layer continues to move forward. Finally, a further increase in the AOA results in a decrease in the value of the C_L. The point where the boundary layer separates has now moved too far forward; the loss of the effective area of the wing is too large to be offset by any increase in the pressure differential that may occur. This is the AOA defined as the stalling AOA. At the AOA for $C_L$max, the slope of the C_L curve has reached zero. Any further increase in the AOA develops a negative slope to the curve–C_L decreases as AOA increases.

7-93. Figure 7-31 shows airfoils that can be compared to the C_L curve in figure 7-30. At an AOA of 12 degrees, the curve slope starts to decrease and boundary-layer separation starts. Separation results when the boundary layer lacks the energy to adhere to the surface of the wing all the way to the trailing edge. In other words, the airflow cannot conform to the sharp bend. When placed at 90 degrees to the airstream, the flat plate shown in figure 7-32 has a turbulent flow behind it. Therefore, the boundary layer will not remain on the back surface of the plate. The same is true of the wing at high angles of attack. There is a limit where the boundary layer no longer remains on the surface of the wing. That limit is the point of boundary-layer separation.

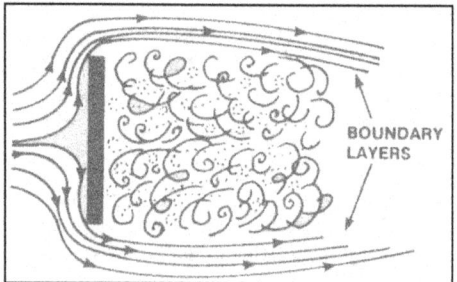

Figure 7-32. Boundary-layer separation

STALL WARNING AND STALL WARNING DEVICES

AERODYNAMIC STALL WARNING

7-94. The turbulent airflow generated when the boundary layer separates is a sign of an impending stall. As this turbulence flows over part of the aircraft, it causes buffeting, which is normally felt in the controls. This notifies the aviator of an approaching stall. Some turbulent flow is generated before the stall actually occurs. Therefore, buffeting, which can occur before the aircraft actually stalls, is a warning.

7-95. Part of the aircraft behind the wing is the horizontal stabilizer. The turbulent flow can pass over it to give the warning. The span of the horizontal stabilizer is less than the wingspan. Therefore, any turbulent flow coming from the wing tips or outer portions of the wing would not flow over the stabilizer. This is one reason to design the wing so the root section stalls before the tip section. The turbulent airflow then creates the aircraft buffet warning before the entire wing is stalled. In addition, when one part of the wing stalls before the other, the stall is not as abrupt as the entire wing stalling at once. Aviators have better lateral control of an approaching stall if the stall progresses from the root of the wing to the tip on aircraft with ailerons located toward the wing tips.

7-96. Although a root-to-tip stall pattern is desirable, it is not always possible to achieve. A rectangular or slightly tapered wing normally stalls root first. However, highly tapered, swept, or delta wings exhibit a strong tendency to stall tip first. Several design techniques can make the root stall before the tip.

Fixed-Wing Aerodynamics and Performance

Geometric Twist

7-97. One method of causing the root to stall is geometric twist (washout); that is, building a twisted wing. The root section angle of incidence is greater than the tip section; the twist is about 3 degrees. If an airfoil section has a stalling AOA of 18 degrees, the root section is at an 18-degree AOA when it stalls. However, the tip section is still at about a 15-degree AOA and is not stalled. Even if there is aerodynamic buffeting from turbulent air around the root section, the aviator can use ailerons for lateral control during recovery.

Aerodynamic Twist

7-98. Another method of stalling the root section before the tip section is aerodynamic twist. A wing with aerodynamic twist is not really twisted as with geometric twist. However, the wing reacts in the same manner and is said to be twisted. In this case, the aircraft designer uses two more types of airfoils. In figure 7-33, page 7-26, the C_L curve for the cambered airfoil and the C_L curve for the symmetrical airfoil have about the same value of $C_L$max; but the AOA at which they attain their $C_L$max is different. In this case, the root section is a cambered airfoil; toward the tip, the wing will gradually transform into a symmetrical airfoil. The angle of incidence is the same for both sections. Therefore, there is no geometric twist to this type of wing. The stall progression from root to tip is controlled aerodynamically by using different types of airfoils. If a wing is constructed with the airfoil sections plotted (figure 7-33) the root will stall at a 15-degree AOA. The tip would not stall until an 18-degree AOA is reached, as indicated by the curves.

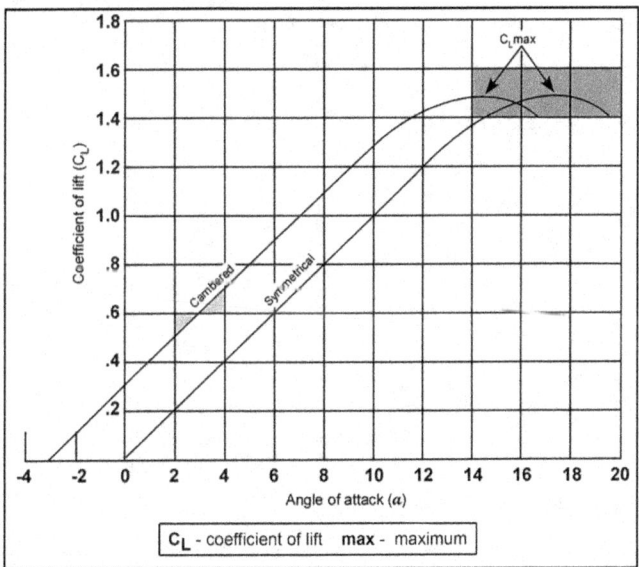

Figure 7-33. C_L curves for cambered and symmetrical airfoils

Stall Strip

7-99. A third method for stalling root sections first, or at least creating a buffet on the aircraft, is to use a stall strip on the wing's leading edge (figure 7-34, page 7-26). This causes the boundary layer to break away from the airfoil at an AOA lower than the stalling AOA for that airfoil. The cruise speed, design load, and general performance requirements of an aircraft determine the airfoil section to be used. These design considerations may preclude the use of twist methods for smooth stall progression. A stall strip, located in the root section, detaches the boundary layer to ensure this section stalls first. This gives adequate warning to allow for a safe recovery with a minimum loss of altitude.

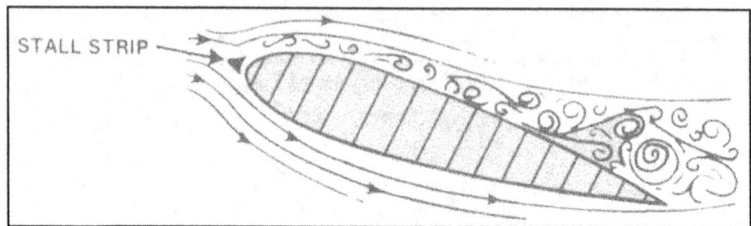

Figure 7-34. Stall strip

MECHANICAL STALL WARNING

7-100. Some aircraft do not have horizontal stabilizers or, as in the C-12, are designed so the horizontal stabilizer is not in the turbulent wake path generated by the wing as it is stalling. These aircraft are usually equipped with a mechanical stall warning. The simplest mechanical stall warning is a flapper switch mounted on the wing's leading edge (figure 7-35). As the wing approaches a stall, the relative wind pushes the flapper up, closing a switch. This, in turn, activates the device to warn the aviator of an impending stall. The flapper can be positioned to vary the attack angle at which the stall warning occurs.

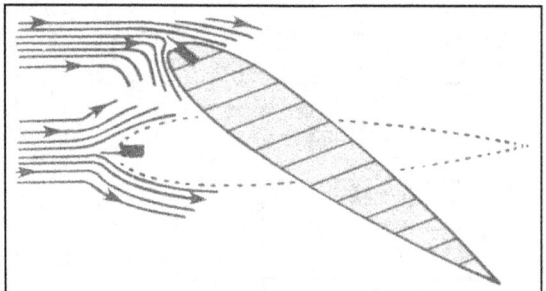

Figure 7-35. Flapper switch

7-101. Some aircraft have AOA sensors on the aircraft that detect when the AOA approaches an attitude known to result in a stall and will activate devices like the stick shaker to warn the aviator.

STALL RECOVERY

7-102. When a stall warning is received, recovery must be immediate. To recover, the aviator corrects the cause, which is a too high attack angle. The only action the aviator must take is to decrease the attack angle. This breaks the stall, stopping the stall warning immediately.

SPINS

7-103. A spin is described is an aggravated stall resulting in what is termed autorotation. In autorotation, the aircraft follows a spiral path in a downward direction. The wings produce some lift, but the aircraft is forced downward by gravity. The aircraft wallows and yaws in this spiral path (figure 7-36, page 7-27). It is assumed many factors contribute to a spin. In fact, the spin is not suited for theoretical analysis.

Fixed-Wing Aerodynamics and Performance

MISHANDLING

7-104. Many aircraft have to be forced to spin; considerable judgment and technique are required to start a spin in these aircraft. However, aircraft forced to spin may accidentally be put into a spin when the aviator mishandles the controls in turns and stalls, and in flight at minimum controllable airspeeds.

YAW

7-105. Once a wing is allowed to drop at the beginning of a stall, the nose attempts to move (yaw) in the direction of the low wing, and the aircraft begins to slip in the direction of the lowered wing. As it does, air meeting the fuselage side, vertical fin, and other vertical surfaces tends to turn the aircraft into the relative wind. This accounts for the continuous yaw present in a spin.

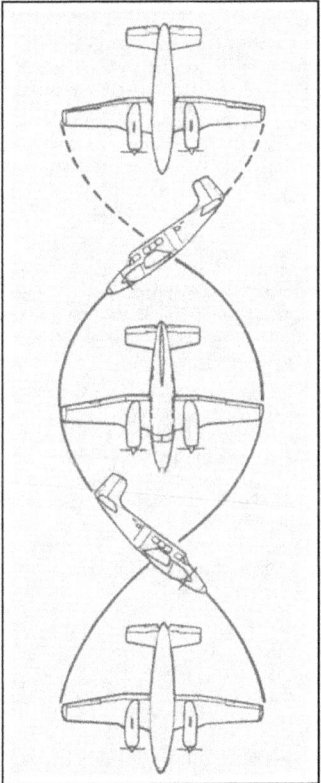

Figure 7-36. Spins

7-106. At the same time, rolling is also occurring about the aircraft's longitudinal axis. This is caused by the lowered wing having an increasingly greater attack angle due to the relative wind's upward motion against its surfaces. This wing is then well beyond the stalling attack angle and, accordingly, has an extreme loss of lift. Because relative wind is striking the lowered wing at a smaller angle, the rising wing has a smaller attack angle. Thus, the rising wing has more lift than the lowered wing, and the aircraft begins to rotate about its longitudinal axis. This rotation, combined with the effects of centrifugal force and different amount of drag on the two wings, then becomes a spin. The aircraft descends vertically, rolling and yawing until recovery is affected.

7-107. The first corrective action taken during any power-on spin is to close the throttles. Power aggravates spin characteristics, causing an abnormal altitude loss in the recovery. As power is reduced, full opposite rudder is applied. Brisk, positive, straight, forward movement of the elevator control (forward of the neutral position) is then applied. Ailerons should be neutral, and controls should be held firmly in this position. The forceful elevator movement decreases the excessive attack angle and breaks the stall. When the stall is broken, spinning stops. This straight, forward position should be maintained, and the rudder neutralized as the spin rotation stops.

7-108. If the rudder is not neutralized at the proper time, the ensuing increased airspeed acting on the fully deflected rudder causes an excessive and unfavorable yawing effect. This yawing effect places tremendous strain on the aircraft. It can cause a secondary spin in the opposite direction.

7-109. Slow and overly cautious control movements during spin recovery must be avoided. In certain cases, such movements have caused the aircraft to continue spinning indefinitely even when full opposite controls have been applied. Brisk and confident operation results in a more positive recovery.

7-110. After the spin rotation has stopped and the rudder has been neutralized, back elevator pressure is applied to raise the nose to level flight. Aviators must be careful not to apply excessive back pressure after rotation stops. To do so causes a secondary stall and may result in another spin that is more violent than the first.

ACCIDENTAL STALLS AND SPINS

7-111. Accidental stalls and spins can result from improperly executed steep turns or increases in the load factor and stalling speed caused by an increase in bank. When the aircraft is close to stalling speed, a slight application of rudder may cause an aircraft to spin. If top (outside) rudder is applied, the aircraft will spin opposite the direction of the turn (over-the-top spin). If bottom (inside) rudder is applied, it will spin in the direction of the turn (under-the-bottom spin).

7-112. Probably the most disastrous of all inadvertent spins occurs when the aviator turns from base to final leg of the traffic pattern. Being close to the ground, the aviator may be dubious about using a steep bank to accomplish the necessary turn rate to align with the runway. The aviator may try to tighten the turn with the bottom rudder without increasing the bank causing a skidding turn that leads to a violent under-the-bottom spin. Conversely, if outside rudder is used to decrease the turn rate, a slip results. If a stall occurs during this slip, an over-the-top spin can result. To conduct a safe turn, airspeed must be kept well above stalling, and controls must be coordinated at all times.

Note. Accidental stalls and spins are not limited to turning situations; they may occur in any flight attitude.

SPIN RECOVERY

7-113. Anytime a spin is encountered, regardless of conditions, the normal spin recovery sequence is as follows:
- Retard power.
- Apply opposite rudder to slow rotation.
- Apply positive forward-elevator movement to break the stall.
- Neutralize the rudder as spinning stops.
- Return to level flight.
- Do not use power.

Fixed-Wing Aerodynamics and Performance

SECTION IV – MANEUVERING FLIGHT

CLIMBING FLIGHT

7-114. Knowledge of climbing performance is essential as climb is encountered in every flight. The type of climb performance used for a certain mission is the aviator's decision. How the desired performance is obtained depends on aviator knowledge of the aircraft and its climb performance. An aircraft in a climb is increasing potential energy by increasing its altitude (potential energy equals weight times height). Generally, potential energy is increased for an aircraft by an expenditure of kinetic energy (airspeed) or chemical energy (propulsion power).

CLIMB POWER

7-115. Like a car going uphill, an aircraft climbs at the cruise power setting with a sacrifice of speed. It can also, within certain limits, climb with added power and no sacrifice in speed. A definite relationship exists among power, attitude, and airspeed.

Available Power

7-116. The amount of excess power available, which is defined as the power available above that required for straight-and-level flight, is the factor most affecting the aircraft's ability to climb. During the climb, lift operates perpendicular to the flight path; it does not directly oppose gravity to support the aircraft's weight. With the flight path inclined, lift acts partially rearward, increasing induced drag; thus adding to total drag. Because weight always acts perpendicular to the earth's surface and drag acts in a direction opposite the aircraft's flight path during a climb, thrust must overcome drag and gravity.

Zoom Climb

7-117. The exchange of kinetic for potential energy is called zoom climb. This is accomplished by flying straight and level to obtain a high airspeed, then increasing pitch to a climbing attitude. Velocity is dissipated as altitude is gained.

Steady-State Climb (Normal Climb)

7-118. The exchange of chemical energy or propulsion power for potential energy produces a steady-state climb, which can then be sustained. This type of climb is used most often.

Sustained Climb

7-119. During a sustained climb, two climbing performance factors concern aviators—the angle and rate of climb. These two factors are discussed later in this section (figure 7-37).

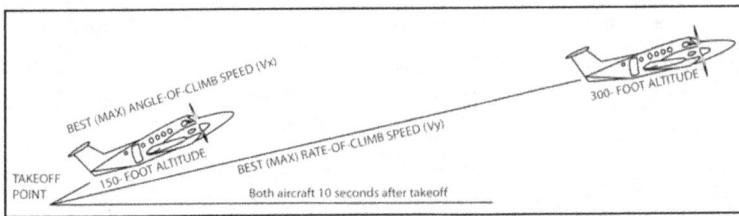

Figure 7-37. Climb angle and rate

Chapter 7

Aerodynamic Forces During Climbing Flight

7-120. All forces acting on the aircraft are resolved into components either perpendicular or parallel to relative wind. In climbing flight, weight is not perpendicular to relative wind. Therefore, weight must be resolved into its two components, one parallel and the other perpendicular, to relative wind (figure 7-38).

7-121. A vector diagram is necessary to understand action of four basic forces (lift, thrust, weight, and drag) that act on the aircraft during a stabilized, steady-state climb. Lift force is less than the weight in a climb. The steeper the climb angle, the less lift required to maintain balanced flight; thrust force supports the portion of the weight not supported by lift. If the aircraft could climb straight up, lift would be zero and thrust would support the entire aircraft weight and overcome drag. Certain assumptions made are the aircraft is climbing at a constant velocity (constant true airspeed [TAS] and straight flight path), and thrust force is considered to be acting along the flight path. Using these assumptions, Newton's first law of motion prevails. The aircraft is in equilibrium; the sum of the forces acting about the aircraft CG equals zero.

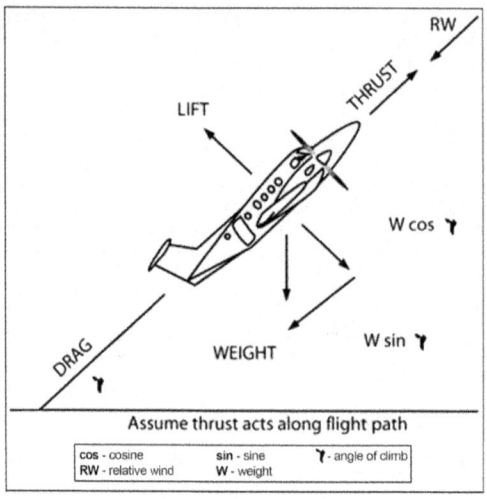

Figure 7-38. Force-vector diagram for climbing flight

TRANSITION TO CLIMB

7-122. Forces acting on an aircraft go through definite changes when the aircraft makes the transition from level flight to a climb. The first change—an increase in lift—occurs when pressure is applied to the elevator control. This initial change is a result of the increase in the attack angle, which occurs when the aircraft's pitch attitude is raised. This results in a climbing attitude. When the inclined flight path and climb speed are established, the attack angle and corresponding lift again stabilize.

NOSE-DOWN TENDENCY

7-123. As airspeed decreases to climb speed, air striking the horizontal stabilizer is reduced. This creates a longitudinally unbalanced condition; the aircraft tends to pitch nose down. To overcome this tendency and maintain a constant climb attitude, additional pressure must be applied to the elevator control.

CLIMBING STALL SPEED

7-124. When an aircraft is in a climb, it will stall at a lower speed. Stalling speed depends on the amount of lift a wing produces. Reducing the amount of lift required of the wing also reduces the aircraft's stalling speed. When an aircraft is in climbing flight, the lift required of the wing is not equal to the weight but only to a portion of the weight. This is due to the vertical component of thrust. No lift force is required when an aircraft is in vertical flight. Therefore, an aircraft cannot aerodynamically stall in vertical flight.

ANGLE OF CLIMB

7-125. The angle of climb (γ) is the angle between the flight path and horizontal plane. The maximum, or best, angle of climb may be required to clear an obstacle after takeoff.

7-126. The amount of excess thrust available (TA) determines the angle of climb that can be maintained. Flight at maximum (best) angle-of-climb speed (V_x) is usually just above stall speed or below minimum control speed in multiengine aircraft. This places the aircraft at a critical flight speed; any increase in attack angle or a loss of power on one engine could result in a stall or loss of control. The recommended airspeed for a maximum climb angle–such as obstacle-clearance airspeed listed in some operator's manuals–is not a true V_x but a safe best angle-of-climb speed. This airspeed is greater than true maximum angle-of-climb speed. It places the aircraft in a safer flight envelope while only slightly sacrificing climb performance. During takeoff when obstacle clearance is primary concern, V_x should be used; the most altitude is gained for the horizontal distance covered.

EFFECTS

7-127. Altitude, weight, and wind each affect angle of climb and are discussed below.

Altitude

7-128. As an aircraft gains altitude, thrust developed by the engine normally decreases. This is true for both turbine and reciprocating engines. The angle of climb must also decrease as a decrease in TA causes a decrease in excess thrust. Thrust required (torque [TR]) remains about constant at all altitudes. As a result, aircraft angle of climb decreases to zero degrees when reaching its absolute ceiling where TA equals TR.

Weight

7-129. A weight increase adversely affects angle-of-climb performance in two ways; it increases both weight and TR. This means there is more weight to be raised with less excess thrust, resulting in a shallower angle of climb.

Wind

7-130. When obstacle clearance is of primary concern, the best angle-of-climb speed for the aircraft must be used. The best angle of climb gains the most altitude for the distance covered. Wind must be considered as it affects the horizontal distance covered to clear an obstacle. With the aircraft climbing at speed for best angle of climb (V_X) (figure 7-39), the horizontal distance covered across the ground in a head wind is less than the horizontal distance covered with no wind or a tail wind. This affects the angle the aircraft climbs over the ground.

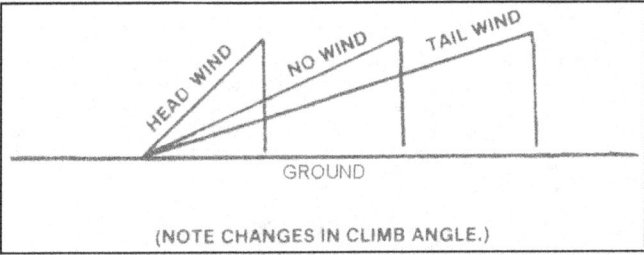

Figure 7-39. Wind effect on maximum climb angle

Chapter 7

ATTACK ANGLE FOR BEST ANGLE OF CLIMB

7-131. To determine angle of climb performance for a propeller aircraft, a TR and TA curve must be constructed. The thrust-required curve is simply the drag curve for the aircraft. As a propeller aircraft increases velocity, thrust force coming from the propeller decreases. The attack angle for a propeller aircraft when it is climbing at its best angle of climb is higher than the attack angle for minimum drag speed (L/D_{max}).

7-132. This high attack angle required of a propeller aircraft is near its takeoff attack angle. If a propeller aircraft must make an obstacle clearance takeoff, it continues to climb at an airspeed close to its takeoff airspeed.

RATE OF CLIMB

7-133. Rate-of-climb performance is the feet-per-minute gain in altitude (vertical velocity). The maximum, or best, rate-of-climb speed (V_y) is flown at a lower climb angle and higher airspeed than V_x. Though the aircraft is flown at a lower climb angle, higher velocity produces a higher rate of climb than could be obtained during a maximum angle of climb (figure 7-37, page 7-29).

EFFECTS

7-134. Altitude and weight affect rate of climb and are discussed below. Since the horizontal and vertical velocities are within the air mass (TAS), wind has no effect on the rate of climb.

Altitude

7-135. Altitude affects engine performance. As with angle of climb, an increase in altitude decreases the rate of climb. The rate of climb at the absolute ceiling of an aircraft is zero. At this altitude, there is no excess power. At the altitude called the service ceiling, an aircraft can maintain a 100-feet-per-minute rate of climb. When operating on a single engine, the aircraft can maintain a 50 feet-per-minute rate of climb.

Weight

7-136. As with angle of climb, weight also affects climbing performance. As weight increases, horsepower required increases. Therefore, a decrease in excess horsepower and increase in weight decreases the rate of climb. As an aircraft burns fuel, its weight decreases. Due to this weight decrease, more excess horsepower is available toward the end of a flight.

ANGLE OF ATTACK FOR BEST RATE OF CLIMB

7-137. The velocity where a propeller aircraft can obtain its best rate of climb is close to the velocity for L/D_{max}. This point is determined from horsepower curves. Measurements are made on those curves; they are not calculated. Maximum excess power produces the best rate of climb.

AIRCRAFT PERFORMANCE IN A CLIMB OR DIVE

PERFORMANCE CAPABILITIES

7-138. The full-power aircraft performance capabilities in a climb or dive can be visualized with a full-power polar diagram. The diagram is plotted as if weight, altitude, and power or thrust are held constant.

Note. If any of the three factors mentioned above change, then curve and performance change.

TYPICAL POLAR DIAGRAM

7-139. Figure 7-40, page 7-33, shows the typical polar diagram for full-power operation at 5,000 feet. This curve represents the plot of vertical and horizontal velocities obtained by the aircraft at full power with different climb and dive angles. Point 1 on the curve represents the maximum aircraft velocity in straight-and-level flight (at full power).

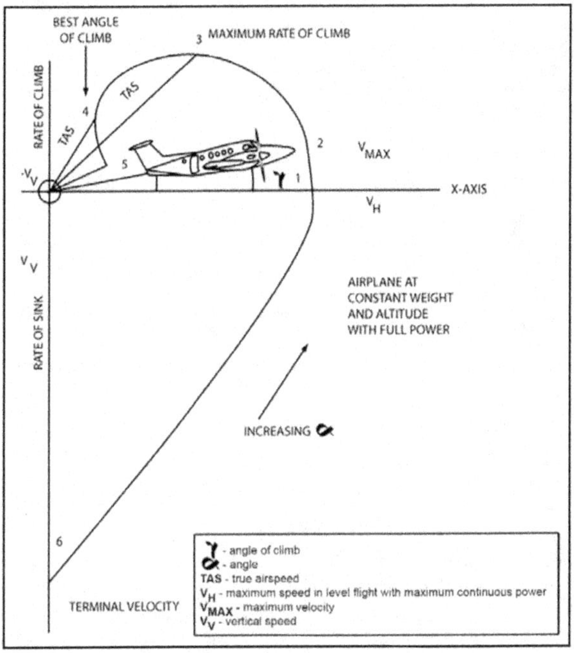

Figure 7-40. Full-power polar diagram

7-140. As the aircraft starts to climb, velocity decreases; the aircraft gains altitude and has a vertical velocity, as shown at point 2. The vertical velocity (rate of climb) can be read on the scale at the left. The angle between the flight path and horizontal velocity line is the angle of climb (γ).

7-141. Point 3 shows the maximum rate of climb (V_y). The curve indicates the maximum vertical velocity obtainable by the aircraft. A line drawn from the origin to the top point of the curve shows the climb angle and TAS when the aircraft is climbing at a maximum rate.

7-142. A line drawn from the origin tangent to the curve indicates the maximum angle of climb (V_x) for the aircraft at point 4. At full power, the aircraft performs somewhere on this curve; there cannot be any steeper climb angle for this aircraft. The TAS and climb angle can be obtained by drawing a line from the origin tangent to the curve.

7-143. If an aircraft stalls with excess power at its stall speed in level flight, then it is in climbing flight when it stalls at full power. Full-power stalling speed is shown at point 5. The climb angle the aircraft is able to attain at its stalling speed can be read from the graph.

7-144. Point 6 is the vertical velocity the aircraft attains if it were diving straight down with full power. Many aircraft would break up before reaching this velocity due to their structural limitations. This point is shown for information purposes only and to complete the curve.

POLAR CURVE

7-145. Any change in altitude, weight, or power setting affects aircraft performance and produces changes in the full-power polar. Curves drawn showing these changes are called the family of polar curves. Figure 7-41 could be the aircraft's polar curve at its absolute ceiling, partial thrust polar, or even sea-level polar in which the aircraft's weight would not allow the aircraft to climb.

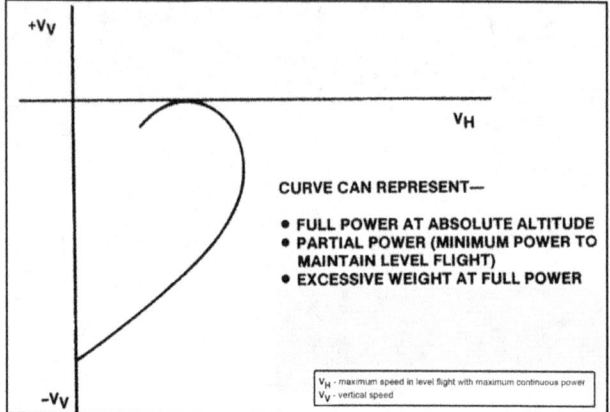

Figure 7-41. Polar curve

7-146. At the absolute ceiling polar, the aircraft cannot climb. Therefore, the full power the power plant can produce at maximum altitude is only adequate to maintain the aircraft in straight-and-level flight.

7-147. Figure 7-41 shows a curve that could also represent partial power. In this case, the aircraft is operating with minimum power to maintain level flight. The aircraft encounters this condition when operating at maximum endurance.

7-148. The third case would show with full power in an overweight condition no climb is possible.

TURNS

PERFORMANCE

7-149. Unlike automobiles or other ground vehicles, an aircraft can rotate about three axes; therefore, it has six degrees of motion. The aircraft can pitch up or down, yaw left or right, and roll left or right. Due to this freedom of motion, an aircraft can perform many maneuvers. All these maneuvers use vertical turns, horizontal turns, or both. This section discusses vertical and horizontal turns separately, as well as the limits imposed on these turns.

7-150. When an aircraft turns, it is not in static equilibrium. Forces must be unbalanced to produce acceleration for turning. When properly performed, the turn does not produce any sideward force pulling the aviator inward or outward from the turn. The net resulting force (lift) acts toward the center of the turn. This turn is called a coordinated turn. The term level turn may be confusing as it refers to a constant-altitude turn, not a wings-level turn; normally, an aircraft is never turned with the wings level. Forces acting on the aircraft must be unbalanced for a turn to occur; this does not happen if wings are level. Incorrect banking during uncoordinated turns causes slips or skids; this is uncomfortable for the crew and passengers. During uncoordinated turns at slow airspeeds, control can inadvertently be lost.

7-151. The force actually turning the aircraft is lift force. The horizontal component of lift is the force that accelerates the aircraft toward the center of the turn. The rudder counteracts adverse aileron effect (yaw). The elevator increases the attack angle to produce the added lift required due to the loss of the vertical lift component and an apparent increase in weight, which is produced by centrifugal force.

Turning Flight

7-152. An aircraft, like any moving object, requires a sideward force to make a turn. In a normal turn, this force is supplied by banking the aircraft so lift is exerted inward as well as upward. The force of lift is then separated into two components at right angles to each other. The lift acting upward combined with opposing weight is called the vertical-lift component. The horizontal-lift component (centripetal force) is the lift acting horizontally combined with opposing inertia or centrifugal force (figure 7-42). Therefore, the horizontal-lift component is the sideward force that forces the aircraft from straight flight, causing it to turn. If an aircraft is not banked, no force is present to make the turn unless rudder application causes the aircraft to skid in the turn. Likewise, if an aircraft is banked, it turns unless it is held on a constant heading with the opposite rudder. Proper control technique assumes an aircraft is turned by banking and that in a banking attitude it should be turning.

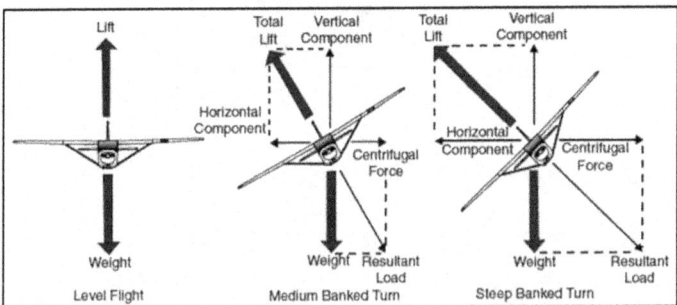

Figure 7-42. Effect of turning flight

Turn Radius

7-153. The aircraft turn radius varies directly with the square of its velocity (TAS) and inversely with the bank angle. Therefore, any two aircraft flying at the same velocity and bank angle can fly in formation, regardless of their weights. However, certain aerodynamic considerations–weight, altitude, load factor, attack angle, and wing area–affect the velocity and also indirectly affect the turn radius. All of these aerodynamic considerations play a part in the lift force produced. To turn an aircraft in the smallest possible radius, an aviator flies at the slowest possible speed and highest possible bank angle. Limits on turn radius performance are aerodynamic, structural, and power limits. The aviator must be constantly aware of these limits while maneuvering the aircraft at or near its design limits.

Aerodynamic Limit of Performance

7-154. Since the horizontal component of lift is the force turning the aircraft, a FW aircraft reaches its aerodynamic turn radius limit when the aircraft turns at its stall velocity. The airspeed at which the minimum turn radius occurs is the stalling speed where maximum Gs can be pulled without exceeding the design load-limit factor. This is maneuvering airspeed. An increase in weight or altitude or a decrease in $C_L$max requires an increase in velocity, which increases the turn radius.

Structural Limit of Performance

7-155. The load factor is purely a function of the bank angle; the aircraft's weight does not affect the G-load imposed on the aircraft. Both the C-12 and C-23 accelerate 2 Gs in a 60-degree bank. The table in figure 7-43, part A shows the load factor at various bank angles. The graph in part B depicts how the increasing load factor affects the aircraft's stalling speed. In the first 60 degrees of bank, the load factor increases by only one. However, in the next 10 degrees (60 to 70 degrees), the load factor increases almost another one (figure 7-43, part A). At higher bank angles, the load factor and stalling speed increase rapidly. A steep turn immediately after takeoff is extremely dangerous due to the load factor imposed and low aircraft velocity.

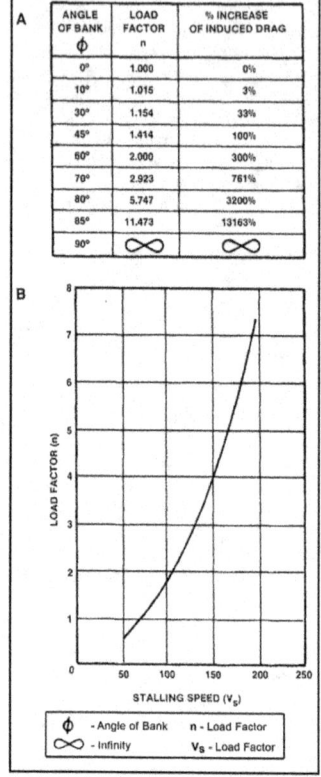

Figure 7-43. Effect of load factor on stalling speed

7-156. The stall speed increases as the bank angle increases. Therefore, a compromise between the bank angle and stalling speed must be made to obtain the minimum aircraft turn radius. Minimum turn radius is found by considering the aircraft's design strength. An aircraft pulling 3 Gs develops a lift force three times its weight. Aircraft are designed to take certain loads; if these load limits are exceeded, the aircraft becomes overstressed. Load limits are published in the appropriate operator's manual. If an aircraft has a load limit of 3 Gs, the minimum turn radius is at its 3-G stalling speed. This occurs at a bank angle of about 73 degrees (figure 7-43, part A). The aircraft's maneuvering speed is the velocity at which the minimum turning radius can be performed at a given altitude without exceeding the load limit.

Power Limit of Performance

7-157. The third limit imposed on the turning performance is the power or thrust limit. The amount of induced drag developed at high-load factors can become quite large. Induced drag is directly proportional to lift squared. In a 73-degree bank, three times more lift is produced than in level flight; therefore, induced

Fixed-Wing Aerodynamics and Performance

drag is nine times that of a level flight at the same velocity. This is a tremendous amount of drag to overcome. The power available from the power plant is the limiting factor.

SLOW FLIGHT

7-158. In aviator training and testing, slow flight is broken down into two distinct elements—
- The establishment, maintenance, and maneuvering of the aircraft at airspeeds and in configurations appropriate to takeoffs, climbs, descents, landing approaches and go-arounds.
- Maneuvering at the slowest airspeed the aircraft is capable of maintaining controlled flight without indications of a stall–usually 3 to 5 knots above stalling speed.

FLIGHT AT LESS THAN CRUISE AIRSPEEDS

7-159. Maneuvering during slow flight demonstrates aircraft flight characteristics and the degree of controllability at less than cruise speeds. The ability to determine characteristic control responses at lower airspeeds appropriate to takeoffs, departures, and landing approaches is a critical factor in stall awareness.

7-160. As airspeed decreases, control effectiveness decreases disproportionately. For instance, there may be a certain loss of effectiveness when airspeed is reduced from 30 to 20 miles per hour above stalling speed, but normally there is a much greater loss as the airspeed is further reduced to 10 miles per hour above stalling speed. The objective of maneuvering during slow flight is to increase aviator confidence and develop their ability to use the controls correctly, thereby improving proficiency in performing maneuvers requiring slow airspeeds.

7-161. Maneuvering during slow flight should be performed using both instrument indications and outside visual reference. Slow flight should be practiced from straight glides, straight-and-level flight, and medium banked gliding and level flight turns. Slow flight at approach speeds should include slowing the aircraft smoothly and promptly from cruising to approach speeds without changes in altitude or heading, and determining and using appropriate power and trim settings. Slow flight at approach speed should also include configuration changes, such as landing gear and flaps, while maintaining heading and altitude.

FLIGHT AT MINIMUM CONTROLLABLE AIRSPEED

7-162. This maneuver demonstrates aircraft flight characteristics and degree of controllability at its minimum flying speed. By definition, the term flight at minimum controllable airspeed means a speed at which any further increase in attack angle, load factor, or reduction in power causes an immediate stall. Instruction in flight at minimum controllable airspeed should be introduced at reduced power settings, with the airspeed sufficiently above the stall to permit maneuvering but close enough to sense flight characteristics at very low airspeed–which are sloppy controls, ragged response to control inputs, and difficulty maintaining altitude. Maneuvering at minimum controllable airspeed should be performed using both instrument indications and outside visual reference. It is important aviators form the habit of frequent reference to flight instruments, especially the airspeed indicator, while flying at very low airspeeds. However, a feel for the aircraft at very low airspeeds must be developed to avoid inadvertent stalls and operate the aircraft with precision.

7-163. This maneuver is performed in the region of reversed command. The FAA definition is a flight regime in which flight at a higher airspeed requires a lower power setting and flight at a lower airspeed requires a higher power setting to maintain altitude. A better description is in normal flight pitch used for altitude control and power for airspeed. In reversed command, pitch controls airspeed and power controls altitude.

7-164. To begin the maneuver, the throttle is gradually reduced from cruising position. While the airspeed is decreasing, the nose position in relation to the horizon should be noted and raised as necessary to maintain altitude.

7-165. When the airspeed reaches the maximum allowable for landing gear operation, the landing gear (if equipped with retractable gear) must be extended and all-gear-down checks performed. As the airspeed reaches the maximum allowable for flap operation, full flaps must be lowered and the pitch attitude adjusted to maintain altitude. Additional power is required as the speed further decreases to maintain airspeed just above a stall. As speed decreases further, the aviator notes the feel of the flight controls, especially the elevator. The aviator also notes the sound of the airflow as it falls off in tone level.

7-166. As airspeed is reduced, flight controls become less effective and normal nose-down tendency is reduced. The elevators become less responsive and coarse control movements become necessary to retain control of the aircraft. The slipstream effect produces a strong yaw so application of rudder is required to maintain coordinated flight. The secondary effect of applied rudder is to induce a roll, so aileron is required to keep the wings level. This can result in flying with crossed controls.

7-167. During these changing flight conditions, it is important to retrim the aircraft as often as necessary to compensate for changes in control pressures. If the aircraft has been trimmed for cruising speed, heavy aft control pressure is needed on the elevators making precise control impossible. If too much speed is lost or too little power is used, further back pressure on the elevator control may result in a loss of altitude or a stall. When the desired pitch attitude and minimum control airspeed have been established, it is important to continually cross-check the attitude indicator, altimeter, and airspeed indicator, as well as outside references to ensure accurate control is maintained.

7-168. The aviator should understand when flying slower than minimum drag speed (L/D_{max}), the aircraft exhibits a characteristic known as speed instability. If the aircraft is disturbed by even the slightest turbulence, the airspeed decreases. As airspeed decreases, total drag also increases resulting in a further loss in airspeed. Total drag continues to rise and speed continues to fall. Unless more power is applied and/or the nose is lowered, speed continues to decay down to the stall. This is an extremely important factor in the performance of slow flight. The aviator must understand at speeds less than minimum drag speed, airspeed is unstable and will continue to decay if allowed to do so.

7-169. When the attitude, airspeed, and power have been stabilized in straight flight, turns should be practiced to determine the aircraft's controllability characteristics at this minimum speed. During turns, power and pitch attitude may need to be increased to maintain the airspeed and altitude. The objective is to acquaint the aviator with the lack of maneuverability at minimum speeds, danger of incipient stalls, and the tendency of the aircraft to stall as the bank is increased. A stall may also occur as a result of abrupt or rough control movements when flying at this critical airspeed.

7-170. Abruptly raising the flaps while at minimum controllable airspeed results in lift suddenly being lost causing the aircraft to lose altitude or perhaps stall.

7-171. Once flight at a minimum controllable airspeed is properly obtained for level flight, a descent or climb at the minimum controllable airspeed can be established by adjusting the power as necessary to establish the desired rate of descent or climb. The inexperienced aviator should note the increased yawing tendency at the minimum control airspeed during high power settings with flaps fully extended. In some aircraft, an attempt to climb at such a slow airspeed may result in a loss of altitude, even with maximum power applied.

7-172. Common errors in performance of slow flight are—
- Failure to adequately clear the area.
- Inadequate back-elevator pressure as power is reduced, resulting in altitude loss.
- Excessive back-elevator pressure as power is reduced, resulting in a climb, followed by a rapid reduction in airspeed and mushing.
- Inadequate compensation for adverse yaw during turns.
- Fixation on the airspeed indicator.
- Failure to anticipate changes in lift as flaps are extended or retracted.
- Inadequate power management.
- Inability to adequately divide attention between aircraft control and orientation.

DESCENTS

7-173. When an aircraft enters a descent, its flight path changes from level to an inclined plane. It is important the aviator know the power settings and pitch attitudes that produce conditions of descent.

Fixed-Wing Aerodynamics and Performance

PARTIAL POWER DESCENT

7-174. The normal method of losing altitude is to descend with partial power. This is often termed cruise or en route descent. The airspeed and power setting recommended by the aircraft manufacturer for prolonged descent should be used. The target descent rate should be 400 to 500 FPM. The airspeed may vary from cruise airspeed to that used on the downwind leg of the landing pattern. But the wide range of possible airspeeds should not be interpreted to permit erratic pitch changes. The desired airspeed, pitch attitude, and power combination should be preselected and kept constant.

DESCENT AT MINIMUM SAFE AIRSPEED

7-175. A minimum safe airspeed descent is a nose-high, power assisted descent condition principally used for clearing obstacles during a landing approach to a short runway. The airspeed used for this descent condition is recommended by the aircraft manufacturer and is normally no greater than 1.3 minimum steady flight spread in the landing configuration (V_{so}). Some characteristics of minimum safe airspeed descent are a steeper than normal descent angle, and the excessive power required to produce acceleration at low airspeed should mushing and/or an excessive descent rate be allowed to develop.

GLIDES

7-176. A glide is a basic maneuver in which the aircraft loses altitude in a controlled descent with little or no engine power; forward motion is maintained by gravity pulling the aircraft along an inclined path, and the descent rate is controlled by the aviator balancing the forces of gravity and lift.

7-177. Although glides are directly related to the practice of power-off accuracy landings, they have a specific operational purpose in normal landing approaches and forced landings after engine failure. Therefore, it is necessary they be performed more subconsciously than other maneuvers as most of the time during their execution the aviator is giving full attention to details other than the mechanics of performing the maneuver. Since glides are usually performed relatively close to the ground, accuracy of their execution and formation of proper technique and habits are of special importance.

7-178. Since the application of controls is somewhat different in glides than in power-on descents, gliding maneuvers require the perfection of a technique somewhat different from that required for ordinary power-on maneuvers. This control difference is caused primarily by two factors—absence of the usual propeller slipstream and difference in the relative effectiveness of various control surfaces at slow speeds.

7-179. The glide ratio of an aircraft is the distance the aircraft will, with power off, travel forward in relation to the altitude it loses. For instance, if an aircraft travels 10,000 feet forward while descending 1,000 feet, its glide ratio is said to be 10:1.

7-180. The glide ratio is affected by all four fundamental forces acting on an aircraft (weight, lift, drag, and thrust). If all factors affecting the aircraft are constant, the glide ratio is constant. Although the effect of wind is not covered in this section, it is a prominent force acting on the aircraft's gliding distance in relationship to its movement over the ground. With a tailwind, the aircraft glides farther because of the higher groundspeed; with a headwind, the aircraft does not glide as far because of the slower groundspeed.

7-181. Variations in weight do not affect the glide angle provided the aviator uses the correct airspeed. Since it is the lift over drag (L/D) ratio that determines the distance the aircraft can glide, weight does not affect the distance. The glide ratio is based only on the relationship of the aerodynamic forces acting on the aircraft. The only effect weight has is to vary the time the aircraft glides. The heavier the aircraft the higher the airspeed must be to obtain the same glide ratio. For example, if two aircraft having the same L/D ratio, but different weights, start a glide from the same altitude, the heavier aircraft gliding at a higher airspeed arrives at the same touchdown point in a shorter time. Both aircraft cover the same distance, but the lighter aircraft takes a longer time.

7-182. Under various flight conditions, the drag factor may change through the operation of the landing gear and/or flaps. When the landing gear or the flaps are extended, drag increases and airspeed decreases unless the pitch attitude is lowered. As pitch is lowered, the glide path becomes steeper and reduces the distance traveled. With the power off, a windmilling propeller also creates considerable drag, thereby retarding the aircraft's forward movement.

7-183. Although the aircraft's propeller thrust is normally dependent on the engine's power output, the throttle is placed in the closed position during a glide so thrust is constant. Since power is not used during a glide or power-off approach, the pitch attitude must be adjusted as necessary to maintain a constant airspeed.

7-184. The best speed for the glide is one at which the aircraft travels the greatest forward distance for a given loss of altitude in still air. This best glide speed corresponds to an attack angle resulting in the least drag on the aircraft and giving the best lift-to-drag ratio (L/D_{max}).

7-185. Any change in gliding airspeed results in a proportionate change in glide ratio. Any speed, other than the best glide speed, results in more drag. Therefore, as glide airspeed is reduced or increased from optimum or best glide speed, glide ratio is also changed. When descending at a speed below best glide speed, induced drag increases. When descending at a speed above best glide speed, parasite drag increases. In either case, the rate of descent increases (figure 7-44).

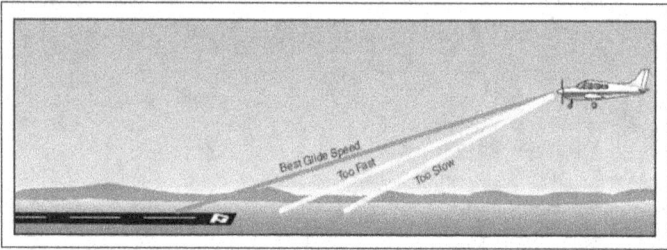

Figure 7-44. Best glide speed

7-186. The aviator must never attempt to stretch a glide by applying back-elevator pressure and reducing airspeed below the aircraft's recommended best glide speed. Attempts to stretch a glide invariably result in an increase in the rate and angle of descent and may precipitate an inadvertent stall.

SECTION V – TAKEOFF AND LANDING PERFORMANCE

PROCEDURES AND TECHNIQUES

7-187. Techniques used in older aircraft during the takeoff and landing phases of flight were essentially the same for each type of aircraft. Because of design differences in today's aircraft, however, procedures and techniques differ from aircraft to aircraft. Some aircraft rotate to their takeoff attitude early in the takeoff roll. Some aircraft make a constant attack angle approach and hold it until landing; others flare just before touchdown, which decreases the rate of sink. Aerodynamic braking is effective in some aircraft during a landing roll. In other aircraft, aerodynamic braking is dangerous.

7-188. The design characteristics of an aircraft determine its takeoff and landing performance. However, a detailed discussion of various aircraft is beyond the scope of this section. This section is concerned primarily with the aerodynamic and physical considerations determining the runway length needed for a successful takeoff or landing.

TAKEOFF

7-189. Takeoff is an acceleration and transition maneuver. During a takeoff run, the aircraft makes the transition from a ground-supported to an air-supported vehicle. At the start of the takeoff roll, the entire aircraft's weight is supported by the wheels. As the aircraft gains velocity, the wing begins to support more of the weight. By the time the aircraft reaches its takeoff velocity, the wing supports the entire aircraft. The aircraft has then completed the transition from ground to air support.

ACCELERATION FORCES

7-190. According to Newton's second law of motion, a body accelerates only if there is an unbalanced force acting on that body. The acceleration takes place in the direction of the unbalanced force. For the aircraft to accelerate during the takeoff run, the sum of the horizontal forces acting on the aircraft must yield an unbalanced force in the thrust direction. Figure 7-45 shows horizontal forces that determine the net accelerating force on the aircraft during its takeoff roll. For a given altitude and RPM, thrust from a propeller-driven aircraft decays as velocity increases during the takeoff roll.

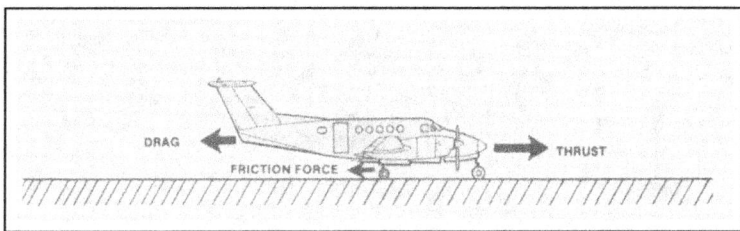

Figure 7-45. Net accelerating force

7-191. The aircraft's wing is close to the ground during the takeoff run. This reduces or cancels out the downwash and wingtip vortexes behind the wing. Reduction of downwash also reduces induced drag. This phenomenon is known as ground effect. It normally reduces induced drag about 1.4 percent at one wingspan, 23.5 percent at one-fourth wingspan, and 47.6 percent at one-tenth wingspan. Parasite drag is directly proportional to the square of the velocity, so this drag force increases as aircraft velocity increases.

7-192. Figure 7-45, which shows friction force, is called rolling friction. It results from the rolling action of the tires against the runway. Like any friction force, this rolling friction is equal to the product of the coefficient of rolling friction and a normal (perpendicular) force. The coefficient of rolling friction varies from 0.02 to 0.3, depending on the runway surface and type of tire. In this case, the normal force is the aircraft weight not supported by the wing. As mentioned before, as the aircraft gains speed during the takeoff roll, the wing supports more of the weight. This weight reduction on the wheels reduces the rolling friction force as the aircraft accelerates. When the aircraft reaches takeoff velocity, the friction force is zero since the normal force is zero.

TAKEOFF DISTANCE

7-193. Takeoff distance is directly proportional to takeoff velocity squared. Because velocity is squared, its effect on takeoff distance is significant. The takeoff velocity is a function of the aircraft's stalling speed and usually 1.1 to 1.25 times the power-off stalling speed. Some aircraft use high-lift devices to lower the stalling speed, which decreases takeoff velocity and, thus, takeoff distance. These high-lift devices are used only to increase C_L.

Altitude

7-194. An aircraft taking off from a field at a 5,000-foot elevation requires a longer takeoff run than the same aircraft taking off at sea level; as altitude increases, air density decreases. This requires an increase in TAS to develop the required amount of lift. This increase in airspeed increases the takeoff distance. Engine performance is also affected as elevation is increased. Increases in altitude decrease the thrust output. Therefore, the net accelerating force decreases at a higher elevation, and takeoff distance increases because of this thrust loss.

Weight

7-195. Weight changes, as well as air density changes, have a compounding effect on the aircraft's takeoff distance. If an aircraft could double its takeoff weight, it might be assumed takeoff distance would double. If weight had no effect on the takeoff velocity, this assumption would be correct. However, doubling the weight doubles the value of the square of takeoff velocity. This value, when combined with the increased value of the weight, yields a takeoff four times the original distance. Another factor is also affected by weight increase–rolling friction force. The increased weight increases normal force and, therefore, rolling friction force. This results in a decrease in net accelerating force, which adds additional distance to the takeoff run.

Wind

Wind Component

7-196. Until now, the discussion of takeoff distance has assumed a no-wind condition; however, wind can be used to decrease the takeoff distance. For example, an aircraft is assumed to be taking off into a head wind equal to its takeoff velocity. While the aircraft is sitting at the end of the runway, the wind velocity over the wing is enough to support the aircraft. The aviator does not need to accelerate because the aircraft does not require a takeoff run or distance. The aircraft's ground speed is zero even if the TAS is equal to the takeoff velocity.

Runway Direction

7-197. Most runways are built in the direction of the local prevailing winds. If a runway must be used that has a tail-wind component, the value of the tail wind must be added to the takeoff velocity, and then the sum of the two must be squared. Thus, a large increase in takeoff distance is necessary.

TAKEOFF PERFORMANCE SUMMARY

7-198. Various factors influence takeoff performance and distance. The interplay between some of these factors makes accurate takeoff distance requirements difficult to predict. In addition, acceleration is assumed to be constant during the takeoff roll. This is not necessarily true, however. The aviator can determine the actual takeoff distance required from the aircraft operator's manual, which has charts that include effects of temperature, PA, aircraft weight, winds, and runway slope. These charts, taken from flight tests, show the expected performance of each aircraft.

LANDING

7-199. In a landing roll, the aircraft must decelerate, not accelerate. As velocity decreases and lift force decays, weight shifts from the aircraft wings to the wheels. This is the reverse of the transition and acceleration scenario mentioned previously.

DISTANCE

7-200. During a landing, the aviator is primarily concerned with dissipating the aircraft's kinetic energy. Any factor affecting the aircraft's mass or velocity must be considered in computing the landing distance. Because velocity is a squared term in the kinetic energy equation, the final approach is always flown at the lowest velocity possible. This condition requires careful planning and execution by the aviator.

Net Decelerating Force

7-201. Compared to takeoff, forces that compose the net accelerating force of the landing are reversed. The acceleration force is now in the direction of the drag and friction forces; therefore, the aircraft slows. For this net decelerating force to develop, drag and friction must be greater than thrust force. However, thrust force must still be considered even if the engine is usually at idle. A residual thrust force must be overcome by drag and friction if the aircraft is to decrease its velocity. Some aircraft are equipped with propellers that can reverse their pitch. Therefore, the propeller can also develop a thrust force in the direction of the retarding forces. This thrust increases the retarding force and decreases the required landing distance.

7-202. The net decelerating force can be increased on some aircraft by aerodynamic braking–increasing the drag on the aircraft during the landing roll. The drag force is proportional to the square of the velocity. Therefore, aerodynamic braking is effective only during the initial landing roll, when the aircraft is at higher velocities.

Fixed-Wing Aerodynamics and Performance

7-203. If a short landing roll is required, it is possible to increase the friction force far above any aerodynamic force that could be applied to the aircraft. This increase in friction force is done with the wheel brakes. Brakes must not be applied too early. During the initial landing roll, most of the aircraft weight is supported by the wings. Using brakes at this time is not effective as the normal force on the wheels is low and the resulting friction force developed is small. Also, the wheels may lock and tires may blow out if the brakes are applied too hard at this time. Velocity should be decreased so enough weight is transferred to the wheels for the brakes to be effective. Some aircraft have wing spoilers used to destroy the lift on the wings. Weight is then transferred to the tires, allowing brakes to be applied earlier in the landing run. Retracting the flaps on some aircraft also reduces lift developed by the wings.

7-204. The condition of the runway surface also affects the aircraft's required stopping distance. On wet or icy runways, the coefficient of friction is small, resulting in a small decelerating force. Therefore, a longer stopping distance is required. A runway condition reading (RCR) is determined by using a decelerometer. The RCR is given in the remarks section of weather sequence reports, which are supplied by the pilot-to-forecaster service and ATC facility. The RCR should always be considered during a landing on a slick runway. The landing distance required may exceed the available runway length.

Deceleration Speed along Landing Roll

7-205. Figure 7-46 shows the speed at various distances from touchdown to a full stop. Again, this assumes a constant deceleration, which is not necessarily true but is easy to visualize. For example, if the total landing distance requires 4,500 feet and touchdown speed is 130 knots, the speed is higher than half the touchdown speed at half the landing distance. With this in mind, the aviator can avoid over-braking.

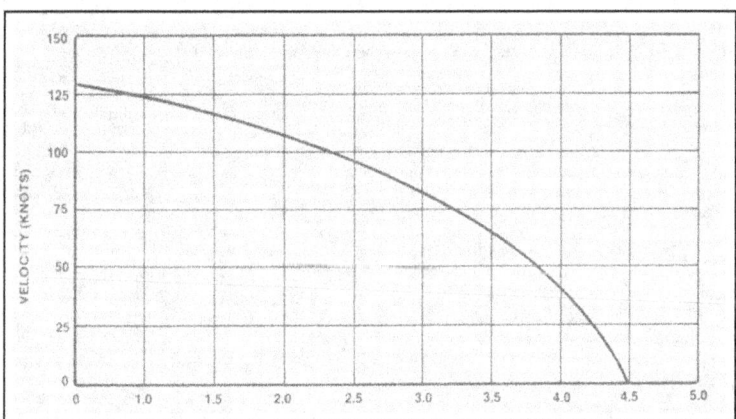

Figure 7-46. Landing roll velocity

Landing Velocity

7-206. The final approach airspeed of an aircraft is about 1.3 times its stalling speed. Therefore, any high-lift device that decreases stalling speed also decreases landing speed. A decreased landing velocity decreases both the kinetic energy and required landing distance. Longer landing distances are necessary for landings made at high altitudes as these higher altitudes result in faster TASs for final approaches.

Wind

7-207. A head wind results in a lower final-approach speed and slower touchdown velocities (ground speed). Reduced kinetic energy, with respect to the ground, decreases landing distance.

Weight

7-208. The aircraft's weight affects landing distance the same as it does takeoff distance. Decreased weight requires less runway, whether an aircraft is landing or taking off.

PERFORMANCE SUMMARY

7-209. As with takeoff performance, landing performance is difficult to predict. Each aircraft operator's manual has landing distance charts, which show results of flight tests conducted under differing conditions. These charts consider the factors affecting landing distance.

Hydroplaning

7-210. Wet or slippery runways can lead to hydroplaning. The hydroplaning aircraft rides on a film of water; tires have little or no contact with the runway surface. The aircraft is supported by a hydrodynamic lift force much like a water skier is supported by skis. The aviator must be aware of conditions that cause hydroplaning and understand how to avoid them.

7-211. For a simplified explanation of hydroplaning, an aircraft can again be compared to a water skier. To support the skier, a hydrodynamic lift force develops that depends on speed. Below speeds where aerodynamic forces dominate, the faster the speed, the easier it is to hydroplane. In the same way, there is a minimum speed at which hydroplaning occurs. Below a certain speed, however, drag is so great and hydrodynamic lift force so small, the skier sinks. Likewise, below the speed where hydroplaning occurs, the tires directly contact the runway.

7-212. The velocity at which total hydroplaning occurs depends on the square root of the tire inflation pressure. Partial hydroplaning may occur at slower speeds. Under-inflated tires can hydroplane at even slower speeds.

7-213. An assumption that heavier aircraft must move at faster speeds than lighter aircraft before hydroplaning occurs is not accurate as experiments and classical hydrodynamic theory show the speed at which hydroplaning occurs is independent of weight. Weight only determines the footprint size the tire makes, however; the ratio of weight per square inch of footprint area is the same. Weight has an indirect effect because a heavier aircraft must fly at a faster approach and touchdown speed. Thus, the possibility of hydroplaning is greater in heavier aircraft.

7-214. Other factors affecting hydroplaning cannot be quantitatively described in a formula. For example, how deep the water must be on a runway before hydroplaning develops is not well defined. Tire tread depth and pattern—as well as the runway surface itself—are also factors. A smooth tire can hydroplane in as little as .15 inches of water on the runway. A tire with deep tread has channels for the water to escape while part of the tire contacts the runway. This tire may need as much as 2 inches of water before hydroplaning occurs. Puddles on the runway can also cause intermittent hydroplaning. A smooth runway surface, in contrast to a coarse surface, may result in earlier hydroplaning.

7-215. Hydroplaning can occur with any aircraft. Tires must be checked during preflight for proper inflation and tread condition. The aviator should avoid crosswind landings when possible and fly at minimum airspeeds when landing on wet or slippery runways. A rule of thumb for determining hydroplaning speeds is to multiply the square root of the tire pressure by nine.

CROSSWIND OPERATIONS

7-216. An aircraft taking off or landing in a crosswind must have a track over the ground parallel to the runway heading. If the aircraft is going to make the desired track over the ground, it must sideslip through the air mass moving across the runway. The rudder produces the required sideslip so landing or takeoff can be made in the direction of the runway heading. The aircraft is traveling at low airspeeds during these phases of flight. Therefore, the directional control problem is again amplified by the lack of high dynamic pressure.

7-217. As mentioned during the directional control requirements discussion, an aircraft must sideslip through the air mass during a crosswind landing or takeoff. Because of the dihedral effect, a sideslip angle produces a roll away from the sideslip. The stronger the dihedral effect of the aircraft (positive lateral stability), the greater the lateral control required during this condition of flight.

SECTION VI – FLIGHT CONTROL

DEVELOPMENT

7-218. After early aircraft designers built surfaces that would yield enough lift to support an aircraft, the greatest problem still remaining was how to gain adequate and positive control of the airborne aircraft. Early gliders flown at the end of the last century were controlled by shifting the CG location in relation to the aerodynamic center. To do this, aviators shifted their body weight. This method of control not only proved inadequate but often was also disastrous. The Wright brothers' greatest contribution was the development of an adequate control system. They developed a method of warping the wings for lateral control. They also added a rudder for use with each wing. Others had already developed the elevator for gliders. This section discusses the theory of control surface operation and control requirements. Also covered are the types of control systems in use.

CONTROL SURFACE AND OPERATION THEORY

CAMBER

7-219. The Wright brothers' method of warping the wings changed the value of the lift coefficient on each wing. In effect, they changed the camber of the airfoils. Today, flaps are used to vary the camber of the airfoils, which varies the lift coefficient. If a flap is deflected downward (figure 7-47) the camber of the airfoil is increased. This results in a higher lift coefficient.

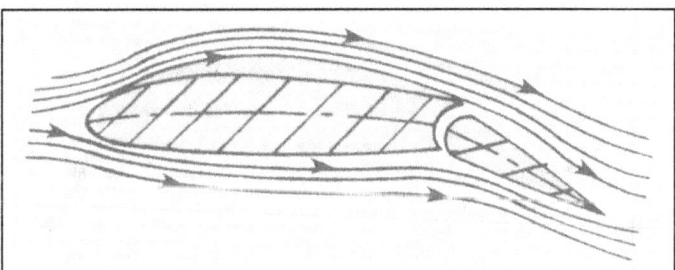

Figure 7-47. Using flaps to increase camber

AILERONS

7-220. Ailerons on a conventional aircraft operate in opposite directions. To bank an aircraft to the right (figure 7-48, page 7-46), the left aileron is lowered while the right aileron is raised. This increases the camber of the left wing and decreases the camber of the right wing. With increased C_L as compared to the right wing, the left wing has a greater lift force; this is assuming both wings are at equal or nearly equal velocity. This unbalanced lift force between the two wings results in a rolling moment about the longitudinal axis. Therefore, the aircraft rolls to the right.

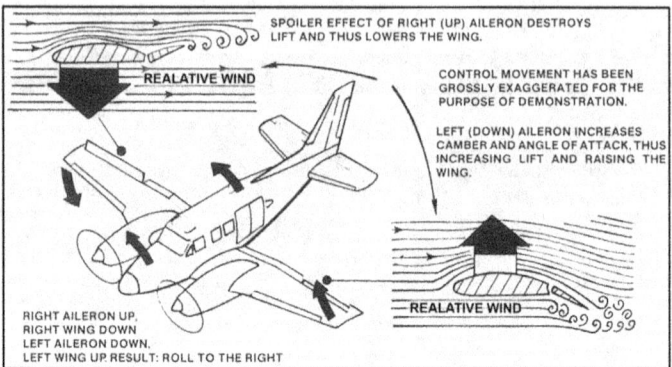

Figure 7-48. Operation of aileron in a turn

Elevators

7-221. The elevators attached to the horizontal stabilizer and the rudder attached to the vertical stabilizer work in the same manner to develop pitching and yawing moments (figure 7-49, parts A and B). The stabilizers are normally symmetrical airfoils. Deflecting the control surface changes the airfoil to either a positive or negative cambered airfoil, depending on direction of the surface movement.

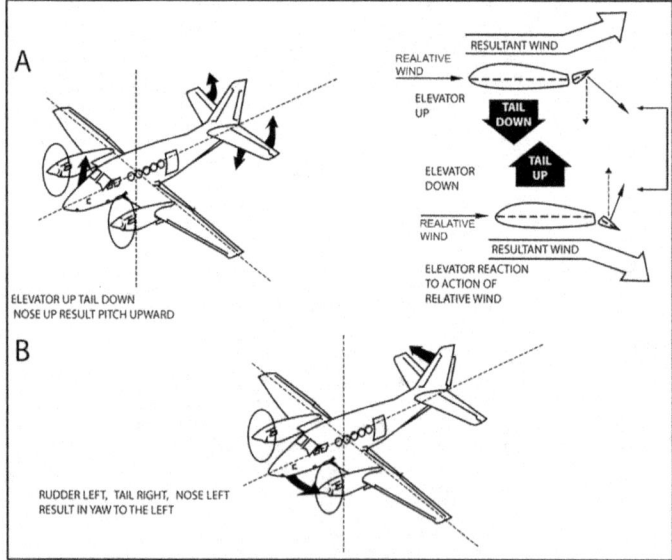

Figure 7-49. Effect of elevator and rudder on moments

CONTROL EFFECTIVENESS

7-222. Control effectiveness is the term used when discussing aircraft control systems. It refers to the amount of change in the lift coefficient for each degree of control-surface deflection rather than the ability of the control surface to maneuver the aircraft. Control effectiveness refers to the change in lift coefficient and not to the amount of lift change produced by deflecting the control surface. Because the lift formula applies to control surfaces, the amount of lift depends not only on the value of C_L but also on velocity, surface area, and air density.

LIFT COEFFICIENT CHANGE

7-223. A control surface deflected 10 degrees produces a greater change in the lift force at 200 knots than at 100 knots. At both airspeeds, the C_L change is the same; therefore, control effectiveness does not change. However, the greater lift change at higher airspeed represents better control response. This should indicate if a definite moment is desired. The control surface deflection must be increased as velocity is decreased.

LONGITUDINAL CONTROL

MANEUVERING CONTROL REQUIREMENTS

7-224. All Army aircraft can attain C_L max. They can fly at any value of lift coefficient designed into the aircraft. Army aircraft can also obtain the maximum lift the airfoil can produce for a given airspeed and altitude.

7-225. Figure 7-50 shows the effect of CG location upon the longitudinal maneuvering capability of an aircraft. The slanted lines represent different locations of aircraft CG. The 30-percent mean aerodynamic chord (MAC) denotes CG is located 30 percent of the MAC length back from the leading edge of the airfoil. The MAC is the chord of a rectangular wing that has the same pitching moments as the wing under consideration. The lower the percentage, the farther forward CG is located. More elevator deflection is required to obtain a certain value of C_L as the CG is moved forward. This increases longitudinal stability and, therefore, would be unsuitable for most military requirements. To correct this situation, the CG must be moved aft or the elevator must be designed so it can produce a larger moment.

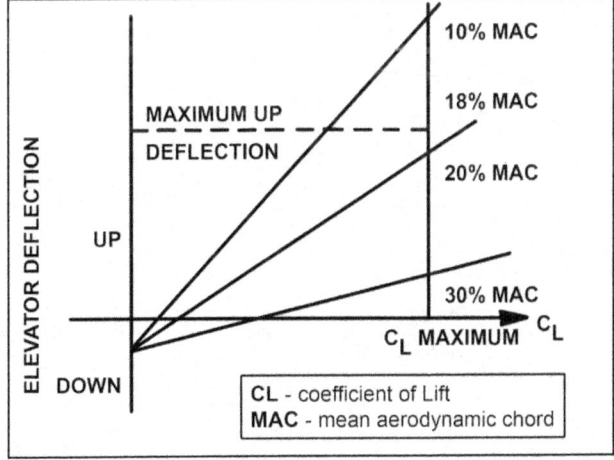

Figure 7-50. Effect of center of gravity location on longitudinal control

Chapter 7

TAKEOFF REQUIREMENT

7-226. An aircraft should be able to rotate to takeoff attitude before reaching takeoff velocity. This rotation occurs about the main landing gear wheels. This is an essential and demanding requirement. Normally, it attains takeoff attitude at about 0.9 V_s. Figure 7-51 represents an aircraft rolling down a runway and shows the forces producing adverse moments. These adverse moments try to pitch the aircraft's nose down. They must be overcome by the elevator control surface as it causes the aircraft's nose to pitch up.

7-227. First, a rolling friction force is generated at the wheels. This force is located below the CG and produces a negative pitching moment. In an aircraft with a tricycle landing gear, the CG must be located ahead of the main wheels. Therefore, if the aircraft is to rotate about the main wheels as it achieves takeoff attitude, this CG location produces an adverse pitching moment that must be overcome. If the lift force of the wing is ahead of the wheels, the lift force assists the elevator in pitching the aircraft's nose up.

7-228. Another condition not evident in figure 7-51 is ground effect or the decrease in downwash as the aircraft is close to the runway. When downwash is decreased, lifting surfaces become more effective and usually more beneficial. In this case, however, the elevator tries to increase the negative lift force of the horizontal stabilizer. The ground effect makes the horizontal stabilizer less effective in producing negative lift. Therefore, greater elevator deflection is necessary when the aircraft is being operated in-ground effect than when it is being operated OGE. Aircraft using flaps during the takeoff run can produce a downwash that may benefit the horizontal stabilizer and decrease the elevator control deflection. However, trailing-edge flaps also generate a negative pitching moment. The net result might be a greater elevator-deflection requirement.

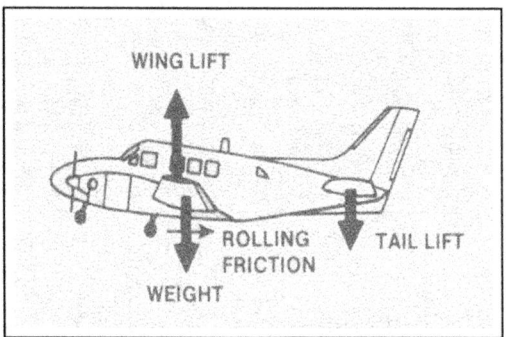

Figure 7-51. Adverse moments during takeoff

7-229. The elevator must be designed to produce a positive pitching moment—below the aircraft flight speed—that can overcome all adverse pitching moments created by some or all of the above mentioned conditions.

LANDING CONTROL REQUIREMENT

7-230. The landing-control requirement is essentially a ground-effect problem. The aircraft should have enough elevator control to maintain a landing attitude to the touchdown point. The elevator must overcome ground effect when the aircraft approaches the runway surface. If an aircraft is able to take off satisfactorily, it usually has enough elevator control to land.

DIRECTIONAL CONTROL

ADVERSE YAW

7-231. Adverse aileron yaw, described previously as yaw, develops when an aircraft is rolled using ailerons. The rudder must develop enough yawing moments to overcome adverse yaw created by an aileron roll. The rudder is used to keep relative wind on the aircraft's nose so a coordinated turn can be performed.

SPIN RECOVERY

7-232. An aircraft has a large sideslip angle as it is spinning. The rudder must be used on all Army FW aircraft to decrease the sideslip angle before the aircraft can recover from a spin. This requirement can be critical in some aircraft.

ASYMMETRICAL THRUST

7-233. A multiengine aircraft has an additional directional-control requirement not required on a single-engine aircraft. This is the yawing moment caused by asymmetrical thrust, which results from a difference in power on each wing. The difference in thrust force developed on each wing produces a yawing moment about the aircraft's CG away from the thrust. This yawing moment must be counteracted by an opposite moment from the rudder. While an engine failure on the critical engine would develop the greatest amount of asymmetrical thrust, it is present any time one engine produces more power than the other.

LATERAL CONTROL

7-234. Each aircraft's roll rate is designed to perform specific types of missions. These designs require different roll rates. A fighter, which is more maneuverable than a transport, must have a fairly high roll rate to perform its mission. To initiate a roll, the aviator deflects the ailerons. One aileron moves up, decreasing the C_L and lift on that wing. The other aileron moves down, increasing the C_L and lift on that wing. This difference in lift between the two wings causes the aircraft to roll toward the wing producing less lift. As the aircraft begins to roll, the attack angle on the down-going wing increases while the up-going wing decreases. This is caused by the component of airflow moving opposite the direction of wing movement. This difference in the attack angle between the two wings produces a rolling moment counter to the rolling moment developed by the ailerons. As the roll rate increases, the counter-rolling moment increases. When the counter-rolling moment produced by the roll equals the rolling moment produced by aileron deflection, a steady-state roll is reached. The roll rate is at maximum for that given aileron deflection. This steady-state roll rate must be rapid enough to be compatible with the mission requirements of the aircraft.

CONTROL FORCES

7-235. Control forces refer to the forces the aviator exerts on the control column to maneuver the aircraft. These forces must be logical and manageable. A logical force means the force must increase as the control surface is deflected or as the speed of the aircraft is increased. Manageable means the magnitude of the control force must be within the comfortable physical capabilities of the aviator.

REQUIRED CONTROL

Overcoming Hinge Moments

7-236. When a control surface is deflected, the pressure differential developed across the control surface creates a force that tends to streamline the control surface (figure 7-52, page 7-50). This force creates a moment about the control surface hinge, which is referred to as the hinge moment. This moment must be overcome by the force applied by the aviator to the control column. The hinge moment is directly proportional to takeoff safety speed (V_2) and the control surface area. If the airspeed is doubled, the hinge moment for a given deflection increases four times. Therefore, the aviator must exert four times the control force on the control column to overcome that hinge moment.

Chapter 7

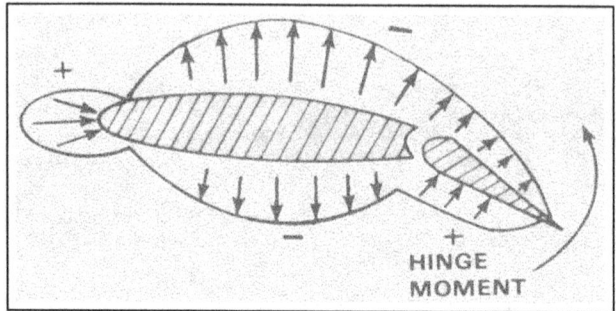

Figure 7-52. Hinge moment

Maintaining Aerodynamic Balance

7-237. As aircraft size increases, larger control surfaces are required. Therefore, as aircraft speed and size increase, control forces become unmanageable. The control system contains many levers and bell cranks that give the aviator a mechanical advantage over the hinge moments developed by the large control areas and dynamic pressures. However, the magnitude of the control force is still too large to overcome. This led to the development of aerodynamic control surface balancing. To achieve this balance, aerodynamic forces aid in deflecting the control surface. Aerodynamic balancing uses the airstream dynamic pressure to reduce the hinge moments of a control surface. One or more of the devices covered below are used.

TYPES OF DEVICES

Horn

7-238. The horn is one the first types of aerodynamic balancing developed. It is an area located ahead of the hinge line. This area creates a moment that opposes the hinge moment developed by the control surface area behind the hinge. The moment developed by the horn is in the same direction as the moment developed by the aviator and aids in displacing the control surface. It decreases the hinge moment the aviator must overcome. The horn can either be exposed (unshielded) or hidden (shielded), depending on the control surface design requirements (figure 7-53, parts A and B). An unshielded horn is normally more effective than the shielded horn; however, it is seldom used in modern high-performance aircraft due to drag considerations.

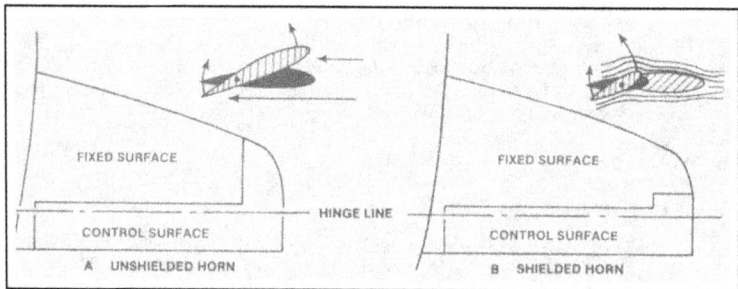

Figure 7-53. Aerodynamic balancing using horns

Balance Board

7-239. Another type of aerodynamic balance is internal balance. With internal balance, a balance board is contained in the fixed portion of the airfoil ahead of the control surface (figure 7-54). A flexible seal separates the balance panel from the wall of the plenum chamber where the balance board is located. Thus, the plenum chamber is divided into two compartments. A slot is located between the control surface and fixed portion of the airfoil. When the control surface is deflected, as shown, the increased velocity caused by the increased camber over the top slot develops a lower static pressure in the compartment at the top of the plenum chamber rather than in the lower compartment. A pressure differential develops across the balance panel moving the panel upward. This assists in deflecting the control surface downward. The decreased pressure over the slot on the top of the airfoil decreases static pressure, which then decreases the hinge moments of the control surface. Therefore, this becomes a form of aerodynamic balancing.

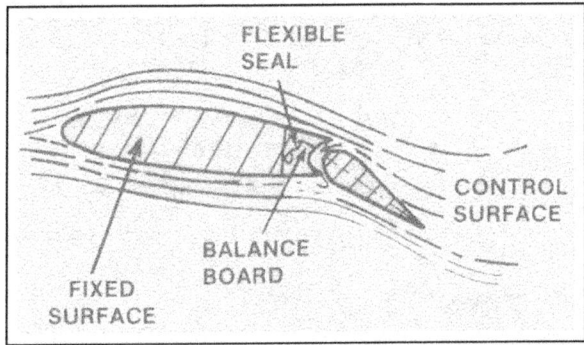

Figure 7-54. Aerodynamic balancing using a balance board

Servo Tabs

7-240. Servo tabs can be considered flaps on flaps (figure 7-55). When it is deflected, the servo tab produces a small aerodynamic force behind the control surface hinge. This small force deflects the control surface, which then moves the aircraft. The servo tab itself does not move the aircraft; it only moves the control surface. The control force required to move a small tab is much less than the force required to move the entire control surface. Servo tabs enable the aviator to use the aerodynamic qualities of the airstream to reduce hinge moments. Jet airliners have servo tabs on many control surfaces.

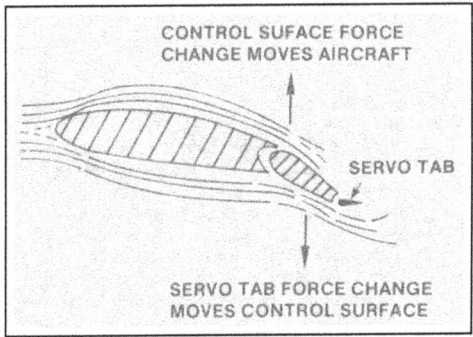

Figure 7-55. Aerodynamic balancing using a servo tab

Balance Tabs

7-241. Another device used to assist with control forces is the balance tab. This small flap is similar to the servo tab shown in figure 7-55. The balance tab moves in the opposite direction of the control flap. If the elevator moves up, the balance tab automatically moves down due to the linkage mechanism controlling its movement. This downward movement produces a moment that assists the movement of the control surface. Sometimes these systems incorporate springs in the linkage, referred to as spring tabs.

Trim Tabs

7-242. Trim tabs are identical in appearance and operation to servo tabs. Trim tabs are connected to trim controls in the cockpit. They are used to trim the aircraft for different weights, airspeeds, and power conditions (for example, when an aircraft has 200 pounds more fuel in one wing than in the other). For the aircraft to remain wings level, the lift force on the heavier wing must be increased to hold up the added weight. This is accomplished by deflecting the ailerons, which requires the aviator to hold the ailerons in that position. To decrease the aviator's workload, trim tabs can be deflected to hold the ailerons.

MASS BALANCING

7-243. Mass balancing of the control surface is important in making the aircraft dynamically stable. If the CG of the elevator is behind the hinge line, a sudden gust of wind moving the aircraft's tail upward causes the elevator to be deflected downward. This downward elevator movement increases tail lift, thereby increasing the upward movement of the tail. If, however, the CG of the elevator is ahead of the hinge line, the pitch-up caused by the air gust deflects the elevator upward, which dampens the upward motion by developing a negative tail lift.

CONTROL SYSTEMS

7-244. Only a few of the various control systems and components are discussed in this section. However, all fall under the categories of conventional, power-boosted, stability-augmenter, full power, flap, or landing gear.

CONVENTIONAL

7-245. The control system discussed in this section is the conventional, mechanical, and reversible control system. Conventional refers to the type of control system employing a rudder, aileron, and elevator. Mechanical refers to the method used in the control system to deflect control surfaces such as cables, pushrods, and bell cranks. Reversible means the control system has feedback. When the aviator moves the stick/wheel, the surface moves; when the surface moves, the stick/wheel moves. The aviator must feel air-loads on control surfaces; it eliminates the aviator's tendency to over control the aircraft, which can produce overstress.

POWER BOOSTED

7-246. As aircraft continue to be designed for travel faster, air loads or hinge moments created on control surfaces become so large, aerodynamic balancing is not effective without creating large increases in drag. Therefore, a new form of control system called the power-boosted conventional, reversible control system was developed. This system is similar to power steering. The aviator supplies part of the control force through a mechanical linkage with the control surface. However, a power system in parallel also supplies part of the force to overcome hinge moments. Usually, a power boosted system is about a 20:1 or 30:1 ratio. If the aviator puts 1 pound of force in the control system, the power system supplies 20 or 30 pounds of force. Normally, the power system is hydraulic, but pneumatic and electrical devices have been used. This type of system is reversible, and the aviator can still feel air loads through the mechanical system. If the power system fails, the aviator controls the aircraft through the mechanical system; however needed control force is greatly increased.

STABILITY AUGMENTER

7-247. When disturbed from equilibrium, an aircraft exhibiting positive static stability naturally oscillates due to the moment of inertia caused by the disturbance. A damping force causes the oscillation to be convergent to equilibrium. In some cases, the aircraft's aerodynamic damping is insufficient to reestablish equilibrium in the desired time. If so, an artificial means of damping must be used. The stability augmenter system is an auxiliary system designed explicitly for this purpose. Pitch and yaw augmenters are common on high performance aircraft. Damping forces are overridden when controls for maneuvering are used, and roll augmenters are normally unnecessary. Most Army aircraft utilize a yaw damper for the yaw axis.

FULL POWER

7-248. A full-power control system is used on supersonic aircraft when the requirement is to deflect large surfaces against extremely high dynamic pressures. This type of system is not necessarily conventional; it can use spoilers rather than ailerons and slab tails instead of elevators or rudders. A control surface of this type is much more effective at supersonic velocities than is a flap control surface (figure 7-56). This system is not employed on Army aircraft.

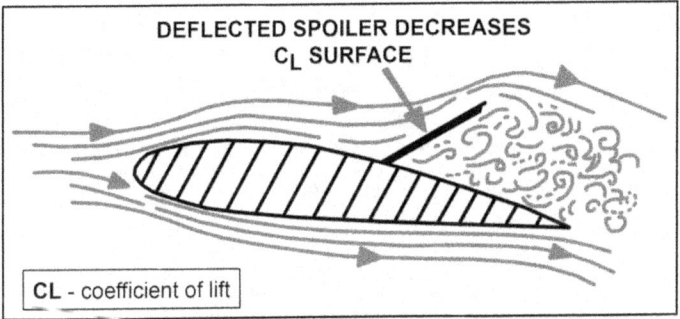

Figure 7-56. Spoiler used as control surface

FLAP

7-249. The wing flap is a movable panel on the inboard trailing edge of the wing. The flap is hinged so it can be extended downward into the airflow beneath the wing to increase lift and drag. The wing flap permits a slower airspeed and steeper descent angle during a landing approach. In some cases, wing flaps are also used to shorten takeoff distance.

7-250. The flap operating control may be an electrical or hydraulic control on the instrument panel or a lever located on the floor or pedestal of the aircraft. The control can be placed in the up position, which raises the flaps if they are in an extended position; neutral position, which allows the flaps to remain in an intermediate position; and down position, which lowers the flaps if they are in the retracted or intermediate position (figure 7-57, page 7-54). In addition to the flap operating control, an indicator usually shows the actual position of the flaps. On most Army aircraft, the maximum extent of flap travel is about 43 to 45 degrees.

Chapter 7

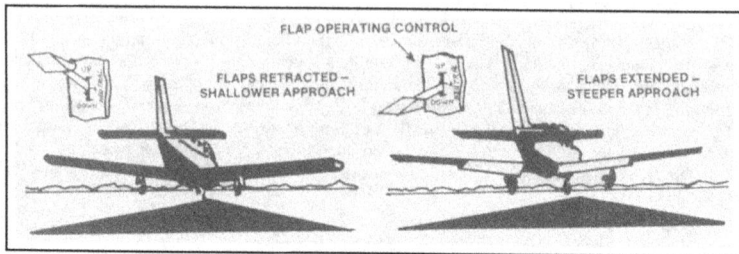

Figure 7-57. Wing flap control

7-251. Performance of an aircraft is noticeably affected by extending and retracting the flaps. For an aircraft with a constant power setting in level flight, airspeed is lower with flaps extended because of the drag created. If power is adjusted to maintain a constant airspeed in level flight, the aircraft's pitch attitude is usually lower with flaps extended.

7-252. When flaps are extended, airspeed must be at or below the aircraft's maximum flap-extended speed (V_{fe}). If flaps are extended above this airspeed, the force exerted by the airflow can damage them. If airspeed limitations are exceeded unintentionally with flaps extended, they must be retracted immediately, regardless of airspeed.

7-253. On the instrument panel, flap control often has the shape of an airfoil. It must be correctly identified before flaps are raised or lowered. This prevents inadvertently operating the landing-gear control and retracting the gear instead of the flaps, particularly when the aircraft is on or near the ground.

LANDING GEAR

7-254. Most all Army aircraft have a retractable landing-gear control. Since the landing gear's only purpose is to support the aircraft on the ground, it becomes excess weight and drag during flight. Although the weight of the gear cannot be reduced during flight, the landing gear can be retracted into the aircraft structure and out of the airflow. This eliminates unnecessary drag.

7-255. The control for operating the landing gear is a switch or lever, often in the shape of a wheel, which differentiates it from the flap control on the instrument panel. When this control is moved to the down position, the gear extends; when moved to the up position, the gear retracts. In addition to the operating control, an indicator or warning light on the instrument panel shows the landing gear's present position.

7-256. The landing gear should be operated only when the airspeed is at or below the aircraft's maximum landing-gear operating speed (V_{LO}); operation at a higher airspeed can damage the operating mechanism. When the gear is down and locked, the aircraft should not be operated above the aircraft's maximum landing-gear extended speed (V_{LE}).

7-257. The landing gear control must be correctly identified before it is raised or lowered. This prevents inadvertently operating the flap control and retracting the flaps instead of the landing gear.

PROPELLERS

OPERATION

7-258. An aircraft propeller converts the power plant shaft horsepower into propulsive horsepower. It provides thrust to propel the aircraft through the air. The propeller consists of two or more blades and the central hub to which the blades are attached. Each propeller blade is an airfoil; therefore, the propeller is a rotating wing.

7-259. The engine furnishes the power to rotate the propeller blades. On high-horsepower turbine engines, the propeller shaft is usually geared to the engine power turbine shaft. The engine rotates the airfoils of the blades through the air at a relatively high velocity. The propeller then transforms the engine's rotary power into thrust.

7-260. Many different factors govern the efficiency of a propeller. Generally, a large-diameter propeller favors a high-propeller efficiency from the standpoint of a large mass flow. However, compressibility affects propeller efficiency, while high-tip speeds adversely affect it. Small diameter propellers favor low-tip speeds. The propeller and power plant must be matched for compatibility of both output and operating efficiency.

TYPES OF PROPELLERS

Fixed Pitch

7-261. A fixed-pitch propeller has blade pitch (blade angle) built into the propeller. Thus, the pitch angle cannot be changed by the aviator as it can be on controllable-pitch propellers. Generally, the fixed-pitch propeller is constructed of aluminum alloy.

7-262. Fixed-pitch propellers are designed for best efficiency at one rotational and forward speed. They fit a specific set of conditions involving both engine rotational speed and forward speed of the aircraft. Any change reduces the efficiency of the propeller and engine.

Constant Speed

7-263. In automatically controllable-pitch propeller systems, a control device adjusts the blade angle to maintain a specific preset engine RPM without the aviator's constant attention. If engine RPM increases as a result of a decreased load on the engine, the system automatically increases the propeller's blade angle. This increases the air load until RPMs return to a preset speed. An automatic control system responds to such small variations in RPM that a constant RPM is usually maintained. These automatic propellers are termed constant-speed propellers.

7-264. An automatic system has a governor unit controlling the blade pitch angle so engine speed remains constant. With cockpit controls, the aviator regulates the propeller governor. Thus, the desired blade angle setting (within its limits) and engine operating RPM can be obtained. This increases the aircraft's operational efficiency in various flight conditions. A low-pitch, high-RPM setting obtains maximum power for takeoff. After the aircraft becomes airborne, a higher pitch and lower RPM setting provide adequate thrust for maintaining proper airspeed (figure 7-58). This can be compared to using low gear in a vehicle to accelerate until high speed is attained and then shifting into high gear for cruising speed.

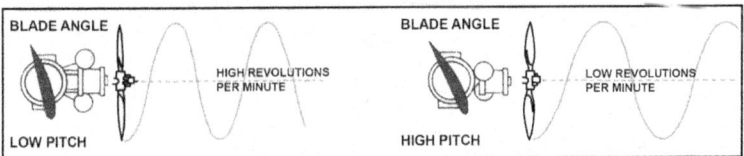

Figure 7-58. Blade angle affected by revolutions per minute

PROPELLER FEATHERING

7-265. If a power plant malfunctions or fails on an aircraft with two or more engines, propeller blades must be streamlined to reduce drag. Flight can then continue on remaining operating engines. This is accomplished by feathering propeller blades, which stops rotation and reduces drag on the inoperative engine. When the propeller blade angle is in a feathered position, parasite drag is at a minimum. On most multiengine aircraft, the added parasite drag from a single feathered propeller adds relatively little to total drag.

7-266. At smaller blade angles near flat-pitch position, the propeller windmilling at a high RPM adds a large amount of drag to the aircraft and may cause the aircraft to become uncontrollable. The propeller windmilling at high speed in the low range of blade angles can produce an increase in parasite drag as great as the parasite drag felt on the rest of the aircraft. An indication of this drag is best shown by an autorotating helicopter. A windmilling rotor can produce autorotation rates of descent approaching that of a parachute canopy with identical disk-area loading. The propeller windmilling at a high speed and small blade angle produces an effective drag coefficient of the disk area that compares to a parachute canopy of the same size. The drag and yawing moments caused by loss of power at high airspeeds are considerable, and the transient yawing displacement of the aircraft may produce critical loads on the vertical tail. For this reason, automatic feathering may be a necessity rather than a luxury.

REVERSE THRUST

7-267. The large amount of drag produced by a rotating propeller can improve the aircraft's stopping performance. Propeller blade rotation to small positive or even negative attack angle values with applied power produces a large amount of drag or reverse thrust. Due to the propeller's high thrust capability at low speeds, reverse thrust alone produces high deceleration.

LIMITATIONS

7-268. Operating limitations of the propeller are closely associated with those of the power plant. Due to the large centrifugal loads and blade twisting moments produced by an excessive rotative speed, overspeed conditions are critical. In addition, propeller blades have various vibratory modes. Certain operating limitations may be necessary to prevent resonance.

SECTION VII – MULTIENGINE OPERATIONS

7-269. Several types and models of twin-engine aircraft are used in performing Army training and operational missions. Primarily, this chapter explains the most prominent flight characteristics of twin-engine, FW aircraft. The appropriate aircraft operator's manual should be consulted for FW, multiengine operations.

TWIN-ENGINE AIRCRAFT PERFORMANCE

7-270. The term twin-engine is used to define Army propeller-driven aircraft having a maximum certified gross weight of more than 12,500 pounds and one engine mounted on each wing. The basic difference between a twin-engine and single-engine aircraft is the potential failure of one of the twin engines.

PERFORMANCE AND OPERATING SPEEDS

7-271. Certain aircraft performance operating limitations are based on airspeed. These airspeeds are called V speeds, (a listing of V speeds is found in the glossary). In addition to performance and operating speeds common to single-engine and twin-engine aircraft, the multiengine aircraft aviator must become familiar with some additional V speeds, defined in the following paragraphs.

Velocity Minimum Control

7-272. Velocity minimum control (V_{mc}) is the minimum airspeed an aircraft is able to be controlled when the critical engine suddenly becomes inoperative and the remaining engine must produce takeoff power. The Federal Aviation Regulation (FAR) (under which the aircraft was certified) states at velocity minimum control, the certificating test pilot must be able to stop the turn that results when the critical engine suddenly becomes inoperative. Using maximum rudder deflection and no more than a 5-degree bank into the operative engine, the test pilot must stop the turn within 20 degrees of the original heading. The FAR also states that after recovery, the certificating test pilot must maintain the aircraft in straight flight with not more than a 5-degree bank (wing lowered toward the operating engine). This means the aircraft must maintain a heading, not that it must be able to climb or hold altitude. The principle displayed here is that at airspeeds less than velocity minimum control, air flowing along the rudder is such that application of the rudder cannot overcome the combined effects of asymmetrical yawing caused by takeoff power on one engine and a powerless windmilling propeller on the other engine.

Fixed-Wing Aerodynamics and Performance

Maximum (Best) Angle of Climb Velocity and Maximum (Best) Single-Engine Angle of Climb Velocity

7-273. V_x is the speed providing the best angle of climb. At this speed, the aircraft gains the greatest height for a given distance. V_x is used for obstacle clearance when all engines are operating. However, when one engine is inoperative, V_x is referred to as best single-engine angle of climb velocity (V_{XSE}).

Maximum (Best) Rate of Climb Velocity and Maximum (Best) Single-Engine Rate of Climb Velocity

7-274. V_y is the speed providing the best rate of climb. This speed provides the maximum altitude gain for a given period when all engines are operating. When one engine is inoperative, V_y is referred to as best single-engine rate-of-climb velocity (V_{YSE}).

SINGLE-ENGINE OPERATION

7-275. Many aviators erroneously believe a twin-engine aircraft will continue to perform at least half as well when only one of its engines is operating. Part 23 of the FAR that governs the certification of twin-engine aircraft does not require that aircraft maintain altitude while in the takeoff configuration with one engine inoperative. In fact, some civilian light twin-engine aircraft are not required to maintain altitude with one engine inoperative in any configuration, even at sea level.

CLIMB PERFORMANCE

7-276. When one of the twin engines fails, aircraft performance is reduced by 80 percent or more. The loss of performance is more than 50 percent because the aircraft's climb performance is a function of thrust horsepower, which is available power in excess of that required for level flight. When more power is added to the engines than needed for straight and level flight, the aircraft climbs. The rate of climb depends on the amount of excess power added, which is power above that required for level flight. When one engine fails, power is lost and drag increased due to asymmetric thrust. The operating engine must carry the full burden by producing 75 percent or more of its rated power. This leaves the engine with very little excess power for climb performance. When one of its engines fails, an aircraft with an all engine climb rate of 1,860 FPM and a single-engine climb rate of 190 FPM loses almost 90 percent of its climb performance. During straight and level flight, the twin offers obvious safety advantages over the single-engine aircraft. However, the aviator must know the options offered by the second engine in the takeoff and approach phases of flight.

ASYMMETRIC THRUST

7-277. The asymmetric thrust, or unequal engine thrust, in multiengine aircraft is the principal flight characteristic that must be counteracted. To achieve the desired stability during power changes, manufacturers position most single-engine aircraft engines so the thrust line passes through or near the CG. In conventional twin-engine aircraft, only the resultant thrust of both engines provides this stability. When both engines are not operating at equal power, asymmetric thrust results and causes movement about the vertical axis or yaw. The rudder is used to prevent this movement. If yaw occurs, the aircraft may also roll or bank. To regain level flight, the aviator must apply both the rudder and aileron.

CRITICAL ENGINE

PROPELLER FACTOR

7-278. P-factor, or asymmetric propeller thrust, is present in twin aircraft just as in single-engine aircraft. It is caused by dissimilar thrust of the rotating propeller blades during certain flight conditions. The P-factor occurs when relative wind striking the blades is not aligned with the thrust line, as it is with a nose-high attitude. Therefore, the downward-moving blade has a greater attack angle than the upward moving blade.

Chapter 7

PROPELLER THRUST

7-279. In Army twin-engine aircraft, both engines rotate clockwise when viewed from the rear and both engines develop equal thrust. With a positive attack angle, low airspeed, and high-power conditions, the downward-moving propeller blade of each engine develops more thrust than the upward-moving blade. In part, this explains why conventional FW aircraft pull to the left on takeoff. The center of thrust of the propeller disk shifts to one side when the thrust line is tilted upward.

ENGINE YAWING THRUST

7-280. As indicated by lines D1 and D2 in figure 7-59, the P-factor results in a center of thrust at the right side of each engine. The yawing force of the right engine is greater than that of the left engine. As indicated by line D2, the center of thrust of the right engine has a longer lever arm and is farther away from the centerline of the fuselage. Therefore, when the right engine is operative and the left engine is inoperative, the yawing force is greater than it is when the left engine is operative and the right engine is inoperative. In an engine-out situation, the greatest demand on the rudder is made when the operative engine is the one on which the downward-moving blade is farther from the fuselage (the right engine). Therefore, the left engine is the critical engine; its loss presents the greatest controllability problem.

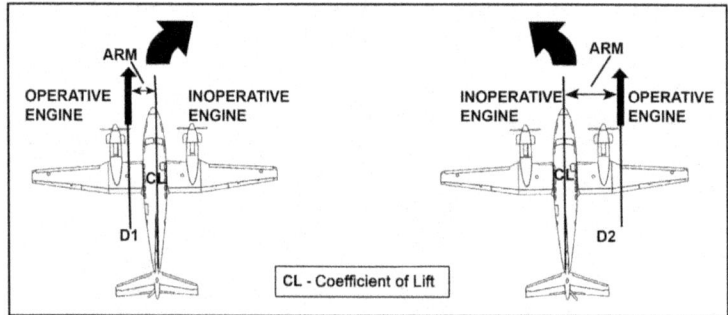

Figure 7-59. Forces created during single-engine operation

MINIMUM SINGLE-ENGINE CONTROL SPEED

ASYMMETRIC THRUST CONTROL

7-281. Maximum asymmetric thrust is created when the critical engine, usually the left engine on Army aircraft, is inoperative and the other engine is operating at takeoff power. If adequate airspeed is maintained, yaw can be prevented by applying rudder. Below this airspeed, directional control can only be maintained by reducing power. Each aircraft's critical airspeed is identified as V_{MC}. When the critical engine is rendered inoperative and the aircraft is in the most unfavorable flight configuration, V_{MC} ensures the aviator can stop the turn and maintain the new heading. V_{mc} applies only to control of asymmetric thrust; it does not ensure altitude can be maintained or a climb can be accomplished. To achieve as low a V_{mc} as possible, a 5-degree bank is always used in flight testing. To determine V_{MC}, the test pilot arrives at an airspeed low enough so that when an engine is cut an immediate bank into the operative engine is required. Full rudder deflection and the 5-degree bank provide the necessary control to keep the aircraft from turning more than 20 degrees into the dead engine. Aviators should refer to the appropriate ATM to practice flight at V_{mc}.

VELOCITY AT MINIMUM CONTROL CERTIFICATION

7-282. The following configuration is required to obtaining a manufacturer's V_{mc} certification:
- Landing gear retracted.

- Aircraft trimmed for takeoff.
- Flaps set to takeoff position.
- Takeoff or maximum available power attainable.
- Rearmost allowable CG exists.
- Maximum sea level takeoff weight maintained.
- Cowl flaps on piston-engine aircraft are in the position normally used for takeoff.
- Propeller windmilling or feathered if aircraft has an autofeather system on the inoperative engine and full power exists on the other engine.

7-283. In addition, rudder control force required to maintain control must not exceed 150 pounds.

BANK ANGLE

7-284. If the aircraft's wings are in a position less than a 5-degree bank angle, V_{MC} is substantially higher than the value shown in the flight manual. On most Army twin-engine aircraft, the difference in V_{MC} between the 5-degree bank condition and wings-level condition may be as high as 15 knots. The complex reasons for this large increase in V_{MC} with varying bank angles are discussed below.

7-285. The effect of bank reduces the amount of rudder power required to overcome the asymmetric thrust condition. As the wings are brought to a level position, more rudder is necessary. At a given rudder deflection, or rudder-pedal force, a higher airspeed is required. Although this characteristic applies to all twin-engine aircraft, it is accentuated in the latest designs due to the amount of power or TA for takeoff. In addition, the thrust lines of the engines are located farther out on the wingspan, which increases the turning moment caused by the unbalanced thrust condition.

7-286. To achieve the best performance when an engine fails during takeoff, climb, or any other flight condition when high power is required, the aviator must keep the aircraft in a 5-degree bank attitude with the inoperative engine on the high side. The normal takeoff procedure ensures the airspeed will be above VMC when the most critical engine is inoperative. However, this is true only if the 5-degree bank angle is maintained.

CONTROL PROBLEMS

7-287. The following paragraphs discuss control problems associated with engine failure. Aviators must be aware of these problems and learn proper procedures to correct them.

7-288. When an engine stops, many aviators instinctively try to center the ball, not understanding how a twin functions with asymmetric thrust.

7-289. Drag normally acts around a point along the centerline of the aircraft fuselage. When the propeller is windmilling or feathered, the center of drag moves toward the dead engine. The operative engine exerts its pull along a line several feet to the side of the center of drag. This causes the aircraft to rotate toward the inoperative engine. The aviator can prevent this rotation in one of two ways—

- The aviator can cut the power on the operative engine and quickly regain control of the aircraft as it is in a symmetrical power-off glide. Unless the aircraft is about to be out of control, this is not a desirable option immediately following takeoff or during low-altitude flight.
- The aviator uses as much power from the operative engine as possible to maintain a safe single-engine flying speed. This requires stopping the rotational movement with the rudder, causing the aircraft to skid toward the inoperative engine. To correct the skid, the aviator must bank into the good engine to maintain the longitudinal axis parallel to the relative wind.

7-290. A variety of rudder and aileron combinations can be used to maintain heading. Most aviators try to center the ball in a wings-level attitude by raising the aileron on the aircraft's operative engine side. This compensates for additional lift produced by the propeller slipstream passing over that wing. When it is viewed from outside the aircraft, however, the fuselage is not aligned with the direction of flight or relative wind; it is yawed toward the operative engine.

Chapter 7

7-291. Many aviators believe during coordinated flight their aircraft flies straight through the air without slipping or skidding. This may be true in a single-engine aircraft or a twin with equal power on both sides. However, when one engine stops and power is off-center, this is not true.

7-292. When the ball is precisely centered during wings-level coordinated flight, a twin-engine aircraft with one engine out will fly with a large sideslip (figure 7-60 part A). If a piece of string were taped to the aircraft's nose or windshield, the string would lean toward the operative engine. Single-engine rate of climb declines or disappears and VMC increases.

7-293. When manufacturers run a performance test, they use precise sideslip indicating instruments to assure zero sideslip and maximum performance. Without these instruments, the aviator has no way of knowing the sideslip angle. Most aviators mistakenly assume zero sideslip occurs when the wings are level and ball is centered.

7-294. Zero sideslip occurs in most twins when the aircraft is banked about 3 to 5 degrees into the operative engine (figure 7-60, Part B). Although it is disturbing to many aviators, the ball will be off-center toward the good engine. A yaw string shows, however, airflow is straight along the nose, which is the proper airflow for minimum drag and maximum performance.

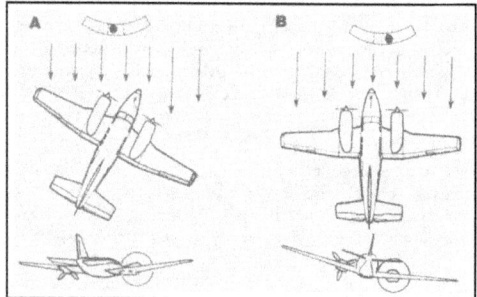

Figure 7-60. Sideslip

SINGLE-ENGINE CLIMBS

CLIMB SPEEDS

7-295. Climbs are made with reserve or excess power. Reserve power is the power available not required to maintain level flight. With one engine shut down, a twin-engine aircraft will not have an abundance of reserve power under the most favorable circumstances. If there are any changes in the best rate speed and angle-of-climb speed which occur above or below the best single engine climb speed, then climb performance can be rapidly decreased. The operator's manual establishes the best angle and rate-of-climb speeds.

DRAG REDUCTION

7-296. To provide adequate power for single-engine climbs, drag should be reduced to a minimum. Drag can be reduced by retracting the landing gear, raising the flaps, and feathering the propeller of the inoperative engine. Single-engine best rate-of-climb speed is the most efficient single engine operating, or V_{YSE}, speed. If altitude cannot be gained at this speed, more power must be obtained, or drag or weight must be reduced.

CLIMB REQUIREMENTS

7-297. After the aircraft reaches the 35-foot height with one engine inoperative, it is required to climb at a specified climb gradient, known as the takeoff flight path requirement. The aircraft's performance must be considered based on a one-engine inoperative climb up to 1,500 feet above the ground. The takeoff flight path profile with required gradients of climb for various segments and configurations is shown in figure 7-61.

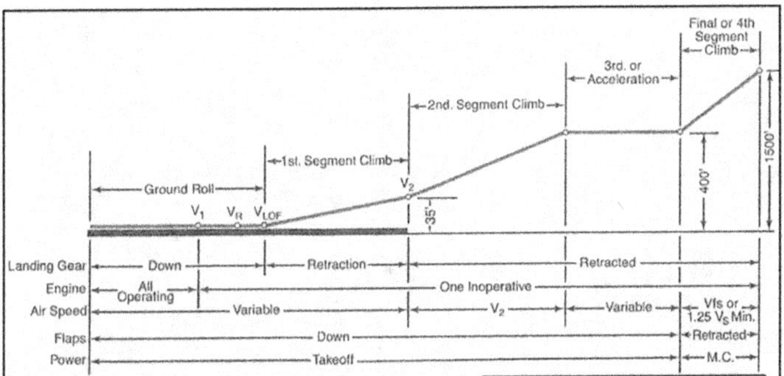

Items	1st T/O Segment	2nd T/O Segment	Transition (Acceleration)	Final T/O Segment
2 Engine	Positive	2.4%	Positive	1.2%
* 3 Engine	3.0%	2.7%	Positive	1.5%
4 Engine	5.0%	3.0%	Positive	1.7%
Wing Flaps	T/O	T/O	T/O	T/O
Landing Gear	Down	Up	Up	Up
Engines	1 Out	1 Out	1 Out	1 Out
Power	T/O	T/O	T/O	M.C.
Air Speed	$V_{LOF} \rightarrow V_2$	V_2	$V_2 \rightarrow 1.25 V_S$ (Min)	1.25 V_S (Min)

* Required Absolute Minimum Gradient of Flight Path

M.C. = Maximum Continuous
V_1 = Critical-Engine-Failure Speed
V_2 = Takeoff Safety Speed
V_S = Calibrated Stalling Speed, or min. steady flight speed at which the airplane is controllable
V_R = Speed at which airplane can start safely raising nose wheel off surface (Rotational Speed)
V_{LOF} = Speed at point where airplane lifts off
T/O = take off

Figure 7-61. One-engine inoperative flight path

First Segment

7-298. This segment is included in takeoff runway required charts and measured from the point at which the aircraft becomes airborne until it reaches the 35-foot height at the end of the runway distance required. Speed initially is speed at point where airplane lifts off (V_{LOF}) and must be at takeoff safety speed (V_2) at the 35-foot height.

Second Segment

7-299. This is the most critical segment of the profile. The second segment is the climb from the 35-foot height to 400 feet AGL. The climb is done at full takeoff power on the operating engine(s), at V_2 speed, and with flaps in the takeoff configuration. The required climb gradient in this segment is 2.4 percent for two-engine aircraft, 2.7 percent for three-engine aircraft, and 3.0 percent for four-engine aircraft.

Second Segment Climb Limitations

7-300. The second segment climb requirements, from 35 to 400 feet, are the most restrictive of the climb segments. The aviator must determine the second segment climb is met for each takeoff. To achieve this performance at higher density altitude conditions, it may be necessary to limit the aircraft's takeoff weight.

7-301. Regardless of the actual available length of the takeoff runway, takeoff weight must be adjusted so second segment climb requirements can be met. The aircraft may well be capable of lifting off with one engine inoperative, but it must then be able to climb and clear obstacles. Although second segment climb may not present much of a problem at lower altitudes, at higher altitude airports and higher temperatures, the second segment climb chart must be consulted to determine effects on maximum takeoff weights before figuring takeoff runway distance required.

Chapter 7

Third or Acceleration Segment

7-302. During this segment, the aircraft is considered to be maintaining 400 feet AGL and accelerating from the V_2 speed to velocity final segment (V_{FS}) speed before the climb profile is continued. Flaps are raised at the beginning of the acceleration segment and power is maintained at the takeoff setting as long as possible (5 minutes maximum).

Fourth or Final Segment

7-303. This segment is from 400 to 1,500-foot AGL altitude with power set at maximum continuous. The required climb in this segment is a gradient of 1.2 percent for two-engine aircraft, 1.55 for three-engine aircraft, and 1.7 percent for four-engine aircraft.

SINGLE-ENGINE LEVEL FLIGHT

LEVEL FLIGHT

7-304. Maintaining level flight with one engine inoperative is possible only below the single-engine absolute ceiling. This ceiling is based on standard atmosphere at sea-level conditions. The operating engine must be at maximum continuous power, while the aircraft is at maximum gross weight. Gear and flaps are up, and the inoperative propeller is feathered. In addition, the aircraft must be at a zero sideslip angle. High-density altitude or failure of the propeller to feather reduces the ceiling. Airspeed is also a factor in maintaining level flight; the best single-engine rate-of-climb speed provides maximum efficiency. The operator's manual contains power settings and ceilings for both normal and single-engine cruise flight.

POWER CHARTS

7-305. Power charts normally provide cruise information for aircraft operating at 45 to 75 percent power. To supply the necessary power for continued flight, one engine may be required to operate above recommended cruise range. The loss of one engine creates an emergency; therefore, flight should not be continued beyond the nearest suitable airfield. Normally, trim controls relieve control pressures when the aircraft is operating with a single engine. Some operator's manuals recommend a bank of no more than 5 degrees toward the operating engine during straight flight. Bank reduces the amount of rudder required to counter drag and asymmetric thrust and reduce sideslip. The degree of bank should be confined to recommended amounts.

SINGLE-ENGINE DESCENTS

7-306. Usually, descent in aircraft powered by reciprocating engines is performed at a specified power setting and airspeed. Correct power setting and airspeed help retain minimum engine operating temperature and reduce plug fouling or engine loading. Aviators should avoid making descents at idle power for prolonged periods. One inch of manifold pressure for each 100 RPM is the general rule for descent power. Due to the low power requirements involved, en route descents with one engine inoperative are not a problem. However, descent for landing is more involved and requires caution. The engine power charts in the operator's manual contain more specific information.

SINGLE-ENGINE APPROACH AND LANDING

7-307. When both engines are operating normally, no special technique or skill is required to perform an approach and landing. However, performing the approach and landing with one engine inoperative demands more skill and judgment. When possible, normal patterns and speeds should be used for single-engine approaches. To preclude loss of directional control and provide for the best rate of climb, speeds above V_{YSE} should be maintained during an approach until landing is assured or during a go-around. Approach and landing speeds vary according to aircraft configuration and type of approach. The operator's manual supplies this information.

PROPELLER FEATHERING

FAILED ENGINE

7-308. When an engine fails in flight, the aircraft's movement through the air keeps the propeller rotating. The failed engine no longer delivers power to the propeller, which produces thrust. Now the propeller is absorbing energy to overcome friction and compression of the engine. The drag of the windmilling propeller is significant. Drag of the windmilling propeller causes the aircraft to yaw toward the failed engine (figure 7-62). To minimize the yawing tendency, all Army multiengine aircraft are equipped with full-feathering propellers.

Figure 7-62. Windmilling propeller creating drag

PROPELLER FEATHERING

7-309. The aviator can position the blades of a feathering propeller to such a high angle they are streamlined in the direction of flight. In this feathered position, blades are streamlined with the relative wind and thus, stop turning. This significantly reduces drag on the aircraft. A feathered propeller creates the least possible drag on the aircraft and reduces its yawing tendency. Therefore, multiengine aircraft are easier to control in flight when the propeller of the inoperative engine is feathered. In addition, a feathered propeller causes less damage to the engine. Feathering a propeller should be demonstrated and practiced in all aircraft equipped with propellers that can be safely feathered and unfeathered during flight.

ACCELERATE-STOP DISTANCE

CRITICAL TIME

7-310. During the 2 or 3 seconds immediately following the takeoff roll, an aircraft accelerates to a safe engine-failure speed. This is the most critical time for a twin-engine aircraft should an engine-out condition occur. Army twin-engine aircraft are controllable at a speed close to engine-out minimum control speed. However, their performance is often so far below optimum, continued flight following the takeoff may be marginal or impossible. A more suitable and recommended speed, which some aircraft manufacturers call minimum safe single-engine speed, is the speed at which altitude can be maintained while landing gear is being retracted and propeller feathered.

ENGINE FAILURE AFTER CRITICAL TIME

7-311. When one engine on a twin-engine aircraft fails on takeoff after having reached the safe single-engine speed, it loses about 80 percent of its normal power. The twin-engine aviator, however, has an advantage over the single engine aviator. If the twin-engine aircraft has single engine climb capability at the existing gross weight and density altitude, the aviator can either stop or continue the takeoff. The single engine aviator has only one choice; land the aircraft.

ENGINE FAILURE AFTER TAKEOFF SAFETY SPEED

7-312. If one engine fails before the aircraft reaches critical engine failure recognition speed (takeoff decision speed [V_1]), the aviator's only choice is to close both throttles and bring the aircraft to a stop. If

Chapter 7

engine failure occurs after the aircraft becomes airborne, the aviator must immediately decide whether to land or continue the takeoff. If the aviator decides to continue the takeoff, the aircraft must be capable of gaining altitude with one engine inoperative. If no obstacles are involved, the aviator must accelerate to V_{YSE}. If obstacles are a factor, the aviator must accelerate the aircraft to V_{XSE}.

ABORT CONSIDERATIONS

7-313. To make a correct decision in this type of emergency, the aviator must consider runway length, field elevation, density altitude, obstruction height, head wind, and the aircraft's gross weight. For simplicity, runway contaminants–such as water, ice, snow, and runway slope–are not discussed here. The flight paths in figure 7-63 indicate the area of decision is bounded by V_1 and V_{LO}. An engine failure in this area demands an immediate decision. If engine failure occurs beyond this decision area, the aircraft can usually be maneuvered back to a landing at the departure airport if it is within the limitations of engine-out climb performance.

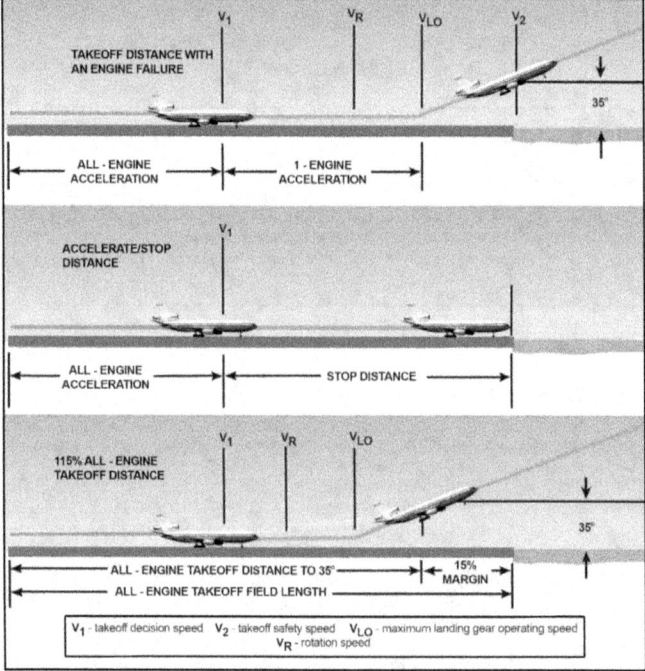

Figure 7-63. Required takeoff runway lengths

ACCELERATION DISTANCE

7-314. The accelerate-stop distance and accelerate-after-lift-off distance are based on the assumption an engine fails at the instant V_1 is attained. The accelerate-stop distance is the total distance required to accelerate the twin-engine aircraft to V_1 and bring it to a stop on the remaining runway.

ACCELERATE-GO DISTANCE

7-315. The accelerate-go distance is the distance required to accelerate to V_1 with all engines at takeoff power, experience an engine failure at V_1, and continue the takeoff on the remaining engine(s). The runway required includes the distance required to climb to 35 feet by which time V_2 speed must be attained. This distance has the same considerations as accelerate-stop, but primary consideration is a safe single-engine takeoff.

BALANCED FIELD LENGTH

7-316. For any given takeoff condition (gross weight, elevation, and temperature), the controlling accelerate-stop distance or accelerate-go distance is shortest when V_1 is chosen so these two distances are equal (balanced). This is a more sophisticated stop-versus-go decision requiring a bit more preflight planning.

7-317. The solid line in figure 7-64 represents accelerate-stop distance. It is plotted as a function of the speed at which the decision is made. Faster speeds require more runway to stop. The curve beyond rotation speed(V_R) represents the decision to abort and land straight ahead.

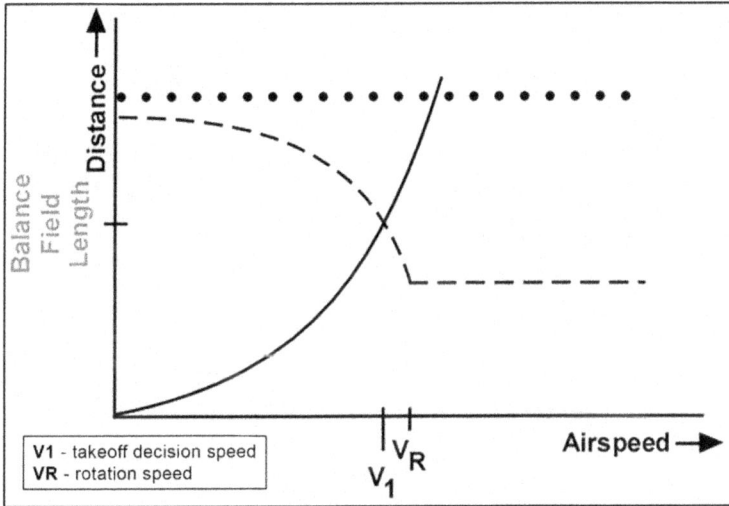

Figure 7-64. Balanced field length

7-318. The dashed line represents accelerate-go distance. The curve has two parts—
- Where it runs horizontally after V_R represents runway usage in a normal takeoff.
- The curved portion shows increased runway requirement during takeoff roll if engine failure occurs due to acceleration impairment.

7-319. If V_1 is chosen at the intersection of accelerate-stop and accelerate-go, this is balanced field length. If an engine failure occurs at this V_1, either option is acceptable.

This page intentionally left blank.

Chapter 8
Fixed-Wing Environmental Flight

This chapter addresses the FW peculiar environmental effects on aircraft performance/mission accomplishment and supplements chapter 3. This overview helps prepare aircrews for mission execution. It does not replace available information; rather, it should supplement unit SOPs and the knowledge of units assigned to and performing missions in these locations. Units tasked to deploy to one of these environments should, in addition to reviewing appropriate FMs and TMs, contact the appropriate units to seek guidance and necessary information to train and prepare. Units operating in these various environments have established training programs and 3000-series tasks not included in individual ATMs yet essential to mission accomplishment. Copies of these tasks and programs should be acquired to train aircrews for operations in unique environments.

SECTION I – COLD WEATHER/ICING OPERATIONS

ENVIRONMENTAL FACTORS

ICING

8-1. One of the hazards to flight is aircraft icing. Pilots should be aware of conditions conducive to icing, types of icing, effects of icing on aircraft control and performance, and use and limitations of aircraft deice and anti-ice equipment.

Forms of Icing

8-2. Aircraft icing in flight is usually classified as being either structural or Induction icing. Structural icing refers to ice forming on aircraft surfaces and components, and induction icing refers to ice in the engine's induction system.

Structural Icing

8-3. Ice forms on aircraft structures and surfaces when super cooled droplets impinge on them and freeze. Small and/or narrow objects are the best collectors of droplets and ice up rapidly. This is why a small protuberance within sight of the pilot can be used as an "ice evidence probe." It will generally be one of the first parts of the aircraft on which an appreciable amount of ice will form. An aircraft's tailplane will be a better collector than its wings, because the tailplane presents a thinner surface to the airstream.

8-4. The type of ice that forms can be classified as rime, clear, or mixed, based on structure and appearance of the ice. The type of ice that forms varies depending on the atmospheric and flight conditions in which it develops. The three types of ice are defined as—

- **Rime.** A rough, milky, opaque ice formed by instantaneous or very rapid freezing of super cooled droplets as they strike the aircraft. Rapid freezing results in the formation of air pockets in the ice, giving it an opaque appearance and making it porous and brittle. Low temperatures, lesser amounts of liquid water, low velocities, and small droplets favor formation of rime ice. This type of ice usually forms on areas such as leading edges of wings or struts.
- **Clear.** A glossy, transparent ice formed by relatively slow freezing of super cooled water. This ice forms from larger water droplets or freezing rain spreading over a surface. This type of ice is denser, harder, and sometimes more transparent than rime ice. Temperatures close to the freezing

point, large amounts of liquid water, high aircraft velocities, and large droplets are conducive to formation of clear ice. This is the most dangerous type of ice since it is clear, hard to see, and can change the shape of the airfoil.
- **Mixed.** This is a mixture of rime and clear ice. It has the bad characteristics of both types and can form rapidly. Ice particles become imbedded in clear ice, building a very rough accumulation.

8-5. Table 8-1 lists temperatures at which types of ice will form.

Table 8-1. Temperature ranges for ice formation

Outside Air Temperature	Icing Type
0°C to -10°C	Clear
-10°C to -15°C	Rime, Clear, and Mixed
-15°C to -20°C	Rime
Celsius (C)	

Induction Icing

8-6. In turbojet aircraft, air drawn into the engines creates an area of reduced pressure at the inlet, which lowers the temperature below that of the surrounding air. In marginal icing conditions, this reduction in temperature may be sufficient to cause ice to form on the engine inlet, disrupting airflow into the engine. Another hazard occurs when ice breaks off and is ingested into a running engine, which can cause damage to fan blades, engine compressor stall, or combustor flameout. When anti-icing systems are used, runback water also can refreeze on unprotected surfaces of the inlet and, if excessive, reduce airflow into the engine or distort the airflow pattern causing compressor or fan blades to vibrate, possibly damaging the engine. Another problem in turbine engines is the icing of engine probes used to set power levels (engine inlet temperature or engine pressure ratio [EPR] probes), which can lead to erroneous readings of engine instrumentation.

8-7. Ice may accumulate on both the engine inlet section and first or second stage of the engine's low-pressure compressor stages. This normally is not a concern with pitot-style engine airflow inlets (straight LOS inlet design). However, on turboprop engines including an inlet section with sharp turns or bird-catchers, ice can accumulate in aerodynamic stagnation points at bends in the inlet duct. If ice does accumulate in these areas, it can shed into the engine, possibly resulting in engine operational difficulties or total power loss. Therefore, with these types of engine configurations, use of anti-icing or deicing systems per the manual is very important.

Intensity

8-8. Ice accumulation is graded at four levels of intensity. Each level is described as—
- **Trace.** Ice becomes perceptible. Rate of accumulation is slightly greater than rate of sublimation. It is not hazardous even though deicing/anti-icing equipment is not used, unless encountered for an extended period of time (over one hour).
- **Light.** The rate of accumulation may create a problem if flight is prolonged in this environment (over one hour). Occasional use of deicing/anti-icing equipment removes/prevents accumulation. It does not present a problem if deicing/anti-icing equipment is used.
- **Moderate.** The rate of accumulation is such that even short encounters become potentially hazardous and use of deicing/anti-icing equipment or diversion is necessary.
- **Severe.** The rate of accumulation is such that deicing/anti-icing equipment fails to reduce or control the hazard. Immediate diversion is necessary.

Effects

8-9. The most hazardous aspect of structural icing is its aerodynamic effects. Figure 8-1, page 8-3, shows how ice often affects the coefficient of lift for an airfoil. At very low angles of attack, there may be little or no effect of the ice on the coefficient of lift. Thus, when cruising at a low AOA, ice on the wing may have little effect on the lift. However, the C_Lmax is significantly reduced by ice, and the AOA at which it occurs (the stall angle) is much lower. Thus, when slowing down and increasing the AOA for approach, a pilot may find ice on the wing, having had little effect on lift in cruise, causes stall to occur at a lower AOA and higher speed. Even a thin layer of ice at the leading edge of a wing, especially if it is rough, can have a significant effect in increasing stall speed. Lift may also be reduced at a lower AOA due to large ice shapes.

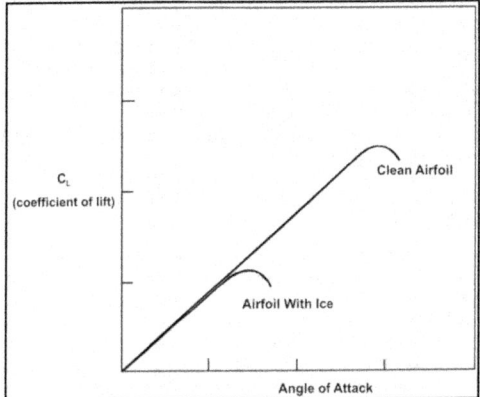

Figure 8-1. Lift curve

8-10. A significant reduction in $C_L$max and the AOA where stall occurs can result from a relatively small ice accretion. A reduction of $C_L$max by 30 percent is not unusual, and a large ice accretion can result in reductions of 40 percent to 50 percent. Drag tends to increase steadily as ice accretes (figure 8-2). An airfoil drag increase of 100 percent is not unusual, and, for large ice accretions, increase can even be 200 percent or higher. Drag effect is significant even at very small AOAs.

8-11. Due to drag, Title 14, Code of Federal Regulations (CFRs), prohibits takeoff when snow, ice, or frost is adhering to wings, propellers, or control surfaces of an aircraft. This clean aircraft concept is essential to safe flight operations.

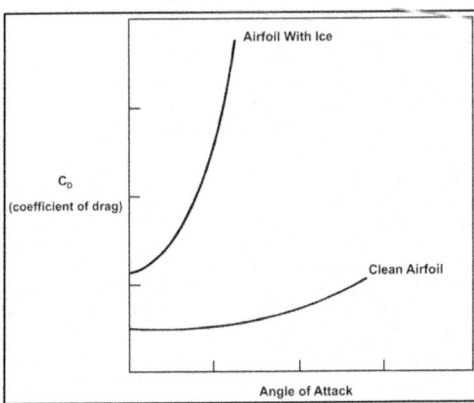

Figure 8-2. Drag curve

Chapter 8

Wings

8-12. The effect of icing on a wing depends on whether the wing is protected and the type and extent of protection provided. There are three types of wings are unprotected, deiced, and anti-iced.

8-13. An aircraft with a completely unprotected wing will not be certificated for flight in icing conditions, but may inadvertently encounter icing conditions. Aircraft certificated for flight in icing would be unprotected if the ice protection system fails. Since a cross-section of a wing is an airfoil, the remarks above on airfoils apply to a wing with ice along its span. The ice, on the wings and other parts of the aircraft, causes increase in drag, which the pilot detects as loss in airspeed. Increase in power is required to maintain the same airspeed. The longer the encounter, the greater the drag increase; even with increased power it may not be possible to maintain airspeed. Ice on the wing also causes a decrease in $C_L max$, possibly on the order of 30 percent, for an extended encounter. The rule of thumb is the percentage increase in stall speed is approximately half the decrease in $C_L max$, so the stall speed may go up by about 15 percent. If the aircraft has relatively limited power (as is the case with many aircraft with no ice protection), it may soon approach stall speed and a very dangerous situation.

8-14. The FAA recommends the deicing system be activated at the first indication of icing. Between and after system activations, some residual ice continues to adhere. Therefore, the wing is never entirely clean. However, if the system is operated properly, the ice buildup on the wing is limited, and drag increase from this buildup should be limited as well. At the AOA typical of cruise, intercycle or residual ice may have very little effect on lift. At the higher AOA characteristic of approach and landing, decrease in $C_L max$ translates into an increase in stall speed. Thus the pilot should consider continuing activation of the deicing system for a period time after exiting icing conditions so the wing will be as clean as possible and any effect on stall speed minimized. If icing conditions cannot be exited until late in the approach or significant icing appears to remain on the wing after activating the system, an increase in the aircraft's stall speed is a possibility and adjustment of the approach speed may be appropriate. Consult the aircraft's manual for guidance.

8-15. An anti-icing system is designed to keep a surface entirely free of ice throughout an icing encounter. Anti-icing protection for wings is normally provided by ducting hot bleed air from engines into the inner surface of the wing's leading edge and thus is found mainly on transport turbojets and business jets, but not on turbopropeller or piston aircraft. Even on transport and business jets, there are often sections along the wing span not protected. An important part of icing certification for these planes is checking the protected sections are extensive enough and properly chosen so ice on unprotected areas will not affect the safety of flight.

Roll Control

8-16. Ice on the wings forward of the ailerons can affect roll control. The ailerons are generally close to the wing tip, and wings are designed so stall begins near the root of the wing and progresses outward. In this way, onset of stall does not interfere with roll control of the ailerons. However, the tips are usually thinner than the rest of the wing, so they efficiently collect ice which can lead to a partial stall of the wings at the tips, affecting the ailerons and thus roll control.

8-17. Ice accumulating in a ridge aft of the boots but forward of the ailerons, possibly due to flight in super-cooled large drop conditions, affects airflow and interferes with proper functioning of the ailerons, even without a partial wing stall at the tip. There are two ways in which the ailerons might be affected by ice in front of them. One has been termed "aileron snatch," in which an imbalance of forces at the aileron is felt by the pilot of an aircraft without powered controls as a sudden change in the aileron control force. Provided the pilot is able to adjust for the unusual forces, the ailerons may still be substantially effective when deflected. The other is ailerons may be affected in a substantial degradation in control effectiveness, although without need for excessive control forces.

> ### American Eagle ATR-72
>
> Of the recent air carrier accidents, the one with arguably the most significant implications regarding in-flight icing is the October 31, 1994, crash of an ATR-72 turbo propeller transport aircraft. The aircraft was on a flight from Indianapolis, Indiana, to Chicago's O'Hare International Airport, flying with autopilot engaged and in a holding pattern, descending to 8,000 feet through super cooled clouds and super-cooled large drops. Later analysis by the National Transportation Safety Board (NTSB) estimated the super-cooled drops in the area ranged between 0.1 mm and 2 mm in size.
>
> Before the aircraft entered the hold, its engine revolutions per minute (RPMs) increased to 86 percent as called for in the ATR-72's aircraft flight manual (AFM) for flight in icing conditions (specified as true air temperature of less than 7 degrees C in the presence of visible moisture). As the aircraft began holding, the flaps were extended to 15 degrees to lower the aircraft's angle of attack (AOA), and the engine RPMs were reduced to 77 percent, presumably because the crew determined they were no longer flying in icing conditions. After holding for over half an hour, the aircraft was cleared to descend to 8,000 feet, and the crew retracted the flaps to avoid a flap overspeed warning.
>
> According to the NTSB, the encounter with the icing conditions in the hold resulted in a ridge of ice accreting aft of the aircraft's wing deicing boots and in front of the aircraft's unpowered ailerons. As the aircraft descended to its cleared altitude, its AOA increased and airflow began to separate in the area of the right aileron. This resulted in a sudden and unexpected aileron hinge reversal exceeding the autopilot's ability to control the aircraft, and it disconnected. This left the flight crew in a full right-wing-down position within a quarter of a second, which was followed by a series of unsuccessful attempts to correct the aircraft's attitude, resulting in a descent that at times reached 24,000 feet per minute (FPM) and precipitated the structural failure of the aircraft's elevators. The aircraft then impacted a soybean field at high speed resulting in the deaths of all 68 passengers and crew.
>
> The NTSB's investigation resulted in several findings, but ultimately, the most important regarding the effects of icing conditions on aircraft was the degree to which conditions affected a properly certificated aircraft and the limited information available to the flight crew with respect to the severity of the conditions they were experiencing.

Tailplane Icing

8-18. Most aircraft have a nose-down pitching moment from the wings because the CG is ahead of the center of pressure. It is the role of the tailplane to counteract this moment by providing "downward" lift. The result of this configuration is CL actions moving the wing away from stall, such as deployment of flaps or increasing speed, may increase the negative AOA of the tail. With ice on the tailplane, it may stall after full or partial deployment of flaps (figure 8-3, page 8-6).

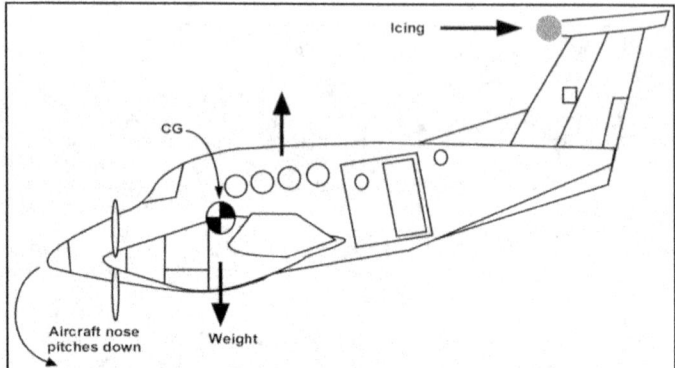

Figure 8-3. Tail stall pitchover

8-19. Since the tailplane is ordinarily thinner than the wing, it is a more efficient collector of ice. On most aircraft the tailplane is not visible to the pilot, who therefore cannot observe how well it has been cleared of ice by any deicing system. Thus it is important the pilot be alert to the possibility of tailplane stall, particularly on approach and landing. For more information, see ice-contaminated tailplane stall (ICTS) later in this chapter.

Propeller Icing

8-20. Ice buildup on propeller blades reduces thrust for the same aerodynamic reasons wings tend to lose lift and increase drag when ice accumulates on them. The greatest quantity of ice collects on the spinner and inner radius of the propeller. Propeller areas on which ice may accumulate and be ingested into the engine normally are anti-iced rather than deiced to reduce the probability of ice being shed into the engine.

Antenna Icing

8-21. Due to their small size and shape, antennas that do not lay flush with the aircraft's skin tend to accumulate ice rapidly. Furthermore, they often are devoid of internal anti-icing or deicing capability for protection. During flight in icing conditions, ice accumulations on an antenna may cause it to begin vibrating or radio signals to become distorted. Besides the distraction caused by vibration (pilots who have experienced the vibration describe it as a "howl"), the antenna may be damaged. If a frozen antenna breaks off, it can damage other areas of the aircraft, and may cause a communication or navigation system failure.

Pitot Tube

8-22. The pitot tube is particularly vulnerable to icing because even light icing can block the entry hole where ram air enters the system. This affects the airspeed indicator and is the reason most aircraft are equipped with a pitot heating system. The pitot heater usually consists of coiled wire heating elements wrapped around the air entry tube. If the pitot tube becomes blocked, the airspeed indicator would still function; however, it would be inaccurate. At altitudes above where the pitot tube became blocked, the airspeed indicator would display a higher-than-actual airspeed. At lower altitudes, the airspeed indicator would display a lower-than-actual airspeed.

Static Port

8-23. Many aircraft also have a heating system protecting the static ports to ensure the entire pitot-static system is clear of ice. If the static port becomes blocked, the airspeed indicator still functions; however, it would be inaccurate. At altitudes above where the static port became blocked, the airspeed indicator would indicate a lower-than-actual airspeed. At lower altitudes, the airspeed indicator would display a higher-than-actual airspeed. The trapped air in the static system would cause the altimeter to remain at the altitude where the blockage occurred. The vertical speed indicator (VSI) would remain at zero. On some aircraft, an alternate static air source valve is used for emergencies. If the alternate source is vented inside the aircraft, where static pressure is usually lower than outside static pressure, selection of the alternate source may result in the following erroneous instrument indications:

- The altimeter reads higher than normal.
- Indicated airspeed reads greater than normal.
- VSI momentarily shows a climb.

Stall Warning Systems

8-24. Stall warning systems provide essential information to pilots. A loss of these systems can exacerbate an already hazardous situation. These systems range from a sophisticated stall warning vane to a simple stall warning switch. The stall warning vane (also called an AOA sensor since it is a part of the stall warning system) can be found on many aircraft. The AOA provides flight crews with a display or feeds data to computers interpreting this information and providing stall warning to the crew when the AOA becomes excessive. These devices consist of a vane, which is wedge-like in shape and has freedom to rotate about a horizontal axis, and are connected to a transducer that converts the vane's movements into electrical signals transmitted to the aircraft's flight data computer. Normally, the vane is heated electrically to prevent ice formation. The transducer is also heated to prevent moisture from condensing on it when the vane heater is operating. If the vane collects ice, it may send erroneous signals to such equipment as stick shakers or stall warning devices. Aircraft using a stall horn may not give any indication of stall if the stall indicator opening or switch becomes frozen. Even when an aircraft's stall warning system is operational, it may be ineffective as the wing will stall at a lower AOA due to ice on the airfoil.

Windshields

8-25. Generally, anti-icing is provided to enable the flight crew to see outside the aircraft in case icing is encountered in flight. On high-performance aircraft requiring complex windshields to protect against bird strikes and withstand pressurization loads, the heating element often is a layer of conductive film or thin wire strands through which electric current is run to heat the windshield and prevent ice from forming.

8-26. Aircraft operating at lower altitudes and speeds have other systems of window anti-icing/deicing. One system consists of an electrically heated plate installed onto the aircraft's windshield to give the pilot a narrow band of clear visibility. Another system uses a bar at the lower end of the windshield to spray deicing fluid onto it and prevent ice from forming.

AIRCRAFT EQUIPMENT

ICE CONTROL SYSTEMS

8-27. Ice control systems installed on aircraft consist of anti-ice and deice equipment. Anti-icing equipment is designed to prevent the formation of ice, while deicing equipment is designed to remove ice once it has formed. Ice control systems protect the leading edge of wing and tail surfaces, pitot and static port openings, fuel tank vents, stall warning devices, windshields, and propeller blades. Ice detection lighting may also be installed on some aircraft to determine the extent of structural icing during night flights. Since aircraft configurations are different, refer to operator's manual, AFM, or pilot's operating handbook (POH) for details.

8-28. Operation of aircraft anti-icing and deicing systems should be checked prior to encountering icing conditions. Encounters with structural ice require immediate remedial action. Anti-icing and deicing equipment is not intended to sustain long-term flight in icing conditions.

Airfoil Ice Control

8-29. Inflatable deicing boots consist of a rubber sheet bonded to the leading edge of the airfoil. When ice builds up on the leading edge, an engine-driven pneumatic pump inflates the rubber boots (figure 8-4, page

8-9). Some turboprop aircraft divert engine bleed air to the wing to inflate the rubber boots. Upon inflation, the ice is cracked and should fall off the leading edge of the wing. Deicing boots are controlled from the cockpit by a switch and can be operated in a single cycle or allowed to cycle at automatic, timed intervals. It is important deicing boots are used in accordance with manufacturer's recommendations. If they are allowed to cycle too often, ice can form over the contour of the boot and render them ineffective.

8-30. Many deicing boot systems use the instrument system suction gauge and a pneumatic pressure gauge to indicate proper boot operation. These gauges have range markings indicating the operating limits for boot operation. Some systems may also incorporate an annunciator light to indicate proper boot operation. Proper maintenance, care, and preflight inspection of deicing boots are important for continued operation of this system. Another type of leading edge protection is the thermal anti-ice system installed on aircraft with turbine engines. This system is designed to prevent the buildup of ice by directing hot air from the compressor section of the engine to the leading edge surfaces. This system is activated prior to entering icing conditions. The hot air heats the leading edge sufficiently preventing the formation of ice.

8-31. An uncommon alternate type of leading edge protection a weeping wing. The weeping-wing design uses small holes located in the leading edge of the wing. A chemical mixture is pumped to the leading edge and weeps out through the holes to prevent formation and buildup of ice.

Windscreen Ice Control

8-32. There are two main types of windscreen anti-ice systems. The first system directs a flow of alcohol to the windscreen. By using it early enough, alcohol will prevent ice buildup on the windshield. The rate of alcohol flow can be controlled by a dial in the cockpit according to procedures recommended by the aircraft manufacturer.

8-33. Another method is the electric heating method. Small wires or other conductive material is imbedded in the windscreen. The heater can be turned on by a switch in the cockpit, at which time electrical current is passed across the shield through the wires to provide sufficient heat to prevent the formation of ice on the windscreen. The electrical current can cause compass deviation errors; in some cases, as much as 40 degrees. The heated windscreen should only be used during flight. Do not leave it on during ground operations, as it can overheat and damage to the windscreen.

Fixed-Wing Environmental Flight

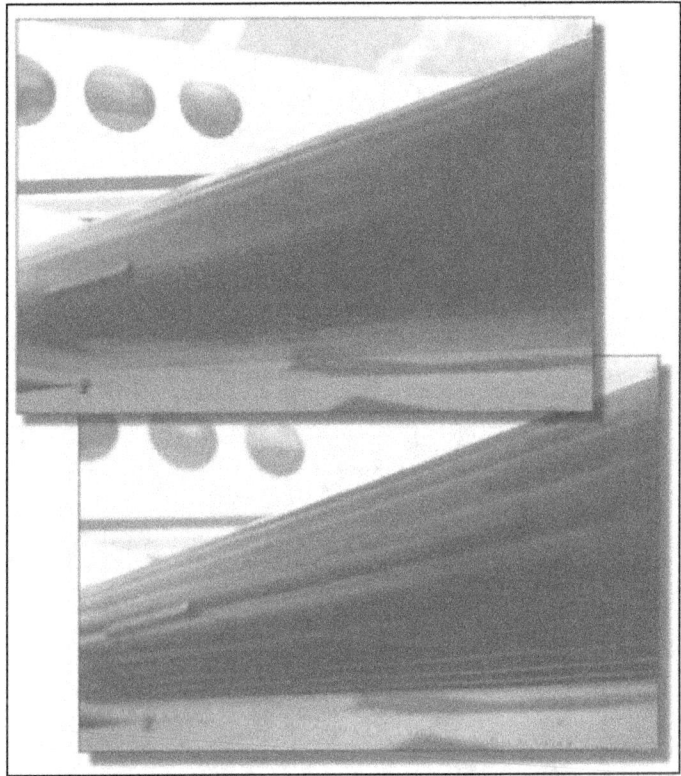

Figure 8-4. Pneumatic boots

Propeller Ice Control

8-34. Propellers are protected from icing by use of alcohol or electrically heated elements. Some propellers are equipped with a discharge nozzle pointed toward the root of the blade. Alcohol is discharged from the nozzles, and centrifugal force makes it flow down the leading edge of the blade preventing ice from forming. Propellers can also be fitted with propeller anti-ice boots (figure 8-5, page 8-10). The propeller boot is divided into inboard and outboard sections. The boots are grooved to help direct the flow of alcohol, and they are also imbedded with electrical wires carrying current to heat the propeller. The prop anti-ice system can be monitored for proper operation by monitoring the prop anti-ice ammeter. During the preflight inspection, check the propeller boots for proper operation. If a boot fails to heat one blade, an unequal blade loading can result, and may cause severe propeller vibration.

Chapter 8

PROP ANTI-ICE AMMETER

When the system is operating, the prop ammeter will show in the normal operating range. As each boot section cycles, the ammeter will fluctuate.

Figure 8-5. Propeller ice control

Other Ice Control Systems

8-35. Pitot and static ports, fuel vents, stall warning sensors, and other optional equipment may be heated by electrical elements. Operational checks of the electrically heated systems are to be checked in accordance with the operator's manual.

FLYING TECHNIQUES

FLAT LIGHT

8-36. In flat light conditions it may be possible to depart but not return to a site. During takeoff, there must be a reference point until a departure reference point is in view. If the departure reference does not come into view, return to the takeoff reference point may be required.

8-37. Flat light is common to snow skiers. One way to compensate for the lack of visual contrast and depth-of-field loss is to wear amber tinted lenses (blue blockers). Eyewear is not ideal for every pilot, personal factors, such as age, light sensitivity, and ambient lighting conditions (LITECON), should be considered. If all visual references are lost—
- Above all, fly the aircraft.
- Trust the cockpit instruments.
- Execute a 180-degree turn around and start looking for outside references.

Landings

8-38. Pilots look for features around the airport or approach path to determine depth perception. Buildings, towers, vehicles, or other aircraft serve well for this measurement. Something that provides a sense of height above the ground, in addition to orienting the runway, should be used.

8-39. Pilots must be cautious of snowdrifts, snow banks, or anything falsely distinguishing the edge of the runway. Look for subtle changes in snow texture or shading to identify ridges or changes in snow depth.

Icing

8-40. Because icing is unpredictable in nature, pilots may find themselves in icing conditions even though they have done everything to avoid it. To stay alert to this possibility while operating in visible moisture, they should monitor the OAT.

8-41. Proper utilization of the anti-icing/deicing equipment is critical to the safety of flight. If the anti-icing/deicing equipment is used before sufficient ice has accumulated, it may not be able to remove all ice accumulation. The operator's manual should be referenced for proper use of anti-icing/deicing equipment.

8-42. Prior to entering visible moisture with temperatures at 4 degrees F or cooler, the appropriate anti-icing/deicing equipment is activated in anticipation of ice accumulation—early ice detection is critical. This may be particularly difficult during night flight. Ice lights or a flashlight is used to check for ice accumulation on the wings.

8-43. At the first indication of ice accumulation, pilots must act to get out of icing conditions. The following are four options for action once ice has begun to accumulate on the aircraft:
- Move to an altitude with significantly colder temperatures.
- Move to an altitude with temperatures that are above freezing.
- Fly to an area clear of visible moisture.
- Change heading and fly to an area of known nonicing conditions.

8-44. Because icing conditions in stratiform clouds often are confined to a relatively thin layer, either climbing or descending may be effective in exiting the icing conditions within the clouds. A climb may take the aircraft into a colder section of cloud consisting exclusively of ice particles. These generally constitute little threat of structural icing as it is unlikely the ice particles will adhere to unheated surfaces. The climb also may take the aircraft out of the cloud altogether to an altitude where ice will gradually sublimate or shed from the airframe depending on conditions. A descent may take the aircraft into air with temperatures above freezing, within or below the cloud, where ice can melt.

8-45. Hazardous icing conditions can occur in cumulus clouds, sometimes having very high water content. Therefore, it is not advisable to fly through a series of such clouds, or to execute holds within them. However, as these clouds normally do not extend very far horizontally, any icing encountered in such a cloud may be of limited duration, and it may be possible to deviate around the cloud.

8-46. Freezing rain forms when rain becomes super cooled by falling through a subfreezing layer of air. Thus, it may be possible to exit the freezing rain by climbing into the warm layer.

8-47. Because freezing drizzle often forms by the collision-coalescence process, a pilot must not assume a warm layer of air exists above the aircraft. A pilot encountering freezing drizzle should exit the conditions as quickly as possible, either vertically or horizontally. Three possible actions are ascend to an altitude where the freezing drizzle event is less intense; descend to an area of warmer air; or make a level turn to emerge from the area of freezing drizzle.

8-48. If none of these options are available, consideration must be given to immediately land at the nearest suitable airport. Anti-icing and deicing equipment is not designed to allow aircraft to operate in icing conditions indefinitely, but provides more time to exit these conditions.

Cruise

8-49. An aircraft certificated for flight in icing conditions whose ice protection system is operating properly will be able to cruise for some time in most icing conditions. However, if it is possible to exit the icing conditions by a change in altitude or a minor change in flight path, it is advisable. During any icing encounter, the behavior of the aircraft should be carefully monitored by the pilot. The aircraft will have some unprotected areas that will collect ice. Although ice in such areas should not compromise the safety of flight, it may cause enough increase in drag to require the pilot to apply more power to maintain flight speed. Residual or intercycle ice on deiced areas can have a similar effect. Typically, adding power is the recommended action, since reduction in flight speed is associated with an increase in AOA, which on many aircraft will expose larger unprotected areas on the underside of the aircraft to the collection of ice. If for any reason the point is reached where it is no longer possible to maintain airspeed through addition of power, the pilot should exit icing conditions immediately. On an aircraft equipped with in-flight deicing systems, there will at all times be residual or some stage of intercycle ice on the wings.

8-50. Airspeed in cruise can have a significant effect on the nature of an icing encounter. An aircraft cruising at a fast airspeed will increase the rate of ice accumulation. However, if airspeed is sufficiently fast, the ram air heat may begin to increase skin temperatures sufficiently to melt some of the ice and prevent accumulation in those areas. Generally, only very high-performance aircraft can attain such speeds. During the flight, pilots periodically verify all anti-icing and deicing systems are working. During the en route portion of the flight, a regularly reevaluated exit plan is necessary.

8-51. Even if the encounter is short and icing is not heavy, the pilot must exercise particular awareness of the behavior of the aircraft. Configuration changes following cruise in icing conditions, such as spoiler/flap deployment, should be made with care. This is because ice on the aircraft which had little effect in cruise may have a much different and potentially more hazardous effect in other configurations. Remember for normal cruise configurations and speeds, both the wing and tailplane are ordinarily at moderate AOAs, making wing or tailplane stall unlikely. After configuration changes and in maneuvering flight, wings or tailplane (especially after flap deployment) may be at more extreme AOAs, and even residual or intercycle ice may cause stall to occur at a less extreme angle than on a clean aircraft.

8-52. Care should be exercised when using the autopilot in icing conditions, whether in cruise or other phases of flight. When autopilot is engaged, it can mask changes in handling characteristics due to aerodynamic effects of icing normally detected by the pilot if the aircraft were being hand flown. In an aircraft relying on aerodynamic balance for trim, autopilot may mask control anomalies otherwise detected at an early stage. If the aircraft has nonboosted controls, a situation may develop in which autopilot servo-control power is exceeded, autopilot disconnects abruptly, and the pilot is suddenly confronted by an unexpected control deflection. Pilots may consider periodically disengaging the autopilot and manually fly the aircraft when operating in icing conditions. If this is not desirable because of cockpit workload levels, pilots should monitor the autopilot closely for abnormal trim, trim rate, or aircraft attitude.

Descent

8-53. Pilots should try to stay on top of a cloud layer as long as possible before descending. This may not be possible for an aircraft using bleed air for anti-icing systems because an increase in thrust may be required to provide sufficient bleed air. This increased thrust may reduce the descent rate of high-performance aircraft whose high-lift attributes already make descents lengthy without use of aerodynamic speed brakes or other such devices. The result may be a gradual descent, extending the aircraft's exposure to icing conditions.

8-54. If configuration changes are made during this phase of flight, they should be made with care in icing conditions, noting the behavior of the aircraft.

Approach and Landing

8-55. During or after flight in icing conditions, when configuring the aircraft for landing, the pilot must be alert for sudden aircraft movements. Often ice is picked up in cruise, when the aircraft's wing and tailplane are likely at a moderate AOA, making a relatively ice-tolerant configuration. If effects in cruise are minor, the pilot may feel comfortable the aircraft can handle the ice it has acquired. Extension of landing gear may create excessive amounts of drag when coupled with ice. Flaps and slats should be deployed in stages, carefully noting the aircraft's behavior at each stage. If anomalies occur, it is best not to increase the amount of flaps or slats and perhaps even to retract them depending on how much the aircraft is deviating from normal performance. Additionally, before beginning the approach, deicing boots should be cycled as they may increase stall speed and it is preferable not to use these systems while landing.

8-56. Once on the runway, pilots also should be prepared for possible loss of directional control caused by ice buildup on landing gear.

8-57. Another concern during approach and landing may be forward visibility. Windshield anti-icing and deicing systems can be overwhelmed by some icing encounters or may malfunction. Pilots have been known to look through side windows or, on small aircraft, attempt to remove ice accumulations with some type of tool (plotter, credit card). Pilot workload can be heavy during approach and landing phases. Autopilots help reduce this load. Advantages of a reduced workload must be balanced against risks associated with using an autopilot during or after flight in icing conditions. An unexpected autopilot disconnect because of icing is especially hazardous in this phase of flight due to the aircraft being flown at a low altitude.

8-58. Accident statistics reveal the majority of icing-related accidents occur in the final phases of flight. Contributing factors are configuration changes, low altitude, higher flight crew workload, and reduced power settings. Loss of control of the aircraft is often a factor. The ice contamination may lead to wing stall, ICTS, or roll upset. Wing stall and roll upset may occur in all phases of flight. However, available statistics indicate ICTS rarely occurs until approach and landing. If an aircraft has accumulated ice on the wings and tailplane, it may be best to perform a no-flap landing at a higher than normal approach speed. However, due to the higher approach speed, longer runways may be needed for this procedure.

Ice-Contaminated Tailplane Stall

8-59. ICTS occurs when a tailplane with accumulated ice is placed at a sufficiently negative AOA and stalls. This angle would not be expected to be reached without at least partial deployment of the flaps. There are few, if any, known incidents of ICTS in cruise (when flaps would not ordinarily be deployed). However, when flaps are deployed, tailplane ice which previously had little effect other than a minor contribution to drag now can put the tailplane at or dangerously close to stall.

8-60. As the pilot prepares for deployment of flaps after or during flight in icing, he or she carefully assesses the aircraft's behavior for buffet or any other signs of wing stall. The initial deployment of flaps is only partial. Vibration or buffeting following deployment is much more likely to be due to incipient tailplane stall than wing stall if there was no vibration buffet before deployment. The reason is after deploying the flaps, the wing will be at a less positive angle, farther from stall, while the tailplane will be at a more negative angle, closer to stall.

8-61. There are a number of specific cues associated with ICTS to which a pilot should be sensitive, particularly during this phase of flight. Most of these cues are less readily detected with the autopilot engaged.

- Elevator control pulsing, oscillations, or vibrations.
- Abnormal nose-down trim change.
- Any other unusual or abnormal pitch anomalies (possibly resulting in pilot-induced oscillations).
- Reduction or loss of elevator effectiveness.
- Sudden change in elevator force (control would move nose down if unrestrained).
- Sudden uncommanded nose-down pitch.

8-62. Pilots observe the following guidelines for action if these cues are encountered:

- Flaps tend to alter airflow reaching the tailplane and must be retracted immediately to the previous setting. Appropriate nose-up elevator pressure must be applied.

- Airspeed is increased appropriately for the reduced flap extension setting.
- Sufficient power is applied for the aircraft's configuration and conditions. High engine power settings may adversely affect the response to ICTS conditions at high airspeed in some aircraft designs. Recommendations in the operator's manual, AFM or POH must be observed regarding power settings.
- Nose-down pitch changes must be made slowly, even in gusting conditions, if circumstances allow.
- If a pneumatic deicing system is used, the system is cycled several times to clear the tailplane of ice.

8-63. Some measures for ICTS recovery are the opposite of those for wing stall recovery. Distinguishing between the two is very important. If for any reason there is a large or rough ice accretion on the wing and tailplane, approach and landing must be managed with great care. Deployment of flaps permits the aircraft to be flown with wings at a less positive attack, decreasing probability of wing stall. However, the AOA at the tailplane is more negative, putting it closer to stall. Similarly, at any particular flap setting, lower speeds put the aircraft closer to wing stall and higher speeds put it closer to tailplane stall. Thus there is a restricted operating window with respect to use of the flaps and airspeed. Pilots must be familiar with guidance provided in the operator's manual.

8-64. When ICTS or wing stall is a possibility, uncoordinated flight such as side or forward slips must be avoided and, to the extent possible, crosswind landings restricted because of their adverse effect on pitch control and the possibility of reduced directional control. Landing with a tailwind component may result in more abrupt nose-down control inputs and should be avoided.

8-65. If an aircraft has ice on the wings and tail, the pilot may be wise to exercise limited or no deployment of flaps, which will likely result in a higher than normal approach speed. Because of the higher speed approach, longer runways may be needed for this procedure.

Roll Upsets

8-66. Roll upsets caused by ice accumulations forward of the ailerons also are possible during an icing encounter, particularly in SLD conditions. During slow speeds associated with approach and landing, such control anomalies can become increasingly problematic. Pilots can remedy roll upsets using the following guidelines:
- Reduce the AOA by increasing airspeed or extending wing flaps to the first setting if at or below VFE (maximum flap extension speed). If in a turn, the wings should be rolled level.
- Set the appropriate power, and monitor airspeed and AOA.
- If flaps are extended, do not retract them unless it can be determined the upper surface of the airfoil is clear of ice. Retracting flaps will increase the AOA at a given airspeed.
- Verify the wing ice protection is functioning normally and symmetrically through visual observation of each wing. If there is a malfunction, follow the manufacturer's instructions.

8-67. These procedures are similar to those for wing stall recovery, and in some respects opposite from those for ICTS recovery. Application of the incorrect procedure during an event can seriously compound the upset. Correct identification and application of the proper procedure is imperative. It is extremely important the pilot maintains awareness of all possibilities during or following flight in icing.

TRAINING

8-68. Units qualifying aviators in cold weather/icing operations are responsible for conducting a well-organized training program. Training programs should be geared to instill confidence and develop skills in all areas. IPs and supervisory maintenance personnel must be highly qualified and skilled.

8-69. Emphasis must be placed on safety and avoidance. Avoiding hazards is the best course and a smart aviator uses all resources at his or her disposal. If hazards are encountered, the crew is prepared to handle them. The professional judgment of the instructor to discontinue training due to unsafe conditions must be accepted and not criticized.

8-70. The flight training program allows each aviator to advance at an individual rate. Initial training is conducted under less challenging conditions. As an aviator's proficiency increases, conditions become more demanding until the most challenging mission can be performed.

8-71. A recommended program of instruction for qualifying aviators is provided. Additional academic subjects may be required, based on specific mission and location of the unit.

Academics

8-72. Suggested topics include—
- Human factors associated with cold weather/icing flying.
- Environmental factors affecting cold weather/icing operations.
- Planning data available on cold weather/icing.
- In-flight equipment and resources to detect, avoid, and continue in these hazards.
- Aircraft operational procedures in cold weather/icing.

Flight

8-73. Flight training may be limited by conditions at the unit's home station as there may not be areas able to replicate conditions adequately. Instructors can demonstrate techniques and procedures to some extent. Crews should be evaluated on these procedures during their APART or no-notice evaluations. Flight simulators are also a great device in training for this environment.

8-74. Suggested maneuvers include—
- Flight (takeoff, en route, and landing) techniques.
- Stall recovery maneuvers.
- Aircraft equipment usage.

SECTION II – MOUNTAIN OPERATIONS

ENVIRONMENTAL FACTORS

MOUNTAIN WAVE

8-75. Mountain waves occur when air is being blown over a mountain range or even the ridge of a sharp bluff area at 15 knots or better at an intersection angle of not less than 30 degrees. As the air hits the upwind side of the range, it starts to climb, thus creating what is generally a smooth updraft which turns into a turbulent downdraft as the air passes the crest of the ridge. From this point, for many miles downwind, there will be a series of downdrafts and updrafts. Satellite photos of the Rockies have shown mountain waves extending as far as 700 miles downwind of the range. Along the east coast area, such photos of the Appalachian chain have picked up the mountain wave phenomenon over a hundred miles eastward.

8-76. To avoid dangerous situations, pilots must understand mountain waves. When approaching a mountain range from the upwind side (generally west), there will usually be a smooth updraft; therefore, it is not quite as dangerous an area as the lee of the range. From the leeward side, it is always a good idea to add an extra thousand feet or so of altitude because downdrafts can exceed the climb capability of the aircraft. Never expect an updraft when approaching a mountain chain from the leeward, and always be prepared to cope with downdraft and turbulence.

MOUNTAIN OBSCURATION

8-77. Mountain obscuration (MTOS) describes a visibility condition distinguished from IFR because ceilings, by definition, are described as AGL. In mountainous terrain clouds can form at altitudes significantly higher than the weather reporting station and at the same time nearby mountaintops may be obscured by low visibility. In these areas, ground level can also vary greatly over a small area. Caution must be used when and if operating VFR-on-top. It is possible to be operating closer to the terrain than thought, as the tops of mountains are hidden in a cloud deck below. MTOS areas are identified daily by the Aviation Weather Center located at www.awc-kc.noaa.gov, under Official Forecast Products, airman's meteorological information (AIRMETs) (WA), IFR/MTOS.

DENSITY ALTITUDE

8-78. Performance figures in the aircraft owner's handbook are generally based on standard atmosphere conditions (59 degrees F [15 degrees C], pressure 29.92 inches of mercury) at sea level. However, pilots may run into trouble when encountering a different set of conditions. This is particularly true in hot weather and at higher elevations. Aircraft operations at altitudes above sea level and higher than standard temperatures are commonplace in mountainous areas. Such operations quite often result in a drastic reduction of aircraft performance capabilities due to changing air density. Density altitude is a measure of air density and is not to be confused with PA, true altitude, or absolute altitude. It is not a height reference, but determines criteria in the performance capability of an aircraft. Air density decreases with altitude. As air density decreases, density altitude increases. The further effects of high temperature and high humidity are cumulative, resulting in an increasing high density altitude condition. High density altitude reduces all aircraft performance parameters. To a pilot, this means normal horsepower output is reduced, propeller efficiency is reduced, and a higher TAS is required to sustain the aircraft throughout its operating parameters. It also means an increase in runway length requirements for takeoff and landings, and a decreased rate of climb. An average small aircraft, for example, requiring 1,000 feet for takeoff at sea level under standard atmospheric conditions will require a takeoff run of approximately 2,000 feet at an operational altitude of 5,000 feet.

Density Altitude Advisories

8-79. At airports with elevations of 2,000 feet and higher, control towers and flight service stations (FSSs) will broadcast the advisory "check density altitude" when the temperature reaches a predetermined level. These advisories will be broadcast on appropriate tower frequencies or, where available, automated terminal information service (ATIS). FSSs will broadcast these advisories as a part of a local airport advisory, and on transcribed weather en route broadcast.

8-80. These advisories are provided by air traffic facilities as reminders to pilots that high temperatures and field elevations cause significant changes in aircraft characteristics. Pilots retain the responsibility of computing density altitude, when appropriate, as part of preflight duties.

FLYING TECHNIQUES

8-81. Proper planning and awareness of potential hazards are a must in mountain-terrain flight. Flat, level fields for forced landings are practically nonexistent. Abrupt changes in wind direction and velocity occur. Severe updrafts and downdrafts are common, particularly near or above abrupt changes of terrain such as cliffs or rugged areas. Clouds even look different and can buildup rapidly. Mountain flight guidelines are the following:

- Plan the route to avoid topography preventing a safe forced landing. The route should be over populated areas and well-known mountain passes. Sufficient altitude must be maintained to permit gliding to a safe landing in the event of engine failure.
- Do not fly a light aircraft when the wind aloft, at your proposed altitude, exceeds 35 MPH. Expect winds to be of much greater velocity over mountain passes than reported a few miles from them. Approach mountain passes with as much altitude as possible. Downdrafts from 1,500 to 2,000 FPM are not uncommon on the leeward side.
- Do not fly near or above abrupt changes in terrain. Severe turbulence can be expected, especially in high wind conditions.
- Do not fly too far up a canyon; ensure a 180-degree turn is possible.
- Approach a ridge at approximately a 45-degree angle to the horizontal direction. This permits a safer retreat from the ridge with less stress on the aircraft should severe turbulence and downdraft be experienced. If severe turbulence is encountered, simultaneously reduce power and adjust pitch until aircraft approaches maneuvering speed, then adjust power and trim to maintain maneuvering speed and fly away from the turbulent area.

- Use the same indicated airspeed used at low elevation fields, when landing at a high altitude field. Due to the less dense air at altitude, this same indicated airspeed actually results in higher TAS, a faster landing speed, and more important, a longer landing distance. During gusty wind conditions which often prevail at high altitude fields, a power approach and landing is recommended. Additionally, due to faster groundspeed, takeoff distance will increase considerably over that required at low altitudes.

SECTION III – OVERWATER OPERATIONS

OCEANOGRAPHIC TERMINOLOGY

8-82. Table 8-2 defines oceanographic terms used in overwater operations.

Table 8-2. Oceanographic terminology

Term	Definition
Sea	Condition of the surface resulting from waves and swells.
Wave/Chop	Condition of the surface caused by local winds.
Swell	Condition of the surface caused by a distance disturbance.
Swell Face	Side of the swell toward the observer. This definition applies regardless of the direction of swell movement.
Backside	Side of the swell away from the observer. This definition applies regardless of the direction of swell movement.
Primary Swell	Swell system having the greatest height from trough to crest.
Secondary Swell	Swell systems of less height than the primary swell.
Fetch	Distance waves have been driven by a wind blowing in a constant direction, without obstruction.
Swell Period	Time interval between passage of two successive crests at the same spot in the water, measured in seconds.
Swell Velocity	Speed and direction of the swell with relation to a fixed reference point, measured in knots. There is little horizontal movement of water. Swells move primarily in a vertical motion, similar to motion observed when shaking out a carpet.
Swell Direction	Direction from which a swell is moving. This direction is not necessarily the result of the wind present at the scene. The swell may be moving into or across the local wind. Swells, once set in motion, tend to maintain their original direction for as long as they continue in deep water, regardless of changes in wind direction.
Swell Height	Height between crest and trough, measured in feet. The vast majority of ocean swells are not more than 12 to 15 feet; swells over 25 feet are uncommon. Successive swells may differ considerably in height.

DITCHING

8-83. If ditching is imminent the following factors, procedures, and techniques aircrews use when an emergency makes further flight unsafe.

PRIMARY FACTORS

8-84. Successful aircraft ditching is dependent on three primary factors. In order of importance, they are—
- Sea conditions and wind.
- Type of aircraft.
- Skill and technique of pilot.

Chapter 8

PROCEDURES AND TECHNIQUES

8-85. To select a good heading when ditching an aircraft, a basic evaluation of the sea is required. Selection of a good ditching heading may well minimize damage and could save your life. It can be extremely dangerous to land into the wind without regard to sea conditions; the swell system, or systems, must be taken into consideration. Remember one axiom—avoid the face of a swell.

Swell Touchdown

8-86. In ditching parallel to the swell, it makes little difference whether touchdown is on the top of the crest or in the trough. It is preferable, however, to land on the top or back side of the swell, if possible. After determining which heading (and its reciprocal) will parallel the swell, select the heading with the most into the wind component (figure 8-6).

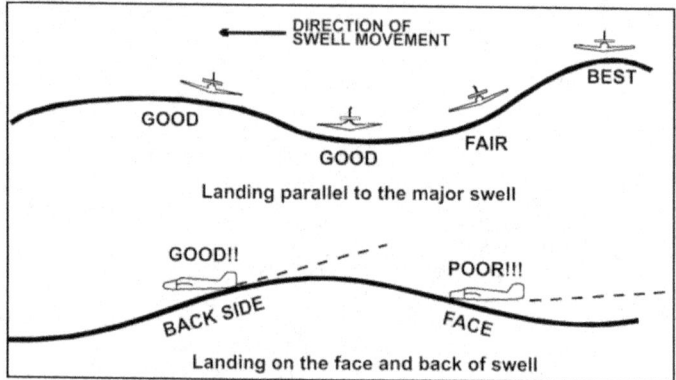

Figure 8-6. Wind swell ditch heading

8-87. If only one swell system exists, the problem is relatively simple even with a high, fast system (figure 8-7, page 8-19). Unfortunately, most cases involve two or more swell systems running in different directions. With more than one system present, the sea presents a confused appearance. One of the most difficult situations occurs when two swell systems are at right angles. For example, if one system is eight feet high and the other is three feet high, plan to land parallel to the primary system (figure 8-8, page 8-19), and on the down swell of the secondary system. If both systems are of equal height, a compromise may be advisable. Select an intermediate heading at 45 degrees down swell to both systems (figure 8-9, page 8-20). When landing down a secondary swell, attempt to touch down on the back side, not on the face of the swell.

Fixed-Wing Environmental Flight

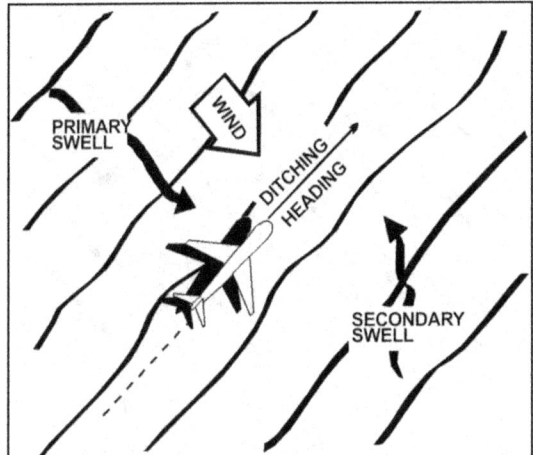

Figure 8-7. Single swell

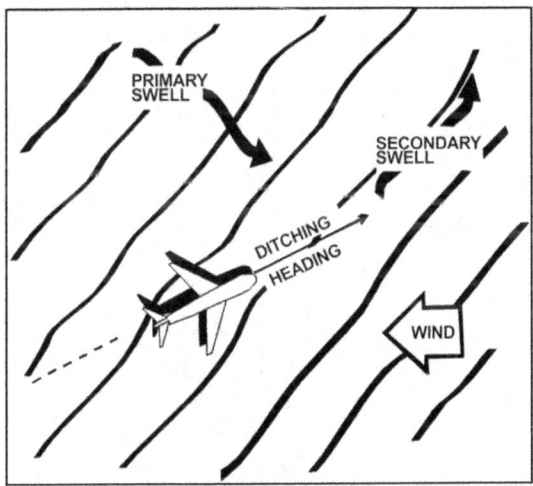

Figure 8-8. Double swell (15 knot wind)

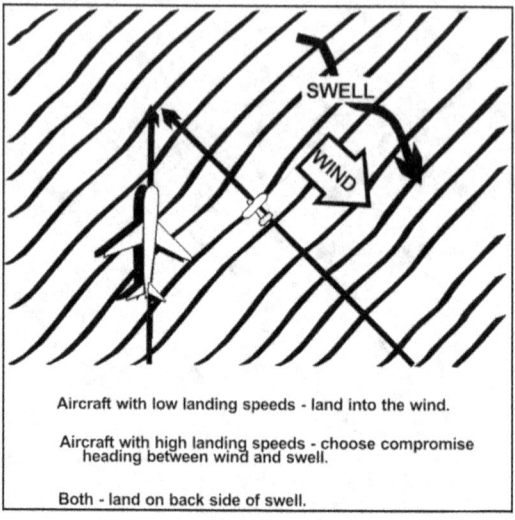

Figure 8-9. Double swell (30 knot wind)

8-88. If the swell system is formidable, it is considered advisable, in landplanes, to accept more crosswind to avoid landing directly into the swell.

8-89. The secondary swell system is often from the same direction as the wind. Here, the landing may be made parallel to the primary system, with the wind and secondary system at an angle. There is a choice of two directions paralleling the primary system. One direction is downwind and down the secondary swell, and the other is into the wind and into the secondary swell. The choice will depend on the velocity of the wind versus the velocity and height of the secondary swell (figure 8-10).

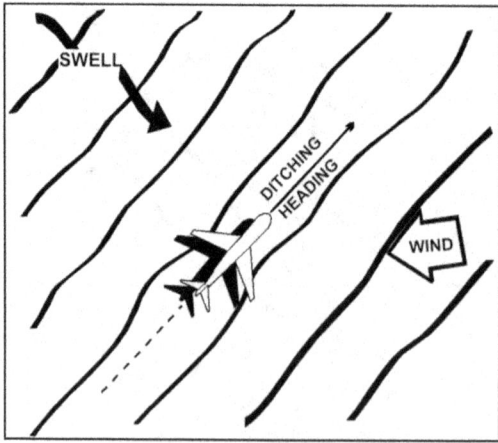

Figure 8-10. Swell (50 knot wind)

Wind Considerations

8-90. The simplest method of estimating wind direction and velocity is to examine windstreaks on the water. These appear as long streaks up and down wind. There may be difficulty determining wind direction after seeing the streaks on the water. Whitecaps fall forward with the wind but are overrun by the waves, this produces the illusion the foam is sliding backward. Knowing this, and by observing the direction of the streaks, wind direction is determined. Wind velocity can be estimated by noting the appearance of whitecaps, foam, and wind streaks.

Water Landing

8-91. Aircraft behavior, on making contact with the water, will vary within wide limits according to the state of the sea. If landed parallel to a single swell system, aircraft behavior may approximate that to be expected on a smooth sea. If landed into a heavy swell or into a confused sea, deceleration forces may be extremely great resulting in breaking up of the aircraft. Within certain limits, a pilot is able to minimize these forces by proper sea evaluation and selection of ditching heading.

8-92. When on final approach a pilot looks ahead and observes the surface of the sea. There may be shadows and whitecaps, signs of large seas. Shadows and whitecaps close together indicate short and rough seas; touchdown in these areas is to be avoided. A pilot selects and touches down in an area (only about 500 feet is needed) where shadows and whitecaps are not as numerous.

8-93. Touchdown is accomplished at the lowest speed and rate of descent permitting safe handling and optimum nose up attitude on impact. Once first impact has been made, there is often little a pilot can do to control a landplane.

8-94. Once preditching preparations are completed, the pilot turns to the ditching heading and commence let-down. The aircraft should be flown low over the water, and slowed down until ten knots or so above stall. At this point, additional power is used to overcome increased drag caused by the nose up attitude. When a smooth stretch of water appears ahead, cut power, and touchdown at the best recommended speed as fully stalled as possible. By cutting power when approaching a relatively smooth area, a pilot will prevent overshooting and touch down with less chance of planning off into a second uncontrolled landing. Most experienced seaplane pilots prefer to make contact with the water in a semi-stalled attitude, cutting power as the tail makes contact. This technique eliminates the chance of misjudging altitude with a resultant heavy drop in a fully stalled condition. Care must be taken to not drop the aircraft from too high altitude or to balloon due to excessive speed.

8-95. The altitude above water depends on the aircraft. Over glassy smooth water or at night without sufficient light, it is very easy to misjudge altitude by 50 feet or more. Under such conditions, a pilot must carry enough power to maintain 9 to 12 degrees nose up attitude, and 10 to 20 percent over stalling speed, until contact is made with the water. The proper use of power on the approach is of great importance. If power is available on one side only, a little power will be used to flatten the approach; however, the engine must not be used to such an extent the aircraft cannot be turned against the good engines right down to the stall with a margin of rudder movement available. When near stall, sudden application of excessive unbalanced power may result in loss of directional control. If power is available on one side only, a slightly higher than normal glide approach speed will be used. This ensures good control and some margin of speed after leveling off without excessive use of power. The use of power in ditching is so important that when it is certain the coast cannot be reached, the pilot should, and if possible, ditch before fuel is exhausted. Use of power in night or instrument ditching is far more essential than under daylight contact conditions.

8-96. If no power is available, a greater than normal approach speed should be used down to the flare-out. This speed margin will allow the glide to be broken early and more gradually, thereby giving the pilot time and distance to feel for the surface—decreasing the possibility of stalling high or flying into the water. When landing parallel to a swell system, little difference is noted between landing on top of a crest or in the trough. If the wings of aircraft are trimmed to the surface of the sea rather than the horizon, there is little need to worry about a wing hitting a swell crest. The actual slope of a swell is very gradual. If forced to land into a swell, touchdown should be made just after passage of the crest. If contact is made on the face of the swell, the aircraft may be swamped or thrown violently into the air, dropping heavily into the next swell. If control surfaces remain intact, the pilot should attempt to maintain the proper nose above the horizon attitude by rapid and positive use of the controls.

After Touchdown

8-97. In most cases drift, caused by crosswind can be ignored. Forces acting on the aircraft after touchdown are of such magnitudes that drift will be only a secondary consideration. If the aircraft is under good control, the "crab" may be kicked out with rudder just prior to touchdown. This is more important with high wing aircraft, as they are laterally unstable on the water in a crosswind and may roll to the side in ditching.

SECTION IV – THUNDERSTORM OPERATIONS

8-98. Turbulence, hail, rain, snow, lightning, sustained updrafts and downdrafts, and icing conditions are all present in thunderstorms. While there is some evidence maximum turbulence exists at the middle level of a thunderstorm, recent studies show little variation of turbulence intensity with altitude.

8-99. There is no useful correlation between the external visual appearance of thunderstorms and the severity, amount of turbulence, or hail within them. The visible thunderstorm cloud is only a portion of a turbulent system whose updrafts and downdrafts often extend far beyond the visible storm cloud. Severe turbulence can be expected up to 20 miles from severe thunderstorms and decreases to about 10 miles in less severe storms.

WIND SHEAR

8-100. One of the more deadly factors associated with thunderstorms is wind shear and has been identified as the cause of many major aircraft accidents and deaths. Wind shear is a sudden, drastic change in wind speed and/or direction over a very small area. Wind shear can subject an aircraft to violent updrafts and downdrafts as well as abrupt changes to horizontal movement of the aircraft. While wind shear can occur at any altitude, low-level wind shear is especially hazardous due to proximity of an aircraft to the ground. Directional wind changes of 180 degrees and speed changes of 50 knots or more are associated with low-level wind shear. Low-level wind shear is also commonly associated with passing frontal systems and temperature inversions with strong upper level winds (greater than 25 knots).

8-101. Wind shear is dangerous to an aircraft for several reasons. The rapid changes in wind direction and velocity alter the wind's relation to the aircraft and disrupt the normal flight attitude and performance of the aircraft. During a wind shear situation, effects can be subtle or very dramatic depending on wind speed and direction of change. For example, a tailwind quickly changing to a headwind will cause an increase in airspeed and performance. Conversely, when a headwind changes to a tailwind, airspeed will rapidly decrease and there will be a corresponding decrease in performance. In either case, a pilot must be prepared to react immediately to maintain aircraft control.

8-102. In general, the most severe type of low-level wind shear is associated with convective precipitation or rain from thunderstorms. One critical type of shear associated with convective precipitation is known as a microburst. A typical microburst occurs in a space of less than 1 mile horizontally and within 1,000 feet vertically. The lifespan of a microburst is about 15 minutes, during which it can produce downdrafts of up to 6,000 FPM. It can also produce a hazardous wind direction change of 45 knots or more, in a matter of seconds. When encountered close to the ground, these excessive downdrafts and rapid changes in wind direction can produce a situation in which it is difficult to control the aircraft (figure 8-11, page 8-23). During an inadvertent takeoff into a microburst, the aircraft experiences a performance-increasing headwind, followed by performance-decreasing downdrafts. The wind then rapidly shears to a tailwind and can result in terrain impact or flight dangerously close to the ground.

Fixed-Wing Environmental Flight

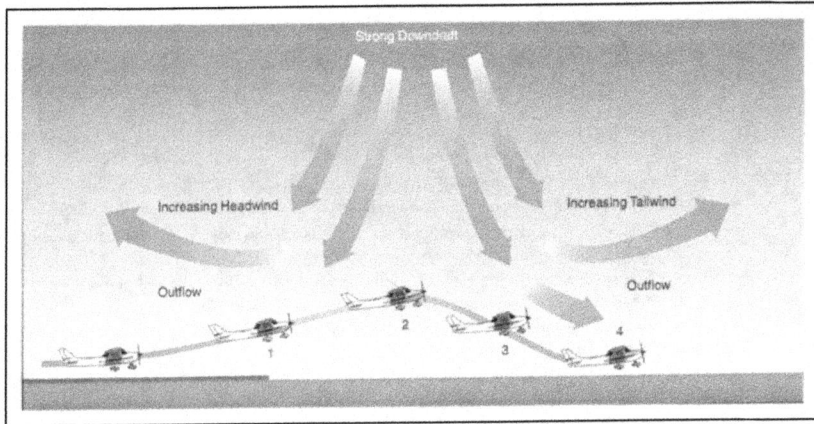

Figure 8-11. Effect of microburst

8-103. Microbursts are often difficult to detect because they occur in a relatively confined area. In an effort to warn pilots of low-level wind shear, alert systems have been installed at several airports around the country. A series of anemometers, placed around the airport, form a net to detect changes in wind speeds. When wind speeds differ by more than 15 knots, a warning for wind shear is given to pilots. This system is known as the low-level wind shear alert system.

8-104. Wind shear can affect any flight and any pilot at any altitude. While wind shear may be reported, it often remains undetected and is a silent danger to aviation. A pilot must always be alert to the possibility of wind shear, especially when flying in and around thunderstorms and frontal systems.

FLYING TECHNIQUES

8-105. Thunderstorms must never be taken lightly, even when radar observers report echoes of light intensity. Avoiding thunderstorms is the best policy. The following are a few guidelines for avoiding thunderstorms:
- Do not land or takeoff in the face of an approaching thunderstorm. A sudden gust front of low level turbulence could cause loss of control.
- Do not attempt to fly under a thunderstorm even with clear visibility through to the other side. Turbulence and wind shear under the storm could be disastrous.
- Do not fly without airborne radar into a cloud mass containing scattered embedded thunderstorms. Scattered thunderstorms not embedded usually can be visually circumnavigated.
- Do not trust visual appearance to be a reliable indicator of turbulence inside a thunderstorm.
- Avoid (by at least 20 miles) any thunderstorm identified as severe or giving an intense radar echo. This is especially true under the anvil of a large cumulonimbus cloud.
- Clear the top of a known or suspected severe thunderstorm by at least 1,000 feet altitude for each 10 knots of wind speed at the cloud top. This should exceed the altitude capability of most aircraft.
- Circumnavigate the entire area if the thunderstorm coverage is—
  - More than 45 percent (DD Form 175-1 [Flight Weather Briefing]).
  - Six-tenths or greater (FAA).
- Remember vivid and frequent lightning indicates probability of a strong thunderstorm.
- Regard as any thunderstorm with 35,000 feet or higher tops as extremely hazardous, whether the top is visually sighted or determined by radar.

Chapter 8

8-106. If you cannot avoid penetrating a thunderstorm, the following guidelines are provided:
- Tighten the safety belt, put on the shoulder harness if one exists, and secure all loose objects.
- Plan and hold course to get through the storm in minimum time.
- Establish a penetration altitude below freezing level or above -15 degrees C, to avoid the critical icing.
- Verify pitot heat is on and turn on carburetor heat or jet engine anti-ice. Icing can be rapid at any altitude and cause almost instantaneous power failure and/or loss of airspeed indication.
- Establish power settings for turbulence penetration airspeed recommended in the AFM.
- Turn up cockpit lights to highest intensity to lessen temporary blindness from lightning.
- If using automatic pilot, disengage altitude hold and speed hold modes. The automatic altitude and speed controls will increase maneuvers of the aircraft thus increasing structural stress.
- If using airborne radar, tilt the antenna up and down occasionally. This will permit detection of any thunderstorm activity at altitudes other than the one being flown.
- Keep visual of the aircraft's instruments. Looking outside the cockpit can increase danger of temporary blindness from lightning.
- Do not change power settings; maintain settings for the recommended turbulence penetration airspeed.
- Do not attempt to maintain constant altitude; let the aircraft ride the waves.
- Do not turn back once in the thunderstorm. A straight course through the storm most likely will get a pilot out of the hazards most quickly. In addition, turning maneuvers increase stress on the aircraft.

WIND SHEAR RECOVERY TECHNIQUE

8-107. The primary recovery technique objective is to keep the aircraft flying as long as possible in hope of exiting the shear. A wide variety of techniques were considered to establish the one best meeting this objective. The best results were achieved by pitching toward an initial target attitude while using necessary thrust. Several factors were considered in developing this technique.

8-108. Studies show wind shear encounters occur infrequently and only a few seconds are available to initiate a successful recovery. Additionally, during high stress situations, pilot instrument scan typically becomes very limited. In extreme cases, this scan may be limited to only one instrument. Lastly, recovery skills will not be exercised on a day-to-day basis. These factors dictated the recovery technique must not only be effective, but simple, easily recalled, and have general applicability.

8-109. Extensive analysis and pilot evaluations were conducted. Although a range of recovery attitudes (including 15 degrees and the range of all-engine initial climb attitudes) provides good recovery capability for a wide variety of wind shears, 15 degrees was chosen as the initial target pitch attitude for both takeoff and approach. Additional advantages of 15-degree initial target pitch attitude are it is easily recalled in emergency situations and prominently displayed on attitude director indicators.

8-110. While other more complex techniques may make slightly better use of aircraft performance, these techniques do not meet simplicity and ease of recall requirements. Evaluations showed the recommended technique provides a simple, effective means of recovering from a wind shear encounter. Proficiency in the techniques for each specific aircraft is critical to wind shear recovery.

OPERATIONAL PROCEDURES

8-111. Weather radar, airborne or ground based, will normally reflect areas of moderate to heavy precipitation (radar does not detect turbulence). The frequency and severity of turbulence generally increases with radar reflectivity which is closely associated with areas of highest liquid water content of the storm. No flight path through an area of strong or very strong radar echoes separated by 20 to 30 miles or less may be considered free of severe turbulence.

Fixed-Wing Environmental Flight

8-112. Turbulence beneath a thunderstorm must not be minimized. This is especially true when relative humidity is low in any layer between the surface and 15,000 feet. Then, the lower altitudes may be characterized by strong out flowing winds and severe turbulence.

8-113. The probability of lightning striking an aircraft is greatest when operating at altitudes where temperatures are -5 degrees C to 5 degrees C. Lightning can strike aircraft flying in the clear within the vicinity of a thunderstorm.

8-114. Meteorological terminal aviation reports do not include a descriptor for severe thunderstorms. However, by understanding severe thunderstorm criteria, information is available in the report to know one is occurring.

8-115. NWS radar systems are able to objectively determine radar weather echo intensity levels by use of video integrator processor equipment. These thunderstorm intensity levels are on a scale of one to six.

TRAINING

8-116. Units qualifying aviators in thunderstorm operations are responsible for conducting a well-organized training program. Training programs should be geared to instill confidence and develop skills in all areas. IPs and supervisory maintenance personnel must be highly qualified and skilled.

8-117. Emphasis must be placed on safety and avoidance; avoiding thunderstorms and all their hazards. The flight training program allows each aviator to advance at an individual rate. Initial training is conducted under less challenging conditions. As an aviator's proficiency increases, conditions become more demanding until the most challenging mission can be performed.

8-118. A recommended program of instruction for qualifying aviators for thunderstorm operations is provided. Additional academic subjects may be required, based on the specific mission and location of the unit.

Academics

8-119. Suggested topics include—
- Human factors associated with thunderstorm flying.
- Environmental factors affecting thunderstorm operations.
- Planning data available on thunderstorms.
- In-flight equipment and resources detecting and avoiding hazards.
- Aircraft operational procedures in thunderstorms.

Flight

8-120. Flight training may be limited by conditions at the unit's home station as there may not be areas able to replicate conditions adequately. Instructors can demonstrate techniques and procedures to some extent. Crews should be evaluated on these procedures during their APART or no-notice evaluations. Flight simulators are also a great device in training for this environment.

8-121. Suggested maneuvers include—
- En route flight techniques.
- Wind shear recovery maneuvers.
- Use of aircraft equipment for avoidance.

This page intentionally left blank.

Chapter 9
Fixed-Wing Night Flight

This chapter discusses night flight, takeoff, and landing considerations. This chapter supplements chapter 4 with information specific to FW operations.

SECTION I – PREPARATION AND PREFLIGHT

9-1. Crewmembers should be familiar with the aircraft, lighting system, and emergency equipment. A thorough preflight of the aircraft and a review of aircraft systems and emergency procedures are important for night operations. The following information supplements, but does not replace, the aircraft checklist.

EQUIPMENT

9-2. Before beginning a night flight, aviators must carefully consider personal equipment needed during a flight. At least one reliable flashlight is required (AR 95-1) as standard equipment on all night flights. A D-cell size flashlight with a bulb switching mechanism used to select white or red light along with a spare set of batteries is preferable. The white light should be used while performing preflight visual inspection of the airplane, and the red light is used when performing cockpit operations. Since the red light is nonglaring, it will not impair night vision. Some aviators prefer two flashlights, a white light for preflight, and a penlight with red light. The latter can be suspended around the neck by a string, ensuring the light is always readily available. If a red light is used for reading an aeronautical chart, red features of the chart will not show up.

9-3. Aeronautical charts are essential for night cross-country flight and, if intended course is near the edge of the chart, the adjacent chart should also be available. The lights of cities and towns can be seen from surprising distances at night. Therefore, if the adjacent chart is not available to identify those landmarks, confusion could result. Regardless of equipment used, organization of the cockpit eases the burden on the aviator and enhances safety.

LIGHTING

AIRCRAFT

9-4. Aviators turn on and check aircraft lights for proper operation. They check position lights for loose connections by tapping the light fixture while the light is on. If the light operation is intermittent, the aviator must determine the cause and correct the deficiency.

9-5. Aviators adjust cockpit lights before takeoff. They adjust the lights to the dimmest level allowing them to read instruments and identify switches without hindering their vision outside the cockpit. Dimming the cockpit lights also eliminates light reflections on the windscreen and windows. Aviators should turn on position and anti-collision lights before starting the engines. These lights remain on during engine operation.

AIRPORT AND NAVIGATION LIGHTING AIDS

9-6. The lighting systems used for airports, runways, obstructions, and other visual aids at night are another important aspects of night flight.

9-7. Lighted airports located away from congested areas can be identified readily at night by lights outlining runways. Airports located near or within large cities are often difficult to identify in the maze of lights. It is important not only to know the exact location of an airport relative to the city, but also to be able to identify these airports by characteristics of their lighting pattern.

Chapter 9

9-8. Aeronautical lights are designed and installed in a variety of colors and configurations, each having its own purpose. Although some lights are used only during low ceiling and visibility conditions, this discussion includes only lights fundamental to VFR night operation.

9-9. Prior to a night flight, particularly a cross-country night flight, the aviator checks the availability and status of lighting systems at the destination airport. This information is found on aeronautical charts and in the airport/facility directory. The status of each facility is determined by reviewing pertinent notices to airmen.

9-10. A rotating beacon is used to indicate the location of most airports. The beacon rotates at a constant speed, thus producing what appears to be a series of light flashes at regular intervals. These flashes may be one or two different colors used to identify various types of landing areas. The following are examples:
- Lighted civilian land airports–alternating white and green.
- Lighted civilian water airports–alternating white and yellow.
- Lighted military airports–alternating white and green, but are differentiated from civil airports by dual peaked (two quick) white flashes, then green.

9-11. Beacons producing red flashes indicate obstructions or areas considered hazardous to aerial navigation. Steady burning red lights are used to mark obstructions on or near airports and sometimes to supplement flashing lights on en route obstructions. High intensity flashing white lights are used to mark some supporting structures of overhead transmission lines stretching across rivers, chasms, and gorges. These high intensity lights are also used to identify tall structures, such as chimneys and towers.

9-12. As a result of technological advancements in aviation, runway lighting systems have become sophisticated in accommodating takeoffs and landings in various weather conditions. Aviators need to be concerned with following the basic lighting system of runways and taxiways.

9-13. This lighting system consists of two straight parallel lines of runway-edge lights defining the lateral limits of the runway. These lights are aviation white, although aviation yellow may be substituted for a distance of 2,000 feet from the far end of the runway to indicate a caution zone. At some airports, the intensity of runway-edge lights can be adjusted to satisfy individual needs of the aviator. The length limits of the runway are defined by straight lines of lights across runway ends. At some airports, runway threshold lights are aviation green, and runway end lights are aviation red.

9-14. At many airports, taxiways are also lighted. A taxiway-edge lighting system consists of blue lights outlining usable limits of taxi paths.

PARKING RAMP CHECK

9-15. Aviators inspect the parking ramp before entering the aircraft. Stepladders, chuckholes, and other obstacles are difficult to see at night. A thorough check of the area can prevent taxiing mishaps.

PREFLIGHT

9-16. Planning for a night flight includes a thorough review of available weather reports and forecasts with particular attention given to temperature/dew point spread. A narrow temperature/dew point spread may indicate possibility of ground fog. Emphasis should also be placed on wind direction and speed, as its effect on the airplane cannot be as easily detected at night as during the day.

9-17. On night cross-country flights, appropriate aeronautical charts are selected, including appropriate adjacent charts. Course lines are drawn in black to be more distinguishable.

9-18. Prominently lighted CPs along the prepared course are noted. Rotating beacons at airports, lighted obstructions, lights of cities or towns, and lights from major highway traffic all provide excellent visual CPs. The use of radio NAVAIDs and communication facilities add significantly to safety and efficiency of night flight.

Fixed-Wing Night Flight

SECTION II – TAXI, TAKEOFF, AND DEPARTURE CLIMB

9-19. Fewer outside visual references are available during night flight. Aviators rely on flight instruments at night for attitude control. This is particularly true for night takeoffs and departure climbs.

TAXI

LIGHTS

9-20. Aviators perform night taxiing slowly and with extreme care. Landing lights can easily disturb the vision of others and overheat due to inadequate airflow to carry heat away. An aviator uses taxi lights whenever possible and the landing light only as necessary.

OTHER AIRCRAFT

9-21. Aviators use extreme caution when taxiing onto an active runway for takeoff. Even at controlled airports, aviators must check the final approach course for approaching aircraft. At uncontrolled airports, aviators execute slow 360-degree turns in the same direction as the flow of air traffic. This will assist in identifying other aircraft in the traffic pattern or the vicinity.

TAKEOFF AND CLIMB

9-22. Night flight is very different from day flying and demands more of the aviator's attention. Flight instruments are used to a greater degree in controlling the airplane. This is particularly true on night takeoffs and climbs.

9-23. After ensuring the final approach and runway are clear of other air traffic, or when cleared for takeoff by the tower, landing lights and taxi lights are turned on and the airplane lined up with the centerline of the runway. If the runway does not have centerline lighting, use the painted centerline and runway-edge lights. After the airplane is aligned, the heading indicator is noted or set to correspond to the known runway direction. To begin takeoff, the brakes are released and the throttle smoothly advanced to maximum allowable power. As the airplane accelerates, it continues moving straight ahead between and parallel to the runway-edge lights.

9-24. The procedure for night takeoffs is the same as normal daytime takeoffs except many of the runway visual cues are not available. Therefore, flight instruments are checked frequently during takeoff ensuring proper pitch attitude, heading, and airspeed are being attained. As airspeed reaches normal lift-off speed, pitch attitude is adjusted to establish a normal climb. This is accomplished by referring to both outside visual references, such as lights, and flight instruments (figure 9-1, page 9-4).

Chapter 9

Figure 9-1. Positive climb

9-25. After becoming airborne, the darkness of night often makes it difficult to note whether the airplane is getting closer to or farther from the surface. To ensure the airplane is continuing a positive climb, verify that climb is indicated on the attitude indicator, VSI, and altimeter. It is also important that airspeed is at best climb speed.

9-26. Necessary pitch and bank adjustments are made by referencing attitude and heading indicators. It is recommended turns not be made until reaching a safe maneuvering altitude.

9-27. Although the use of landing lights provides help during takeoff, they become ineffective after the airplane has climbed to an altitude where the light beam no longer extends to the surface. The light can cause distortion when reflected by haze, smoke, or fog existing in the climb. Therefore, when the landing light is used for takeoff, it may be turned off after climb is well established provided other traffic in the area does not require its use for collision avoidance.

SECTION III – ORIENTATION AND NAVIGATION

9-28. During night flight, aircrew members must be alert for other aircraft. The relative position of other aircraft is determined by color, position, and movement direction of their position lights.

VISIBILITY

9-29. Clouds and visibility restrictions may be difficult to see. This is more pronounced during periods of low moon illumination or overcast sky conditions. While flying VFR, aviators must avoid flying into clouds or a fog layer. Usually, the first indication of restricted visibility is the gradual disappearance of ground lights. If a halo appears surrounding the lights, aviators avoid further flight in that direction. If descent through fog, smoke, or haze is necessary, horizontal visibility is considerably less than vertical visibility. Aviators must avoid night VFR flight during poor or marginal weather conditions.

MANEUVERS

9-30. Night maneuvers are practiced in designated areas or at least in an area known to be comparatively free of other air traffic. Aviators practice and acquire competency in straight-and-level flight, climbs and descents, level turns, climbing and descending turns, and steep turns. Recovery from unusual attitudes is practiced, but only as covered in the appropriate ATM. They must also practice these maneuvers with all cockpit lights turned off. This blackout training is necessary if aviators experience an electrical or instrument light failure. Training also includes use of navigation equipment and local NAVAIDs.

DISORIENTATION AND REORIENTATION

9-31. Disorientation can happen to the most experienced aviator. An orderly plan for reorientation must be developed in advance. Thorough knowledge of the area, current navigation charts, identified radio NAVAIDS, and assistance from ATC agencies and other aircraft may be used for reorientation.

CROSS-COUNTRY FLIGHTS

9-32. Cross-country night flights do not present unusual problems if aircrew members complete adequate preplanning. NAVAIDs, if available, are used to assist in monitoring en route progress.

OVERWATER FLIGHTS

9-33. Crossing large bodies of water during night flights is potentially hazardous. Emergency procedures, such as ditching, are primary hazards and briefed to aviators. Aircraft control is also a potential problem that must be prepared for. The horizon may be difficult to see under certain atmospheric conditions and can essentially disappear, creating conditions conducive to spatial disorientation. On clear low-moon illumination nights, stars can reflect on the water's surface leading to a variation of the visual illusion called ground light misinterpretation.

ILLUSIONS

9-34. Lighted runways, buildings, or other objects may cause illusions when seen from different altitudes. At an altitude of 2,000 feet, an aviator may see a group of lights individually. At 5,000 feet or higher, the same lights may appear to be one solid mass of light. These illusions may become acute with altitude changes and could cause problems when an aviator approaches a lighted runway.

SECTION IV – APPROACHES AND LANDINGS

DISTANCE

9-35. Distance may be difficult to judge at night. This is due to limited lighting, lack of visual references on the ground, and an aviator's inability to compare size and location of ground objects. Altitude and airspeed are also difficult to estimate at night. Therefore, the aircrew must closely monitor flight instruments, particularly the altimeter and airspeed indicator.

AIRSPEED

9-36. Inexperienced aviators often tend to make night approaches and landings at excessive airspeeds. A night approach uses the same techniques as a day approach; however, it is important to frequently crosscheck the flight instruments with particular attention to the altimeter and airspeed indicator.

DEPTH PERCEPTION

9-37. Even the most experienced aviators make mistakes in depth perception. Using power during the landing flare reduces rate of descent and assists in maintaining a safe airspeed before touchdown. The use of power is essential during landings to unlighted airfields when the surface is not visible. This is also an effective technique in preventing errors in judgment and perception on lighted airfields. An effective night technique is maintaining a slight amount of power and airspeed above stall until the wheels make ground contact.

APPROACHING AIRPORTS

9-38. When approaching an airport, an aviator must identify runway lights and other airport lighting as soon as possible. If aircrew members are unfamiliar with the airport, they may have difficulty sighting the runway because of light congestion in the area. Figure 9-2 illustrates the difficulty of sighting a runway surrounded by lights. Because airport beacons are hard to see when directly overhead, aircrew members identify the airport through beacon identification while at a distance. Once identified, the aircrew continues to fly toward the beacon until they can identify the runway lights and environment.

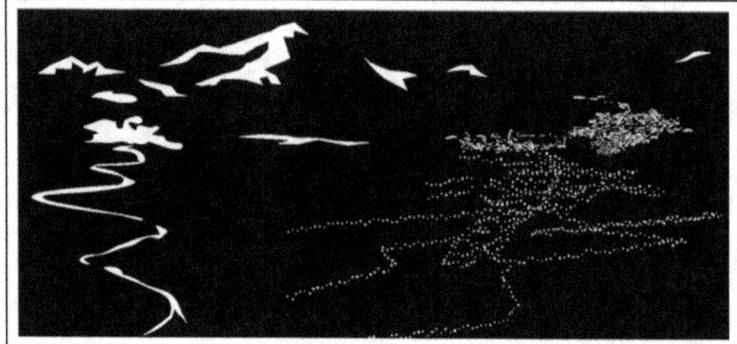

Figure 9-2. Typical light pattern for airport identification

ENTERING TRAFFIC

9-39. When aircrew members identify runway lights, they keep the approach threshold lights–including a visual approach slope indicator (VASI), if available–in sight throughout the traffic pattern and approach.

FINAL APPROACH

9-40. After turning onto the final approach leg, all available lighting, including a VASI or precision approach path indicator, are used to maintain a proper approach angle. Obstruction or runway lights also assist in judging the proper approach angle, especially when the runway environment is on level ground and lights are spaced at a known interval. Uneven terrain or nonstandard light spacing makes the angular reference unreliable. In such cases, an aviator takes advantage of other reference points near the approach area. Figure 9-3, page 9-7, depicts a VASI.

Fixed-Wing Night Flight

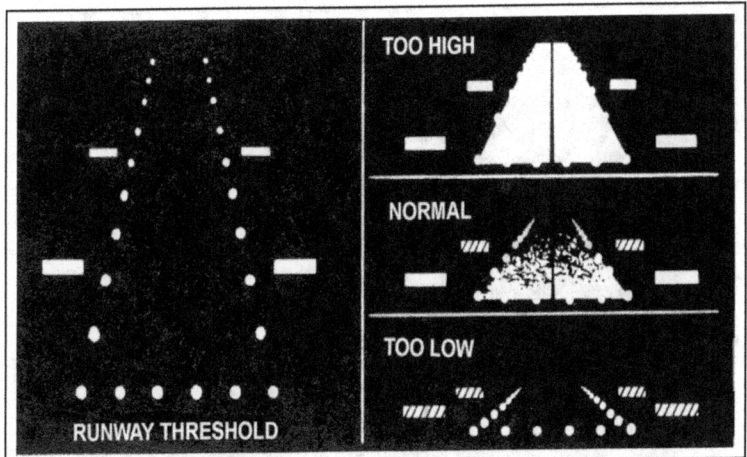

Figure 9-3. Visual approach slope indicator

EXECUTING ROUNDOUT

9-41. Inexperienced aviators may tend to roundout too high until they become familiar with the apparent height for the correct roundout position. To aid in determining the proper roundout point, an aviator continues a constant approach descent until the landing light reflects off the runway and tire marks or expansion joints on the runway can be clearly seen. The aviator then smoothly starts the roundout for touchdown and continues to apply standard daytime procedures outlined in the appropriate ATM. Figure 9-4, page 9-8, illustrates the proper roundout point. During landings without landing lights or where marks on the runway are not visible, an aviator starts the roundout when runway lights at the far end of the runway first appear to be rising higher than the aircraft. This demands a smooth and timely roundout and requires an aviator to feel the runway surface using necessary power and pitch changes to settle the aircraft softly onto the runway.

Chapter 9

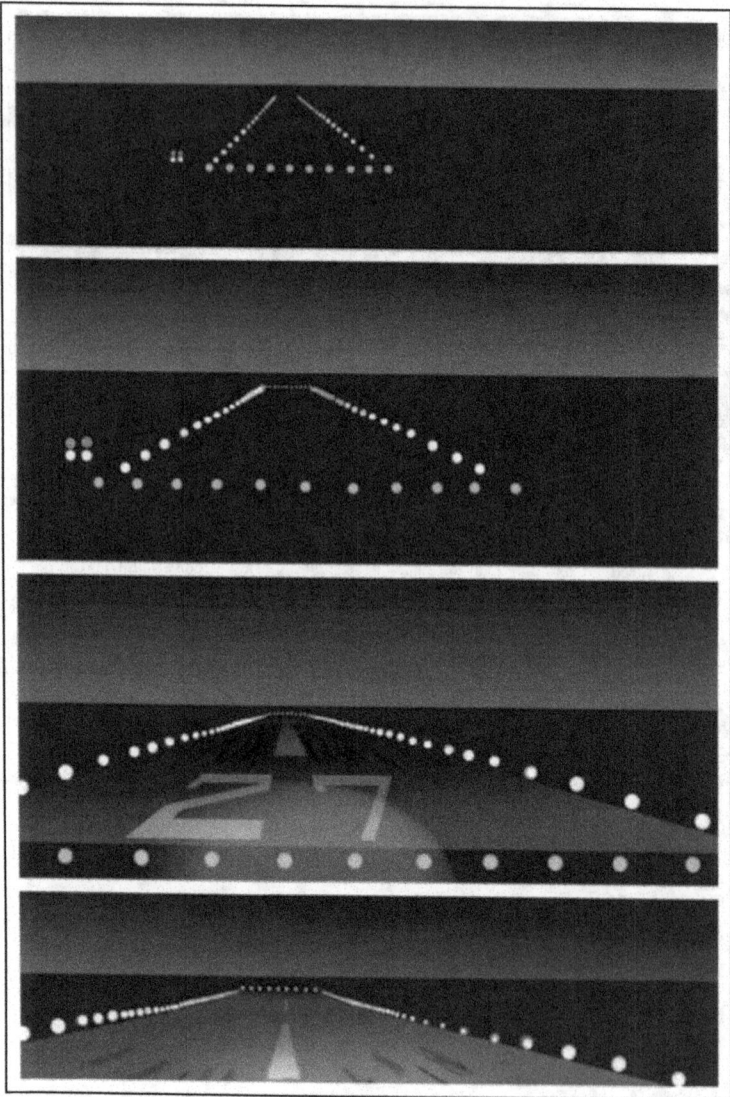

Figure 9-4. Roundout (when tire marks are visible)

Fixed-Wing Night Flight

SECTION V – NIGHT EMERGENCIES

9-42. An aviator's greatest concern about night flight is the possibility of an emergency and the subsequent landing. This is a legitimate concern, even though continuing flight into adverse weather and poor aviator judgment account for most serious accidents.

9-43. If an emergency occurs at night, keep the following important procedures and considerations to keep in mind:

- Focus on flying. If an aviator gets distracted by the emergency, a flyable aircraft could crash.
- Dual engine failure or unable to maintain single engine flight. Maintain positive control of the airplane and establish the best glide configuration and airspeed. Turn the airplane toward an airport or away from congested areas. If possible, maintain orientation with the wind to avoid a downwind landing.
- Check to determine the cause of malfunction, such as position of fuel selectors, switch, or circuit breaker. If possible, the cause of malfunction must be corrected immediately.
- Announce the emergency situation to ATC or universal integrated communication. If already in radio contact with a facility, do not change frequencies, unless instructed.
- Forced landing. If the condition of the nearby terrain is known turn towards an unlighted portion of the area. Plan an emergency approach to an unlighted portion.
- Consider an emergency landing area close to public access if possible. This may facilitate rescue or help, if needed.
- Complete the before-landing checklist. Check landing lights for operation at altitude and turn on in sufficient time to illuminate terrain or obstacles along the flight path. Complete the landing in the normal landing attitude at the slowest possible airspeed. If landing lights are unusable and outside visual references are not available, the airplane must be held in level-landing attitude until ground is contacted.

This page intentionally left blank.

Glossary

β	beta – sideslip angle
ρ	rho - density of the air
AAA	antiaircraft artillery
AATF	air assault task force
AATFC	air assault task force commander
ABC	automatic brightness control
ACL	allowable cargo load
ACO	airspace control order
ACP	air control point
ADA	air defense artillery
AFM	aircraft flight manual
AGL	above ground level
AH	attack helicopter
AHO	above highest obstacle
AIM	Aeronautical Information Manual
AIRMET	airman's meteorological information
AKO	Army Knowledge Online
ALSE	aviation life support equipment
AMC	air mission commander
AMPS	aviation mission planning system
ANVIS	aviator's night vision imaging system
AO	area of operations
AOA	angle of attack
APART	annual proficiency and readiness test
AR	Army regulation
ASE	aircraft survivability equipment
ATA	actual time of arrival
ATC	air traffic control
ATIS	automated terminal information service
ATM	aircrew training manual
AWS	area weapons system
BCM	basic combat manuver
BLC	boundary-layer control
BMCT	begin morning civil twilight
BMNT	begin morning nautical twilight
BP	battle position
BSP	bright source protection
C_l	coefficient of the rolling moment
CAS	close air support

Glossary

CASEVAC	casualty evacuation
C_D	coefficient of drag
CG	center of gravity
CH	cargo helicopter
CHUM	chart update manual
C_L	coefficient of lift
C_{L-MAX}	maximum value of the coefficient of lift
C_M	coefficient of pitching moment
C_N	coefficient of yawing moment
CONUS	continental United States
CP	checkpoint
CRT	cathode ray tube
DA	Department of the Army
DC	direct current
DD	Department of Defense
DEP	design eye point
DOD	Department of Defense
DTS	data transfer system
EECT	end evening civil twilight
EENT	end evening nautical twight
EPR	engine pressure ratio
ETA	estimated time of arrival
ETE	estimated time en route
ETL	effective translational lift
FAA	Federal Aviation Administration
FAR	Federal Aviation Regulation
FARP	forward arming and refueling point
FLIP	Flight Information Publication
FLIR	forward-looking infrared
FM	field manual
FOD	foreign object damage
FOV	field of view
FPM	feet per minute
FSS	flight service station
FW	Fixed-wing
G	gravitational
GPS	global positioning system
HAATS	high altitude aviation training site
HDU	helmet display unit
Hg	Mercury
HUD	heads-up display

Glossary

I2	image intensifier
IATA	International Air Transport Association
ICTS	ice-contaminated tailplane stall
IFR	instrument flight rules
IGE	in ground effect
IIMC	inadvertent instrument meteorological conditions
IMC	instrument meteorological conditions
IP	instructor pilot
IR	infrared
ITO	instrument takeoff
JCDB	joint common data base
JOG	joint operations graphic
JP	joint publication
KIAS	knots indicated airspeed
L/D	lift over drag
L/D$_{MAX}$	minimum drag speed
LCU	lightweight computer unit
LED	light emitting diode
LITECON	lighting conditions
LOS	line-of-sight
LZ	landing zone
MAC	mean aerodynamic chord
MCP	microchannel plate
METT-TC	mission, enemy, terrain and weather, troops and support available, time available, civil considerations
MPH	miles per hour
MPNVS	modernized pilot night vision system
MRT	minimum resolvable temperature
MSL	mean sea level
MTADS	modernized target acquisition device system
MTOS	mountain obscuration
MTRA	maximum torque rate attenuator
N	north
NATO	North Atlantic Treaty Organization
NAVAID	navigational aid
NCM	nonrated crewmember
NM	nautical miles
NOE	nap-of-the-earth
NTSB	National Transportation Safety Board
NVD	night vision device
NVG	night vision goggle
NVS	night vision system

NWS	National Weather Service
OAT	outside air temperature
OGE	out of ground effect
OH	observation helicopter
OPORD	operations order
OPSEC	operations security
P	pilot not on the controls
P*	pilot on the controls
PA	pressure altitude
PC	pilot in command
P-factor	propeller factor
POH	pilot's operating handbook
PPC	performance planning card
PZ	pickup zone
RCR	runway condition reading
RP	release point
RPM	revolutions per minute
SA	situational awareness
SLS	sea level standard (+15 degrees C and 0 feet PA)
SOP	standing operating procedures
SP	start point
STANAG	standardization agreement
TA	thrust available
TAF	total aerodynamic force
TAS	true airspeed
TC	training circular
TM	technical manual
TPC	tactical pilotage chart
TR	torque
TTP	tactics, techniques, and procedures
U.S.	United States
UH	utility helicopter
USN	United States Navy
UTM	universal transverse mercator
V_1	takeoff decision speed (same as V_R)
V_2	takeoff safety speed
VASI	visual approach slope indicator
V_{FE}	maximum flap extended speed
VFR	visual flight rules
V_{FS}	final segment climb speed
V_{LE}	maximum landing gear extended speed

Glossary

V_{LO}	maximum landing gear operating speed
V_{LOF}	lift-off speed (V_R + 3 knots)
VMC	visual meterological conditions
V_{MC}	minimum control speed with the critical engine inoperative
V_R	rotation speed
V_S	stalling speed or the minimum steady flight speed at which the airplane is controllable
VSI	verticle speed indicator
V_X	speed for best angle of climb
V_{XSE}	speed for best single engine angle of climb
V_{YSE}	speed for best single engine rate of climb

This page intentionally left blank.

References

REQUIRED PUBLICATIONS

These documents must be available to intended users of this publication.

ADRP 1-02. *Terms and Military Symbols.* 16 November 2016.

Department of Defense Dictionary of Military and Associated Terms. 15 October 2016.

RELATED PUBLICATIONS

ARMY PUBLICATIONS

Most Army publications are available online at http://www.apd.army.mil/.

AR 50-6. *Nuclear and Chemical Weapons and Materiel Chemical Supply.* 28 July 2008.

AR 95-1. *Flight Regulations.* 11 March 2014.

AR 95-2. *Air Traffic Control, Airfield/Heliport, and Airspace Operations.* 31 March 2016.

AR 95-27. *Operational Procedures for Aircraft Carrying Hazardous Materials.* 11 November 1994. AR 700-68. *Storage and Handling Of Liquefied and Gaseous Compressed Gasses and Their Full and Empty Cylinders.* 16 June 2000.

ATP 4-35.1. *Techniques for Munitions Handlers.* 31 May 2013.

FM 1-564. *Shipboard Operations.* 29 June 1997.

FM 3-04. *Army Aviation.* 29 July 2015.

TC 3-04.11. *Commander's Aviation Training and Standardization Program.* 03 August 2016.

TC 3-04.93. *Aero Medical Training for Flight Personnel.* 31 August 2009.

TC 4-13.17. *Cargo Specialist's Handbook.* 12 May 2011.

TM 4-48.09. *Multiservice Helicopter Sling Load: Basic Operations and Equipment.* 23 July 2012.

TM 38-701. *Packaging of Material Packing.* 27 October 2015.

AIR FORCE PUBLICATIONS

The following publication is available online at http://www.e-publishing.af.mil/.

AFMAN 24-204. *Preparing Hazardous Materials for Military Air Shipments.* 03 December 2012.

NATIONAL AGREEMENT TREATY ORGANIZATION PUBLICATIONS

The following publication is available online at NSO.NATO.INT/NSO/

STANAG 3854 (Edition 3). *Policies and Procedures Governing the Air Transportation of Dangerous Cargo.* 13 May 2008.

OTHER PUBLICATIONS

The following publication is available online at www.ecfr.gov.

Electronic Code of Federal Regulation (CFR), Title 14. *Aeronautics and Space, Chapter 1-Federal Aviation Administration, Department of Transportation (Continued), Subchapter G-Air Carriers and Operators for Compensation or Hire: Certification and Operations, Part 121-Operating Requirements: Domestic, Flag, and Supplemental Operations.* 16 November 2016.

PRESCRIBED FORMS

None

REFERENCED FORMS

These documents must be available to the intended users of this publication. Unless otherwise indicated, DA Forms are available on the Army Publishing Directorate (APD) web site: www.apd.army.mil. DD Forms are available on the Office of the Secretary of Defense (OSD) web site: http://www.dtic.mil/whs/directives/forms/index.html.

DA Form 2028. *Recommended Changes to Publications and Blank Forms.*

DD Form 175-1. *Flight Weather Briefing.*

DD Form 365-3. *Weight and Balance Record, Chart C-Basic.*

DD Form 365-4. *Weight and Balance Clearance Form F-Transport/Tactical.*

Index

Entries are by paragraph number unless indicated otherwise

A

absolute ceiling, 7-31, 7-32, 7-34, 7-62
acceleration, 1-31, 1-32, 7-41, 7-62, 7-65
action-reaction, 1-3
adverse yaw, 7-14, 7-38, 7-49
aerodynamic
    angle, 1-10
    balance, 7-50, 7-51
    braking, 7-40, 7-42
    center, 1-7, 1-69, 7-4, 7-5, 7-6, 7-7, 7-8, 7-10, 7-45
    force, 1-2, 1-3, 1-6, 1-7, 1-10, 1-11, 1-28, 1-44, 7-30
    load, 2-22
    properties, 1-6
    stall, 7-10, 7-22, 7-24
    twist, 7-25
ailerons, 7-14, 7-45, 7-49, 8-4
aircraft, 1-17
    component, 3-11, 3-19, 3-24, 7-10
    design, 4-2
    equipment, 8-7
    lighting, 4-19, 4-27, 9-1
    reaction, 1-17
    station, 2-3
    towing, 3-12
airflow, 1-1, 1-3, 1-8, 1-11, 1-33, 1-36, 1-43, 1-59, 1-66, 1-68, 3-28, 7-24, 8-2
airfoil
    airflow, 1-3
    characteristics, 1-6
    terminology, 1-6
    types, 1-7, 7-25
airspeed, 1-59
    best angle of climb, 7-31, 7-57
    best rate of climb/maximum endurance, 1-50
    bucket, 1-50
    differential, 1-12
    drag relationship, 1-27
    forward, 1-37
ambient, 3-6, 4-2, 4-3, 4-11, 5-1, 5-14
angle
    anhedral, 7-12, 7-13
    approach, 4-32, 4-33, 6-8
    climb, 7-30
    critical, 1-61
    dihedral, 7-12
    moon, 4-9
    of attack, 1-7, 1-11, 1-14, 1-40
    of bank, 1-52, 7-59
    of climb, 1-59, 7-31, 7-32, 7-57
    of descent, 1-60, 7-40
    of incidence, 1-7, 1-10, 7-25
    offset, 4-24
    sideslip, 7-9, 7-11, 7-12, 7-13
angular acceleration, 1-52
antitorque rotor, 1-30
approach
    from a hover, 3-9, 3-18
    icing, 8-13
    mountain, 3-39
    paths, 3-38
    procedure, 2-23
    single engine, 7-62
    terrain flight, 3-44
    to a T, 4-33
    to a Y, 4-32
    to the ground, 3-9, 3-18
arm, 2-3, 2-4, 2-24, 7-58
asymmetric
    loading, 7-15
    thrust, 7-57, 7-58
attitude
    fuselage, 1-22, 1-23
    indicator, 3-11, 9-4
    landing, 7-48
    shift, 1-25
    takeoff, 7-40
autorotation, 1-43, 1-44
    blade regions, 1-43
    phases, 1-46
    spin, 7-26

B

balance, 2-1, 2-2
    angular, 1-66
    calculation, 2-6
    definitions, 2-3
    lateral, 2-2
    of forces, 1-31
    point, 2-14
    tab, 7-52
balanced field length, 7-65
bank angle, 1-50, 4-18, 6-8, 7-35, 7-59
Bernoulli's Principle, 1-2
blade
    actions, 1-12, 1-13
    pitch, 1-40, 7-55
    span, 1-7
    speed, 1-12
    twist, 1-8

C

centrifugal
    force, 1-22
climbing, 7-30, 7-32
    flight, 7-29
    performance, 7-29, 7-32
    stall speed, 7-31
climbs, 1-55, 3-33, 6-11, 7-29, 9-5
    single-engine, 7-60
coefficient
    of drag, 1-27, 8-3
    of lift, 1-26, 7-5, 7-13, 7-16, 7-17, 7-47, 8-2
    of rolling moment, 7-12
    yawing moment, 7-9
collective, 1-11, 1-12, 1-17, 1-33, 1-46, 1-48, 1-51, 1-62, 1-65, 3-8
compressibility, 1-67, 1-68, 1-69, 7-55
coning, 1-28, 1-50
conservation
    of angular momentum, 1-15, 1-52
    of energy, 1-2
control, 1-16, 1-17, 1-18, 1-20, 1-24, 1-30, 1-61, 3-12, 3-20, 4-21, 6-31, 7-37, 7-38, 7-43, 7-45, 7-46, 7-47, 7-49, 7-50, 7-51, 7-52, 7-53, 7-54, 7-59
Coriolis, 1-15, 1-52
crest, 3-26, 8-18
critical
    engine, 7-56, 7-57
    mach, 1-68
    rate, 1-60
    time, 7-63
crossover, 4-22, 4-25, 6-14

Index

crosswind, 3-35, 3-38, 6-31, 7-44
crown, 3-32
cyclic, 1-13
    feathering, 1-39
    pitch, 1-19

D

damping, 7-3, 7-53
deceleration, 1-47
demarcation, 3-26, 3-39
density, 1-55, 1-56, 2-24, 8-16
descent, 7-38, 8-12
    rate, 1-48
dihedral, 7-12, 7-13
dissymmetry of lift, 1-13, 1-38, 1-39, 1-40
divergence, 7-15
downwash, 1-9, 1-41, 1-51, 7-48
drag, 1-25, 1-26, 1-46, 7-38, 7-60
driven region, 1-43
driving region, 1-43
Dutch roll, 7-15
dynamic
    pressure, 1-3, 7-53
    rollover, 1-61
    stability, 7-2, 7-4

E

effective translational lift, 1-42
elevator, 1-14, 1-24, 7-28, 7-35, 7-46
emergency, 2-23, 3-40, 4-35, 7-64, 9-5
engine failure, 1-46, 7-49, 7-64
equilibrium, 7-1, 7-3, 7-30

F

feathering, 1-12, 7-55, 7-63
flapping, 1-13, 1-14, 1-39
flaps, 7-19, 7-21, 8-13
flow, 1-2, 1-3
    compressible, 1-67
    incompressible, 1-67
    supersonic, 1-68
    transonic, 1-68
fog, 3-3, 3-26, 4-9, 4-19, 4-23, 5-5, 9-5
force, 7-49, 7-52
    centrifugal, 1-28, 7-27
    centripetal, 7-35
    damping, 7-4
friction, 7-41, 7-43

frise ailerons, 7-14
fuselage, 1-24, 7-7, 7-8, 7-10

G

g
    force, 1-54, 2-18
    load, 7-35
    loading, 1-52
    maneuver, 1-51
glide, 7-39, 9-9
go-around, 1-63, 3-10, 3-40, 3-44, 6-8
ground effect, 1-33, 1-34, 7-41, 7-48
gyroscopic precession, 1-16, 1-20

H

heading control, 1-30
high-lift, 7-16, 7-19, 7-43
hinge, 1-13, 7-49
horn, 7-50, 8-7
    pitch-change, 1-20
hovering, 1-11, 1-33, 1-40, 1-58, 3-8
humidity, 1-56, 3-22, 8-16
hydroplaning, 7-44

I

induced
    drag, 1-26, 7-14, 7-36
    flow, 1-7, 1-9, 1-33, 1-34, 1-39, 1-59
    roll, 6-31
inertia, 1-1
in-ground effect, 1-34
inverted, 4-32

J

join-up, 6-12

K

kinetic energy, 1-43, 7-29, 7-42

L

landing, 3-9, 3-10, 3-18, 3-24, 3-39, 4-32, 7-22, 7-42, 8-13, 9-5
    formation, 6-8
    gear, 7-54
    light, 4-28, 4-30
    velocity, 7-43
    water, 8-20
lateral, 1-62, 2-2, 7-11, 7-13
lead and lag, 1-14
leading-edge, 1-7, 7-21
lift, 1-31, 6-30

equation, 1-26
force, 7-17
pattern, 1-64
linear, 5-7, 7-5
load factor, 7-4, 7-28, 7-35
longitudinal, 1-23, 1-24, 2-2, 7-1, 7-4, 7-7
    control, 7-47

M

maneuvering, 1-50, 1-55, 6-7, 6-18, 7-35, 7-37
map, 5-8, 5-9, 5-10
mass, 1-52, 1-67, 7-52
maximum
    endurance, 1-50
    glide, 1-48
mean
    camber line, 1-7
mean aerodynamic chord, 7-47
minimum
    control speed, 7-31, 7-37, 7-56
    drag speed, 7-38
    resolvable temperature, 4-20
    turn radius, 7-36
moment, 2-3, 2-5, 7-1, 7-4, 7-12, 7-15

N

no-lift, 1-37, 1-64
nonoscillatory, 7-2
nonsymmetrical, 1-8

O

obstacle
    clearance, 4-25
    detection ability, 5-15
obstructions, 4-18, 9-2
operations
    external load, 2-9
    slope, 1-61
oscillatory, 7-3
out-of-ground effect, 1-34, 7-48
overcontrolling, 1-24

P

parallelogram method, 1-4
parasite drag, 1-27, 1-32, 7-40, 7-41
pendular action, 1-24
performance, 1-55, 1-58, 7-32, 7-44, 7-56
    charts, 1-58
    climb, 1-58, 7-29, 7-57

Index

FLIR, 4-22
hovering, 1-58
limits, 7-35
planning, 2-22
slow flight, 7-38
terrain flight, 5-15
turns, 7-34
weight, 2-2
P-factor, 7-16, 7-57, 7-58
phase lag, 1-20
photocathode, 4-15
pinnacle, 3-32, 3-39
pitch
    angle, 1-7, 1-9, 1-22, 1-35, 1-40
    change, 1-19
    control, 1-17, 1-22
    cyclic, 1-19
    stability, 7-3
    variation, 1-22
polar diagram, 7-32
polygon method, 1-4
positive
    cambered, 7-7
    dynamic stability, 7-3
    lift, 1-37
    stall, 1-37
    static stability, 7-2
pressure
    altitude, 1-55
    atmospheric, 1-56
    center of, 1-7
    contact, 2-11
    differential, 1-25, 6-30, 7-17, 7-23
    formulas, 2-12
propeller, 7-54
    factor, 7-57
    feathering, 7-55, 7-63
    icing, 8-6
    region, 1-43
proverse roll, 7-14

R

rate
    of accumulation, 8-2
    of climb, 7-32
    of closure, 3-39, 4-31, 6-8
    of sublimation, 8-2
reconnaissance, 3-3, 3-36, 3-39
relative wind, 1-7, 1-8, 1-39, 7-9, 7-14
resonance, 1-65
restraint, 2-9, 2-18, 2-19
resultant relative wind, 1-7, 1-10

retreating blade stall, 1-64
reverse flow, 1-37
ridge, 3-29, 3-30, 3-34, 3-41, 5-7, 8-16
roll
    control, 8-4
    landing, 7-42
    rate, 7-49
    stability, 7-11
    takeoff, 7-41
    upsets, 8-14
rolling
    friction, 7-41
    motion, 1-61
rotational
    relative wind, 1-9
rotor
    blade actions, 1-11
    blade angles, 1-10
    clouds, 3-29
    head control, 1-17
    streaming, 3-29
    system, 1-28, 1-36
rotor efficiency, 1-34
roundout, 9-7
route, 3-5, 3-34, 3-41, 5-6, 5-7, 5-8, 5-10, 5-11, 8-16
rudder, 7-16, 7-35, 7-49
runway, 4-32, 6-32, 7-41, 7-42, 7-43, 7-44, 7-64, 9-3

S

safety, 2-2, 2-23, 2-26, 2-27, 3-18, 4-35, 5-17, 6-29, 7-64
scalars, 1-4
settling with power, 1-59
shoring, 2-9, 2-11
sideslip, 7-9, 7-60
single-engine
    climb, 7-60
    flight, 7-62
    operation, 7-57
skidding, 7-60
sling, 2-22, 2-23
slipstream, 7-15, 7-38
spin, 7-26, 7-49
stability, 7-1
    augmenter, 7-53
stall, 1-8, 1-51, 1-65, 7-22, 7-23, 7-28, 7-37, 8-4, 8-5
    negative, 1-37
    recovery, 7-26
    region, 1-44
    speed, 7-17
    strip, 7-25
    tailplane, 8-13

warning, 7-24, 7-26, 8-7
static
    electricity, 2-9
    eletricity, 3-6
    equilibrium, 2-5
    leaks, 3-12
    port, 8-6
    pressure, 1-67
    rollover angle, 1-61
    stability, 7-2
steady-state
    climb, 7-29
    descent, 1-47
strain, 7-28
structural
    components, 2-10
    failure, 1-65, 1-69
    icing, 8-1
supersonic, 1-67
switch, 7-26, 7-54, 8-8
symmetrical, 1-7, 7-7, 7-25

T

tab, 7-51
tail, 1-30, 1-39
takeoff, 1-58, 1-62, 2-2, 2-23, 3-8, 3-17, 3-33, 3-41, 4-30, 6-7, 6-34, 7-31, 7-40, 7-48, 7-61, 9-3
taxiing, 3-7, 3-17, 3-23, 4-36, 9-3
terrain, 3-5, 3-10, 3-15, 3-22, 3-30, 4-10
    flight, 3-8, 3-18, 3-23, 3-40, 4-25, 5-1
    interpretation, 4-5
thrust, 1-31, 1-35, 7-8, 7-49, 7-56, 7-58
torque, 1-30, 1-50, 1-51, 1-54, 1-55
total aerodynamic force, 1-25, 1-43, 1-53, 1-54
traffic, 3-39, 6-34, 9-3, 9-6
trailing, 1-69, 6-7, 6-31, 7-19, 7-20, 7-24
translating, 1-22, 1-35
translational, 1-40, 1-42, 1-58, 6-7
transverse, 1-41
trim, 1-55, 7-4, 7-5, 7-52
trunnion, 1-23
turbulence, 1-40, 3-28, 3-34, 3-44, 6-7, 6-30, 8-16, 8-22, 8-25
turns, 1-55, 7-34

Index

formation, 6-8
  tactical, 6-23
twist, 7-25

U

unbalanced
  forces, 7-34

V

vector, 1-4, 1-5, 7-30
velocity, 1-7, 1-36, 1-52, 7-42

Venturi, 1-2, 1-3, 3-30
vertical, 1-19, 1-43, 1-44, 6-5, 6-14, 6-27, 7-9, 7-10
visual, 1-63, 3-7, 3-9, 3-35, 3-40, 4-1, 4-4, 4-5, 4-9, 4-25, 4-32, 4-35, 6-13
vortex, 1-33, 1-34, 1-59, 6-30, 6-31, 6-32, 7-18

W

weight, 1-57, 2-1, 2-3, 2-5, 2-11, 4-16, 7-30, 7-31, 7-41, 7-44
windmilling, 7-40, 7-55, 7-59, 7-63
wires, 3-42, 4-18, 5-7, 5-14, 8-8

Y

yaw, 7-27

By Order of the Secretary of the Army:

MARK A. MILLEY
*General, United States Army
Chief of Staff*

Official:

GERALD B. O'KEEFE
*Administrative Assistant to the
Secretary of the Army*
1634701

DISTRIBUTION:
Active Army, Army National Guard, and United States Army Reserve: Distributed in electronic media only (EMO).

PIN: 201153-000

www.ingramcontent.com/pod-product-compliance
Lightning Source LLC
Chambersburg PA
CBHW050046230526
45470CB00004B/1427